Engineering Materials

This series provides topical information on innovative, structural and functional materials and composites with applications in optical, electrical, mechanical, civil, aeronautical, medical, bio- and nano-engineering. The individual volumes are complete, comprehensive monographs covering the structure, properties, manufacturing process and applications of these materials. This multidisciplinary series is devoted to professionals, students and all those interested in the latest developments in the Materials Science field, that look for a carefully selected collection of high quality review articles on their respective field of expertise.

Indexed at Compendex (2021) and Scopus (2022)

Raju Khan · Neeraj Kumar ·
Mohd. Abubakar Sadique · Arpana Parihar
Editors

Electrochemical Exfoliation of Graphene and Its Derivatives

Commercial Applications

Springer

Editors
Raju Khan
CSIR-Advanced Materials and Processes
Research Institute (AMPRI)
Bhopal, India

Neeraj Kumar
CSIR-Advanced Materials and Processes
Research Institute (AMPRI)
Bhopal, India

Mohd. Abubakar Sadique
CSIR-Advanced Materials and Processes
Research Institute (AMPRI)
Bhopal, India

Arpana Parihar
CSIR-Advanced Materials and Processes
Research Institute (AMPRI)
Bhopal, India

ISSN 1612-1317 ISSN 1868-1212 (electronic)
Engineering Materials
ISBN 978-981-97-2127-6 ISBN 978-981-97-2128-3 (eBook)
https://doi.org/10.1007/978-981-97-2128-3

This Springer imprint is published by the registered company Springer Nature Singapore Pte Ltd.
The registered company address is: 152 Beach Road, #21-01/04 Gateway East, Singapore 189721, Singapore

The purpose of learning is growth, and our minds, unlike our bodies, can continue growing as we continue to live.
—Mortimer Adler

Preface

Among the intricate fabric of modern materials, few strands are as luminous as those crafted from graphene and its extraordinary derivatives. These remarkable materials possess an exceptional combination of strength, conductivity, and adaptability, which has attracted researchers and technologists. They hold the potential to revolutionize various industries, including electronics, energy storage, biosensing, and medicinal applications. However, fully harnessing the complete capabilities of graphene requires more effort than simply exploiting its known extraordinary qualities. It requires the development of efficient, scalable, and environmentally sustainable methods for its synthesis and processing. The book extensively describes the technique of electrochemical exfoliation, which possesses the remarkable ability to bring about transformation. Among various known synthesis methods, the electrochemical exfoliation approach eliminates the use of harsh chemicals and energy-intensive methods commonly linked to the synthesis of graphene. Electrochemical exfoliation utilizes electrical energy to gently remove layers of graphene from its original source, providing a more environmentally friendly method. This precise and careful synchronization heralds a new era in the field of materials science, where the principles of sustainability converge with unmatched performance. However, the benefits extend beyond environmental excellence. This book thoroughly examines the complexities of electrochemical exfoliation, highlighting its clear advantage over traditional techniques. The approach demonstrates exceptional mastery in manipulating the structure and properties of graphene, allowing for the customization of specific capabilities to suit a wide range of applications.

The chapters of this book explore an enthralling way of the intricate realm of graphene and its derivatives. Every chapter reveals a different aspect of this remarkable 2D material, which has been carefully created by electrochemical exfoliation. The text explores the theoretical framework of electrochemical exfoliation, its environmental benefits, and its potential applications. It highlights the various methods of synthesizing graphene derivatives, their unique properties, and their interaction with electrons. The process promotes sustainability and resourcefulness by converting waste streams into valuable resources. The text also explores the unique features of electrochemically exfoliated multi-dimensional graphene, its potential in energy storage applications, and its role in biosensing and bioimaging. It also discusses the therapeutic possibilities of electrochemically exfoliated graphene and its derivatives, which could revolutionize medical treatments and health care. The text concludes by navigating the complexities of large-scale manufacturing and commercialization and addressing the obstacles and possibilities in the field of electrochemical exfoliation.

This book transcends being a mere technical treatise; it serves as an enticement to go on a scientific voyage. Through the progression of each chapter, you will explore the captivating realm of graphene and its derivatives, gaining direct insight into the profound impact of electrochemical exfoliation. The limitless potential that resides inside the atomic structure of these extraordinary substances will inspire, challenge, and ultimately invigorate you.

Raju Khan
CSIR-Advanced Materials
and Processes Research Institute
(AMPRI)
Bhopal, India

Neeraj Kumar
CSIR-Advanced Materials
and Processes Research Institute
(AMPRI)
Bhopal, India

Mohd. Abubakar Sadique
CSIR-Advanced Materials
and Processes Research Institute
(AMPRI)
Bhopal, India

Arpana Parihar
CSIR-Advanced Materials
and Processes Research Institute
(AMPRI)
Bhopal, India

Acknowledgments

The editors are sincerely thankful to every contributor to the book. Their heartfelt efforts, hard work, and rational approach are greatly acknowledged. At the same time, the editors would like to acknowledge the publisher and the associated team for their constant support, guidance, and inspiration, which regularly thrust us to the fore to complete this book.

The beneficial suggestions and support given by Dr. Avanish Kumar Srivastava, Director, CSIR-Advanced Materials and Processes Research Institute, Bhopal, MP, India, are duly acknowledged.

Raju Khan expresses his special thanks to his parents, wife Shazia M. Siddiqui, daughter Sara Khan, and son Aayan Khan for their everlasting love, enthusiasm for science, and encouragement to pursue every task successfully.

Neeraj Kumar would like to acknowledge with gratitude the support and love of his family parents, Pan Kunwar and Bhikhari Lal, and his brothers Vinod, Ashok, Santosh, and Amit. Special thanks to wife Pratiksha Prajapati and son Aarav for their everlasting love. They all kept him going, and the work done would not have been possible without their continual support.

Mohd. Abubakar Sadique expresses heartfelt gratitude to God for his blessings and guidance. He extends his deepest appreciation to his parents, Mohammad Sadique and Nishat Sadique, for their unwavering love and constant support; and to his siblings, Adiba, Fazila, and Nabiha, for their unwavering belief and encouragement. He is also grateful to his colleagues, Shalu Yadav and Ayushi Singhal, for their tireless teamwork and insightful contributions.

Arpana Parihar is thankful to Almighty and acknowledges the support of Ashish, Adhiraj, and Aviraj during the tenure of this work. The fellowship offered by the Department of Science and Technology, Government of India under the DST-WoS-B scheme (DST/WOS-B/HN-4/2021) is duly acknowledged by her.

Additionally, the editors are sincerely appreciative of all who have directly or indirectly offered important input to this book.

Raju Khan
Neeraj Kumar
Mohd. Abubakar Sadique
Arpana Parihar

Contents

Editors and Contributors

About the Editors

Dr. Raju Khan is currently working as a Senior Principal Scientist and Professor at Industrial Waste Utilization, Nano and Biomaterials (IWUNB), CSIR-Advanced Materials Processes and Research Institute, Bhopal, Madhya Pradesh, India. He has a background in Chemistry and did his M.Sc. in Chemistry and Ph.D. in Chemistry from Jamia Millia Islamia, Central University, New Delhi, India. He has worked as a Post Doctorate at the University of the Western Cape, Cape Town, South Africa, and CSIR-National Physical Laboratory, New Delhi, India as a Young Scientist, Visiting Scientist at the University of Texas and Sanantonio, UTSA, USA. He has been awarded Fellow of the Royal Society of Chemistry (FRSC) and also awarded a reputed BOYSCAST Fellowship, Government of India. He is the recipient of many prestigious achievements like being a Scientific Advisor for FSSAI, Government of India, a Life member of the Materials Research Society of India, and a Member of the International Advisory Committee, DST, Government of India. His work has also been highlighted in Nature India twice in 2008 for tracking food toxicity and recently in 2021 for the fabrication of electrochemical biosensors from carbon waste materials. He has published more than 180 articles with a high citation score of more than 4700 in several renowned journals including 40 book chapters in esteemed publishers, editing 25 books with various reputed publishers, and patent file filled. He has several ongoing projects and completed including international scientific collaborations with the USA, Czech Republic, and Russia.

Mr. Neeraj Kumar is currently working as a Ph.D. candidate (CSIR-Senior Research Fellow) under the supervision of Dr. Raju Khan at IWUNB, CSIR-Advanced Materials and Processes Research Institute (AMPRI), Bhopal 462026, Madhya Pradesh, India. His current research work focuses on the investigation of carbon nanomaterial-based composites for electrochemical biosensor applications. He has been awarded CSIR-UGC-JRF (NET) and GATE 2 times. He has more than five years of research and development experience in the Council of Scientific and Industrial Research at CSIR-AMPRI, Bhopal, Madhya Pradesh, India (Government of India). He received a Bachelor of Science (Chemistry) from Bundelkhand University, Jhansi (Uttar Pradesh), 2013, a Master of Science in Chemistry from Bundelkhand University, Jhansi (Uttar Pradesh), 2015, and is currently pursuing a Doctor of Philosophy in Chemical Sciences from the Academy of Scientific and Innovative Research (AcSIR), CSIR-AMPRI, Bhopal, M. P., India. He has published more than 10 articles in SCI journals, one article in Indian journals, and 2 book chapter in the reputed book Springer and IOP.

Mr. Mohd. Abubakar Sadique is currently working as a research scholar at CSIR-AMPRI, Bhopal, India. He has worked on a DST-SERB project entitled "Development of Rapid Electrochemical-Based Diagnostics for the Detection of SARS-CoV-2 Infection." He has worked as a Senior Project Fellow at CSIR-CEERI, Pilani, Rajasthan, India, on a mission-mode project, "Design and Fabrication of Microfluidic-Based Bio-Sensors for Biochemical Detection." His recent research includes the use of nanotechnology for healthcare applications. He was a gold medalist in M.Tech., and he received his post-graduate degree in nanotechnology from Aligarh Muslim University, India, in 2019. He was a branch topper in his bachelor's and received a B.Tech. degree in Material Science Engineering with specialization in Nanotechnology from the University of Petroleum and Energy Studies, Dehradun, India, in 2016. Mr. Sadique's research interests include the green synthesis of nanomaterials, their characterization, and their healthcare applications. Mr. Sadique has a particular interest in carbon-based nanostructures for their effectiveness in the fields of biosensing, diagnostics, therapeutics, and healthcare applications. He has much experience in experimental research with numerous internships at ARCI-Hyderabad, UGC-DAE-CSR, Indore, RRCAT, Indore, etc. His academics are evident in his skills and qualities for the research domain, even in a brief time. He has been associated with various research and development projects funded by different agencies, which include the Department of Science and Technology (DST, India) and the Council of Scientific and Industrial Research (CSIR), India. Mr. Sadique has published several reviews and research articles, book chapters, and book editing, and one patent.

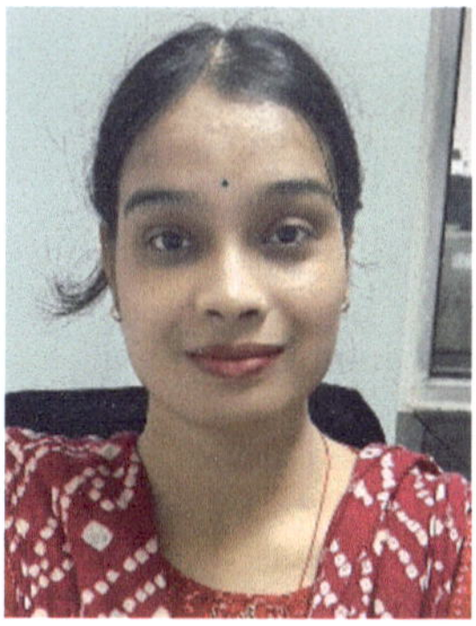

Dr. Arpana Parihar is working as a DST Women Scientist B at IWUNB, CSIR-Advanced Materials and Processes Research Institute (AMPRI), Bhopal (462026), MP, India. She did her Ph.D. from the Raja Ramanna Centre for Advanced Technology, Indore, under the aegis of HBNI, Mumbai. Her doctoral research work involves the evaluation of the tumor selectivity and photodynamic therapy (PDT) efficacy of chlorin p6 through receptor-mediated targeted delivery. After her Ph.D., she worked as a post-doctoral researcher at CBME, IIT Delhi. At IITD, her research work was to enhance the osteoinductive and osteoconductive properties of various implants made up of metals, ceramics, and polymers. Dr. Parihar has been awarded GATE, CSIR-NET, DBT RA, and DST-WoS B fellowships. She has more than eight years of research and teaching experience at various prestigious institutes. Her current research interests include the fabrication of biosensors for early diagnosis of cancer, infectious diseases, stem cell therapy, tissue engineering, photodynamic therapy of cancer, receptor-mediated targeting of tumors, biosensors for early diagnosis of diseases, 3D cell culture, expression of receptors and cytokines, molecular docking, molecular dynamic simulations, and immunoinformatics. At present, she has published seven books with renowned publishers such as Elsevier, CRC-Press, IOP, and has written several book chapters for esteemed publishers such as Springer, Wiley, Elsevier, and Taylor & Francis. Her research work has gained considerable attention in various international peer-reviewed journals and conferences.

Contributors

Soumya Basu Cancer and Translational Research Centre, Dr. D. Y. Patil Biotechnology and Bioinformatics Institute, Tathawade, Pune, Maharashtra, India

Diksha Bhatt Rajendra Singh Nanoscience and Nanotechnology Lab, Department of Chemistry, Kumaun University, Nainital, Uttrakhand, India

Kalpesh V. Bhavsar National Centre for Nanosciences and Nanotechnology, University of Mumbai, Kalina Campus, Kalina, Mumbai, Santacruz (E), India

Prakash Chandra Department of Chemistry, Institute of Basic Science, Bundelkhand University, Jhansi, Uttar Pradesh, India

Atul Chaskar National Centre for Nanosciences and Nanotechnology, University of Mumbai, Mumbai, India;
Department of Chemistry, Institute of Chemical Technology Mumbai, Mumbai, India

Deepali Chaudhary Department of Bio-Sciences and Technology, Maharishi Markandeshwar (Deemed to Be) University, Mullana, Ambala, India

Hardik S. Churi Sonopant Dandekar Shikshan Mandali (SDSM) College, Tembhode, Palghar, Maharashtra, India

Aishik Das Cancer and Translational Research Centre, Dr. D. Y. Patil Biotechnology and Bioinformatics Institute, Tathawade, Pune, Maharashtra, India

Sudeepa Devi Defence Materials and Stores Research and Development Establishment, Kanpur, India;
Janta Maha Vidyalaya (CSJM University), Ajitmal, Auraiya, Uttar Pradesh, India

Nkurunziza Florien Department of Bio-Sciences and Technology, Maharishi Markandeshwar (Deemed to Be) University, Mullana, Ambala, India

Kamal Garwal Rajendra Singh Nanoscience and Nanotechnology Lab, Department of Chemistry, Kumaun University, Nainital, Uttrakhand, India

Shagun Gupta Department of Bio-Sciences and Technology, Maharishi Markandeshwar (Deemed to Be) University, Mullana, Ambala, India

Uday P. Jagtap National Centre for Nanosciences and Nanotechnology, University of Mumbai, Kalina Campus, Kalina, Mumbai, Santacruz (E), India

Rimpa Jaiswal High Pressure and Synchrotron Radiation Physics Division, Bhabha Atomic Research Centre, Mumbai, India;
Defence Materials and Stores Research and Development Establishment, Kanpur, India

Kevin Jose National Centre for Nanosciences and Nanotechnology, University of Mumbai, Kalina Campus, Kalina, Mumbai, Santacruz (E), India

Nandakumar Kalarikkal International and Inter University Centre for Nanoscience and Nanotechnology (IIUCNN), Mahatma Gandhi University, Kottayam, Kerala, India;
School of Pure and Applied Physics, Mahatma Gandhi University, Kottayam, Kerala, India

Deepshikha Pande Katare Proteomic and Translational Research Lab, Centre for Biomedical Biotechnology, Amity Institute of Biotechnology, Amity University, Noida, India

Ankur Kaushal Department of Bio-Sciences and Technology, Maharishi Markandeshwar (Deemed to Be) University, Mullana, Ambala, India

Raju Khan Industrial Waste Utilization, Nano and Biomaterials, CSIR-Advanced Materials and Processes Research Institute (AMPRI), Bhopal, MP, India;
Academy of Scientific and Innovative Research (AcSIR), Ghaziabad, India

Neeraj Kumar Industrial Waste Utilization, Nano and Biomaterials, CSIR-Advanced Materials and Processes Research Institute (AMPRI), Bhopal, MP, India;
Academy of Scientific and Innovative Research (AcSIR), Ghaziabad, India

Mantu Modak High Pressure and Synchrotron Radiation Physics Division, Bhabha Atomic Research Centre, Mumbai, India

Tebello Nyokong Insititute of Nanotechnology Innovation (INI), Rhodes University, Makhanda, South Africa

Vijayeta Pal Materials Science and Technology, Indian Institute of Technology (BHU), Varanasi, Uttar Pradesh, India

Arpana Parihar Industrial Waste Utilization, Nano and Biomaterials, CSIR-Advanced Materials and Processes Research Institute (AMPRI), Bhopal, MP, India

Mayank Pathak Rajendra Singh Nanoscience and Nanotechnology Lab, Department of Chemistry, Kumaun University, Nainital, Uttrakhand, India

Sharon J. Paul Photonic Materials Metrology Sub Division, Advanced Materials and Device Metrology Division, CSIR-National Physical Laboratory, New Delhi, India

Rachna Poria Department of Bio-Sciences and Technology, Maharishi Markandeshwar (Deemed to Be) University, Mullana, Ambala, India

Renu Poria Department of Bio-Sciences and Technology, Maharishi Markandeshwar (Deemed to Be) University, Mullana, Ambala, India

Trishala Rawat Department of Bio-Sciences and Technology, Maharishi Markandeshwar (Deemed to Be) University, Mullana, Ambala, India

Melkamu Biyana Regasa Department of Chemistry, College of Natural and Computational Sciences, Wollega University, Nekemte, Ethiopia

Debmalya Roy Directorate of Nanomaterials, DMSRDE, Kanpur, Uttar Pradesh, India

Mohd. Abubakar Sadique Industrial Waste Utilization, Nano and Biomaterials, CSIR-Advanced Materials and Processes Research Institute (AMPRI), Bhopal, MP, India;
Academy of Scientific and Innovative Research (AcSIR), Ghaziabad, India

Nanda Gopal Sahoo Rajendra Singh Nanoscience and Nanotechnology Lab, Department of Chemistry, Kumaun University, Nainital, Uttrakhand, India

Shikha Saxena Amity Institute of Pharmacy, Amity University, Noida, India

Faiz Ullah Shah Chemistry of Interfaces, Luleå Technology University, Luleå, Sweden

Dhriti Shenoy National Centre for Nanosciences and Nanotechnology, University of Mumbai, Mumbai, India;
Department of Chemistry, Institute of Chemical Technology Mumbai, Mumbai, India

Gurmehar Singh Amity Institute of Pharmacy, Amity University, Noida, India

Gaurav Tatrari Chemistry of Interfaces, Luleå Technology University, Luleå, Sweden

Upendra K. Tripathi Janta Maha Vidyalaya (CSJM University), Ajitmal, Auraiya, Uttar Pradesh, India

Fadeela Chundekatt Ummer International and Inter University Centre for Nanoscience and Nanotechnology (IIUCNN), Mahatma Gandhi University, Kottayam, Kerala, India

Saneeya Vichare Cancer and Translational Research Centre, Dr. D. Y. Patil Biotechnology and Bioinformatics Institute, Tathawade, Pune, Maharashtra, India

Devyani Wakchoure National Centre for Nanosciences and Nanotechnology, University of Mumbai, Mumbai, India;
Department of Chemistry, Institute of Chemical Technology Mumbai, Mumbai, India

Pravin Walke National Centre for Nanosciences and Nanotechnology, University of Mumbai, Mumbai, India;
Department of Chemistry, Institute of Chemical Technology Mumbai, Mumbai, India

Neha Yadav Department of Biological Science & Bioengineering, Indian Institute of Technology, Kanpur, Uttar Pradesh, India

Chapter 1
Overview of Electrochemical Exfoliation Approaches

Rimpa Jaiswal, Mantu Modak, and Sudeepa Devi

Abstract In recent time, researchers and scientists have extensively studied electrochemical exfoliation (ECE) techniques to achieve defect-free, highly pure forms of 2D nanomaterials. It is eco-friendly, cost-effective, time-saving, simple, single-step, facile, and overall a highly efficient technique. Moreover, this method is also beneficial for the bulk production. The ECE involves an intercalation followed by exfoliation, upon applying a potential bias to the electrodes (anode or cathode) in an aqueous or organic electrolyte. However, the present chapter emphasizes the role of electrolytes with their potential applications to the quality and quantity of exfoliated 2D materials, especially graphene, and its composite materials. Additionally, the advantages and challenges associated with the electrochemical exfoliation technique over the existing top-down and bottom-up methods, along with applications of ECE, are also discussed.

Keywords Electrochemical exfoliation · 2D materials · Nanomaterials · Electrolyte

1.1 Introduction

Various nanomaterials, categorized as 0D (Fullerene), 1D (carbon nanotube, CNT), 2D (graphene, metal chalcogenides, MXene), and 3D (graphite) nanomaterials (1–100 nm), have been widely studied (Fig. 1.1A) and employed in several applications such as solar cell, photovoltaic, supercapacitor, biomedical, stealth technology, opto-electronics, and so on, owing to their promising characteristic properties and high surface area. It is worthy to mention that, nanomaterials always have a tendency to

R. Jaiswal (✉) · M. Modak
High Pressure and Synchrotron Radiation Physics Division, Bhabha Atomic Research Centre, Mumbai 400 085, India
e-mail: rimpajai@gmail.com

R. Jaiswal · S. Devi
Defence Materials and Stores Research and Development Establishment, Kanpur 208013, India

R. Khan et al. (eds.), *Electrochemical Exfoliation of Graphene and its Derivatives*, Engineering Materials, https://doi.org/10.1007/978-981-97-2128-3_1

agglomerate and make them non-dispersible in aqueous or organic solvent. Consequently, this drawback restricts their characteristic properties and ultimately limits their potential applications (Danial et al. 2021). There were two methods employed to synthesize nanomaterials. In the bottom-up approach, the atoms/ions/molecules interacted to yield nanomaterials (e.g., hydrothermal, precipitation, chemical vapor deposition (CVD), chemical wet method). In contrast, the top-down process (Danial et al. 2021) involves breaking of the bulk material into several nano dimensional structures by applying forces like mechanical milling, sputtering; etching methods developed via mechanical, electrochemical, solution based exfoliation of graphite. It reduces the van der Waal interaction between the layers and gets 2D nanomaterials without compromising their hierarchical structure and properties and it also helps to avoid aggregation. The synthesis of nanomaterials via CVD is a promising method that yields highly pure materials but is not cost-effective because it consumes more time, requires high temperatures, and compromises mass production (Danial et al. 2021). On the other hand, the wet chemical method uses some toxic and hazardous reagents and produces the material with some structural defects. It is also a time-consuming, complex, and tedious procedure. Among these methods or techniques, the electrochemical method is found to be excellent, owing to its environment-friendly, inexpensive, good controllability, simple, and facile single-step process that allows mass production of materials with minimum defect or defect-free lattice structure (Danial et al. 2021).

This chapter mainly focuses on the ECE of graphite to graphene owing to their significant role in basic researches and their potential applications in industries. Ideally, graphene is a single-atom-thick, hexagonal honeycomb structure, sp^2 hybridized carbon sheet. First time, the free-standing exfoliation of graphene from graphite was successfully accomplished via micromechanical exfoliation, employing scotch tapes in 2004 by Geim and Novoselov (2007). For this achievement they were bestowed the Nobel Prize in Physics in 2010. The presence of free π-electrons and

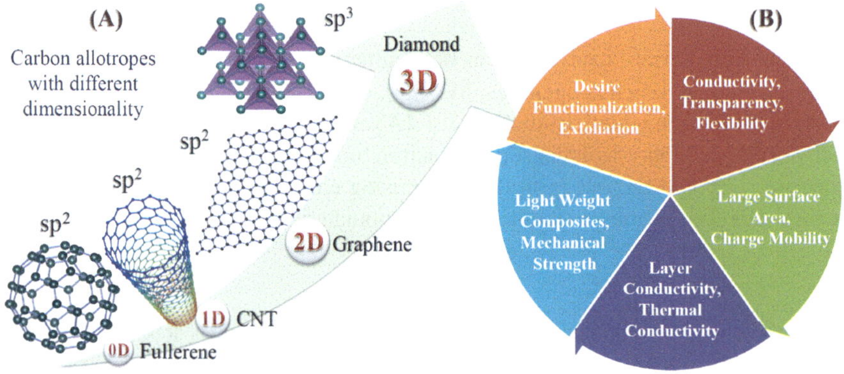

Fig. 1.1 **A** Carbon allotropes with different dimensionality and hybridization, **B** Graphene nanomaterials possess wide range of potential applications

their delocalization in graphene are responsible for its fascinating physical and chemical properties, such as thermal conductivity (~5000 Wm^{-1} K^{-1}), electrical conductivity (~6000 S/cm), optical properties (optical transmittance ~97.7% with insignificant reflectance), and mechanical strength (Young's module ~1 TPa,) (Danial et al. 2021; Xia et al. 2019; Prekodravac et al. 2021). Furthermore, graphene possesses a very high theoretical specific surface area of 2630 m^2 g^{-1}, intrinsic mobility ~200 000 cm^2V^{-1} s^{-1} at the Dirac point, and charge carrier mobility of 10^6 cm^2V^{-1} s^{-1} at room temperature for dispersed graphene (Prekodravac et al. 2021). These extraordinary properties of graphene nanomaterials enable extensive possibilities for their use in many promising applications ranging from electrochemical catalysis to environmental remediation, energy storage (supercapacitors) to conversion (photovoltaics) (Sood et al. 2015; Jaiswal et al. 2023, 2018; Saha et al. 2022; Li et al. 2016; Markovic et al. 2011, 2017; Kepic et al 2014, 2018), fuel cells, electrochemical and bio sensors (Ambrosi et al. 2014), as presented in Fig. 1.1B.

However, the angle of stacking, number of layers, sheet size, and quality (Chen et al. 2021a, b) are deciding parameters that govern the properties of graphene as mentioned above. Unlike other 2D materials, graphene is considered to be a zero-band gap semiconductor (Gui et al. 2008; Sahu and Rout 2017) owing to symmetrically distributed valence and conduction bands on both sides of the Fermi level. On the other hand, graphene oxide (GO) is the most frequently used derivative of graphene, synthesized from graphite via oxidative chemical wet methods (e.g., Hummers method) followed by ultrasonication to minimize the Van der Waals force of attraction (5.9 kJ/mol) among graphene layers of graphite oxide. The stable dispersion of exfoliated GO sheets could be achieved in some of the solvents having matching surface energies, like water, N-methyl-2-pyrrolidone (NMP), and dimethyl formamide (DMF). Chemically tuned graphene possesses oxygen-containing functional groups like hydroxyl (–OH) and epoxy (C–O–C) on their basal planes while carboxylic or carbonyl (–COOH/–C = O) at their edges with lower C/O atomic ratios.

The sp^2-bonded carbon atoms of graphene get transformed to sp^3 upon chemical oxidation treatment, which consequently compromises the desired properties of graphene such as electrical conductivity due to loss of π-electron and inhibition of its delocalization. Therefore, GO is subjected to a reduction method involving chemical or thermal treatment to obtain reduced graphene oxide (rGO) with higher C/O atomic ratios while retaining the graphitic structure. However, the properties of graphene are still compromised and cannot fully match with the requirement (Pei and Cheng 2012). Thus, the synthesis of graphene often faces critical issues between controlled properties (quality) and the yield (quantity) of the resultant graphene materials (Lin et al. 2019). For example, exfoliation of chemically modified graphite oxide has a much higher yield and is being explored commercially. In contrast, mechanical exfoliation of graphite produces a poor yield of graphene, and its dispersion concentration is generally less than 0.5 mg/L. Therefore, it remains a major challenge to synthesize graphene materials in bulk with well-controlled properties that are cost-effective and environmentally friendly.

In the recent past, electrochemical exfoliation has been widely explored as a technique to produce single, bi-, or few layers (up to 10 layers) of 2D nanomaterials from bulk layered materials. This green technique takes place owing to its simplicity, quality/purity, ambient synthesis temperature, cost efficient mass production, and ease of processing in a single step within few minutes. The high production of graphene nanomaterials acquired via electrochemical anodic or cathodic exfoliation of graphite in an appropriate electrolyte provides power in terms of positive and negative potentials or biases applied on anodic and cathodic electrodes, respectively. That means, it has been done by providing an external electric field to bring about exfoliation of the layers and consequently leading to separation of mono layers or few stacks of atomic layers. Figure 1.2A, B shows the electrochemical exfoliation of graphene and graphene quantum dots by applying suitable power. In Fig. 1.2A, two graphite rods are immersed in an aqueous solution of ammonium sulfate $(NH_4)_2SO_4$ with 10 ml of hydrogen peroxide and connected to a positive and negative source of DC power supply of 10 V for 2 h. These two graphite electrodes are placed 4 cm apart from each other. The whole electrochemical system is placed in a water bath, where the temperature changes from 25 to 95 °C. The electrochemical process starts at the positive electrode, and the exfoliated graphene is filtered, washed with DI water, and dried in an oven at 75 °C for 12 h (Hossain and Wang 2016). Figure 1.2B demonstrates the electrochemical exfoliation of graphene quantum dots (GQDs) from a small piece of coke. In this two-electrode method, platinum is employed as the counter electrode and coke as the working electrode, which are dipped in a mixture of methanol and water electrolyte containing 0.1 mol/L of ammonium persulfate ($(NH_4)_2S_2O_8$) at pH 7 for 1 h. The exfoliated graphene quantum dots are collected via centrifugation (rpm 12,000), followed by purification. The obtained GQDs exhibit different fluorescence colors depending upon the current intensities supplied during the electrochemical process. Figure 1.2C shows the progression of graphene based material development with a historical timeline.

1.2 Electrochemical Exfoliation

There are various electrochemical exfoliation methods that have been developed and employed in different ways. An overview of some of the frequently used techniques is mentioned as follows:

1.2.1 Liquid-Phase Exfoliation

The liquid-phase exfoliation method is widely used by researchers and scientists in laboratories as well as in industries for cost-effective, high-yield production of 2D materials (viz., graphene) with balance qualities (Li et al. 2020a, b). It involves submerging a layered material, typically graphite, in an appropriate solvent, followed

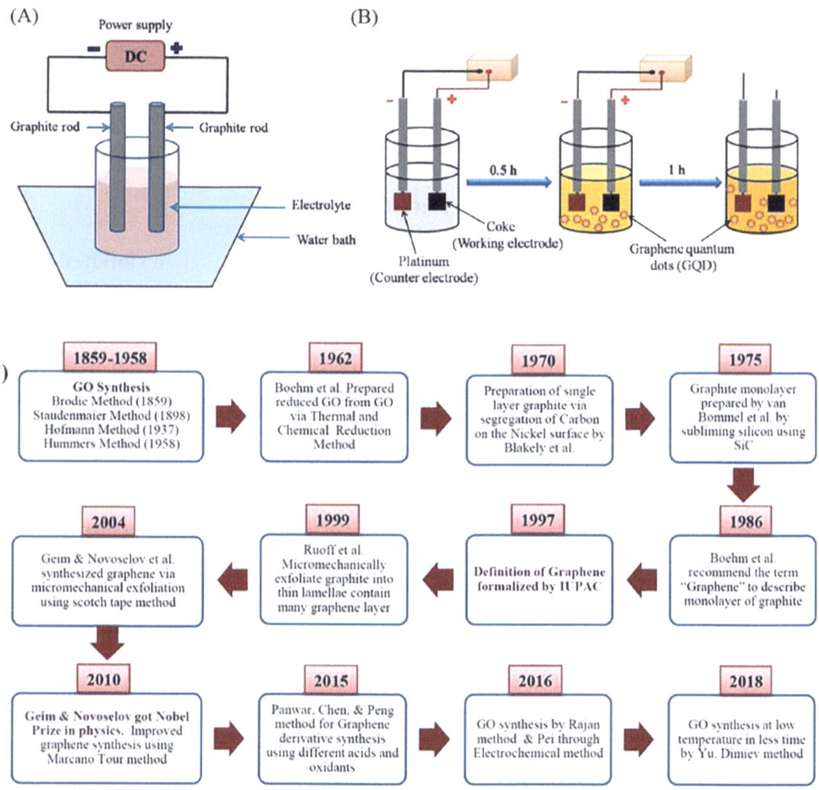

Fig. 1.2 Schematic diagram illustrating the electrochemical exfoliation of **A** graphene, **B** graphene quantum dots, and **C** timeline flow chart representing the evolution of graphene based materials

by extended sonication, which causes fragmentation of graphite leading to its interaction or intercalation with solvent molecules and results in the exfoliation of graphite to graphene. The quality and thickness of exfoliated 2D materials could be controlled and optimized by the duration of ultrasonication. Figure 1.3 illustrates the formation of scalable nanosheets by providing energy from ultrasonication in a controlled manner.

1.2.2 Electrochemical Bubbling

In this approach, the layered material is dispersed in an electrolyte solution, where it acts as the cathode. When an electric field is applied as voltage, it facilitates the ionic molecule or species towards the graphite electrode and gets intercalated with it, producing gas bubbles at the surface of the electrode and causing a localized pressure that induces the layers to exfoliate (Achee et al. 2018).

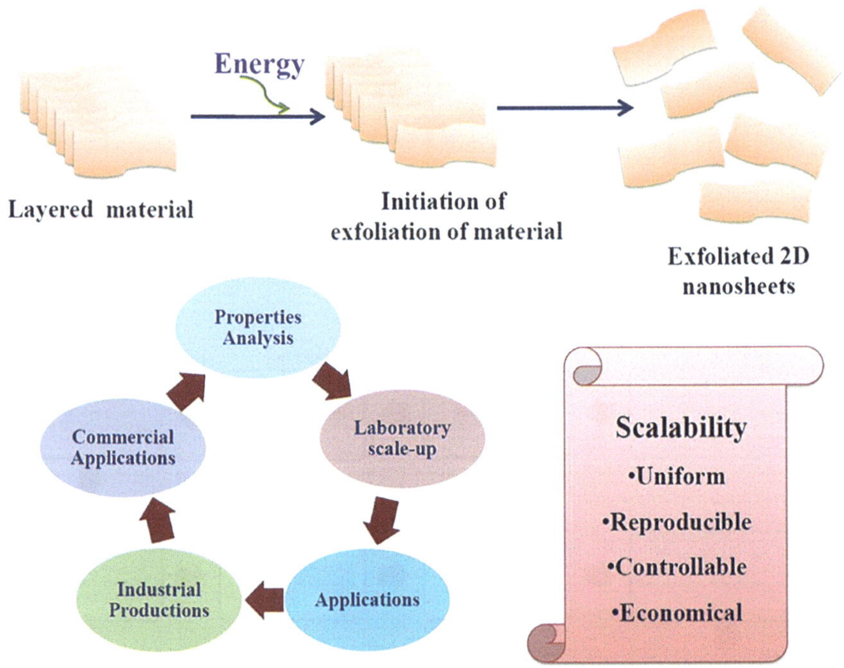

Fig. 1.3 Schematic diagram representing the fabrication of scalable nanosheets by means of energy provided by sonication and its application in laboratory and industry

1.2.3 Electrochemical Delamination

This method involves the use of a sacrificial layer, generally metallic (for example, a copper oxide layer), deposited onto the surface of the layered material. The sacrificial layer serves as an anode, while the layered material acts as the cathode. An electric field is applied, causing dissolution of the sacrificial layer and subsequent exfoliation of the laminated material. For example, Loh and co-workers illustrated that a single-layer graphene sheet is separated from copper (Cu) foil utilizing electrochemical delamination. During the delamination processes, it produces hydrogen bubbles due to water electrolysis, leading to mechanical force on the Cu substrate. This method is better than the chemical etching method because it is devoid of harmful chemicals during the whole procedure and the grown substrate could be available for reuse for further processing (Tang et al. 2020). The repeated growth and transport of WS_2 monolayer and graphene from gold (Au) and platinum (Pt) substrates, respectively, was demonstrated by electrochemical delamination (Gao et al. 2015, 2012). However, this method is explored mainly for the transfer of inorganic layers of 2D material.

Yang et al. demonstrated and explored the novel electrochemical delamination method, i.e., the dual intercalation/exfoliation method, for the production of excellent-quality graphene in large amounts (ca. 80%), by applying alternating current (AC) in an aqueous organic sulphate salt solution. In this method, they observed that exfoliation took place simultaneously at both electrodes (anode and cathode), facilitating an excellent graphene formation rate (20 g/h) with 1–3 layers of thick graphene flakes having a lateral size in the range of 1–5 μm. Moreover, in-situ reduction facilitated by the switch of voltage polarity during the exfoliation process yields graphene sheets of low defect density (Dali et al. 2019).

1.2.4 Pulse Electrochemical Exfoliation

In this process, a small pulse of the electric field applied rather than a continuous one (i.e., DC current) to the graphite electrode submerged in the suitable electrolyte (acidic, neutral or alkaline) lead to exfoliation of graphite in a controlled manner over the quality and thickness (Pogacean et al. 2019).

1.2.5 Electrochemical Intercalation

This technique is of great importance for electronic property tuning. The intercalation process involves the introduction of ions or molecules between the layers or the stratified material, which ultimately causes the layers to expand into a single layer.

Overall, it has been observed that the properties of exfoliated 2D materials depend on the appropriate selection of electrolyte solution, current density, applied potential, and duration of exfoliation. To attain desired results in terms of thickness, flake size, and quality, it is essential to optimize these parameters. Therefore, electrochemical exfoliation techniques are found to be more feasible methods than conventional methods. It produces 2D materials on a large scale and integrates them into various applications, such as sensing, energy storage, and electronics. Research and development in this area continues to explore new approaches and improve the effectiveness and extensibility of the exfoliation process.

1.3 Electrochemical Exfoliation Mechanism

Numerous electrochemical exfoliation methods based on a top-down approach have been developed. In this method, layered raw materials used as a working electrode in a two- or three-electrode system undergo exfoliation in an aqueous (water) or non-aqueous (organic) electrolyte. Strong electric fields created by applying a bias to the electrode system can cause the insertion of the ions generated as a result of electrolysis or solvated anions or cations from the electrolytes into the layered bulk raw material. Thus, it leads to interlayer spacing and breaks the bonds out-of-plane. The electrochemical exfoliation of graphene can be carried out by the following three methods based on applied bias and intercalation of ions: anodic, cathodic, and double electrode exfoliation (Li et al. 2020a, b). Besides these three methods, alternative and multistep intercalation techniques are also utilized for the production of high-quality 2D materials. Various techniques, such as bipolar intercalation, alternative intercalation, and sequential intercalation, have also been developed based on these electrochemical exfoliation mechanism principles to improve the exfoliation/delamination efficiency and production of 2D materials. The obtained 2D nanosheets are used as building blocks for the preparation of hybrid materials in the meantime (Ambrosi and Pumera 2018; Zhong and Swage 2012).

1.3.1 Anode Exfoliation Mechanism

In anodic exfoliation, two electrodes are used to carry out anion intercalation. Here, platinum (Pt) is employed as a negative electrode and layered materials as a positive electrode. The anions are induced to intercalate into the layered materials by applying current or positive voltage. These anions come from electrolytes or are developed as a result of electrolysis during the exfoliation process. Thus, ultrathin-exfoliated nanosheets are obtained either by ultrasonication or a mild shake.

The anode exfoliation is easily understood by taking the following examples: In anode exfoliation, graphite is used as an anode for graphene synthesis. The most frequently explored anions involved in anodic exfoliation are sulfate (SO_4^{2-}), organic sulfonate (RSO_3^-), nitrate (NO_3^-), phosphate (PO_4^{2-}), halide (X^-), perchlorate (ClO_4^-), tetrafluoroborate (BF_4^-), tungstate (WO_4^{2-}), surfactant, and some alkaline electrolytes (Li et al. 2020a, b). On applying an electric field, the anions present in the electrolyte migrate towards the graphite layers and subsequently expand and increase the interlayer spacing with gas production. Hence, graphene is obtained as a result of the weakening of the van der Waals force of attraction between graphite layers, leading to their peeling off.

In anodic exfoliation, the hydroxyl ion (OH^-) is formed by the reduction of water, and the peroxide ion (O_2^{2-}) is produced by adding hydrogen peroxide (H_2O_2), as presented in Eqs. 1.1 and 1.2 (Parvez et al. 2014). These ions attack the edge of the graphite and grain boundaries, followed by oxidation, which causes depolarization along with increasing the interlayer spacing of the graphite layer. Thus, facilitating the insertion of SO_4^{2-} ions across the graphite layer, where SO_4^{2-} ions produce SO_2 gas and water generates O_2 gas as consequences of the reduction of sulfate ions and the self-oxidation of water, respectively, as shown in Eqs. 1.3 and 1.4 (Parvez et al. 2014). The formation of these gases promotes the expansion and exfoliation of graphite layers, as shown in Fig. 1.4 (Hossain and Wang 2016).

Reduction of water at cathode:

$$4\,H_2O + 4\,e^- \rightarrow 4\,OH^- + 2\,H_2 \tag{1.1}$$

Reduction of H_2O_2:

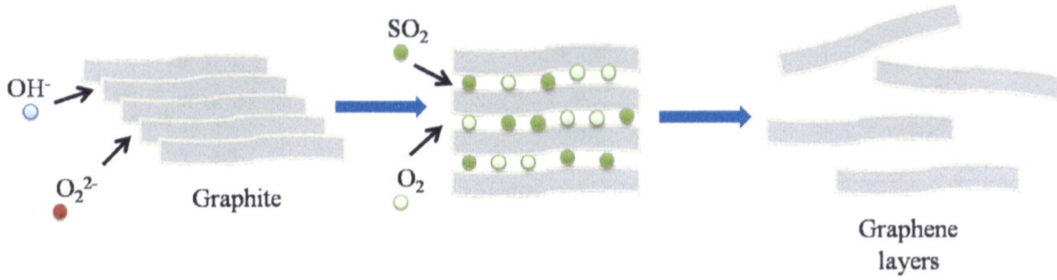

Fig. 1.4 Schematic diagram for proposed anodic exfoliation mechanism

$$H_2O_2 + 2OH^- \rightarrow O_2^{2-} + 2H_2O \tag{1.2}$$

Reduction of sulfate ion:

$$SO_4^{2-} + 4H^+ + 4e^- \rightarrow SO_2 + 2H_2O \tag{1.3}$$

Self-oxidation of water:

$$2H_2O \rightarrow 2OH^- + 2H^+ \rightarrow 2O + 4H^+ \rightarrow O_2 \tag{1.4}$$

Since the oxide-free and high-quality graphene produced by the anode exfoliation method possesses oxygen-containing functional groups with structural defects owing to the electrochemical oxidation of anode graphite.

1.3.2 *Cathodic Electrode Exfoliation Mechanism*

Anodic exfoliation is a good technique for exfoliating multilayer materials, however, it is not a very successful way to delaminate layered compounds since the anion intercalation requires a high positive voltage to oxidize the delaminated nanosheets. Cathodic exfoliation is an alternate process to create crystalline structures devoid of defects in answer to this problem. Cations from the electrolytic solution travelled into the interlayer gaps of multi-layered materials, causing them to expand when a negative voltage or current was supplied to the layered material. Co-intercalation of solvent molecules can occasionally occur along with this process, increasing the interlayer gap further and aiding in exfoliation. The negative voltage inhibits the material's oxidation when it is explored as a cathode electrode (Lu et al. 2023).

In the mechanism involved in cathode exfoliation, cations move towards the cathode graphite and reach the graphite layer in the electrolyte. Hydrogen gas is produced as a result of the cathodic reduction of the water electrolyte, leading to the expansion of the graphite cathode. Finally, graphene is produced from the peeling of graphite layers under the influence of both hydrogen and anions. The most commonly used electrolytes for cathodic electrode exfoliation are metallic ions (for example,

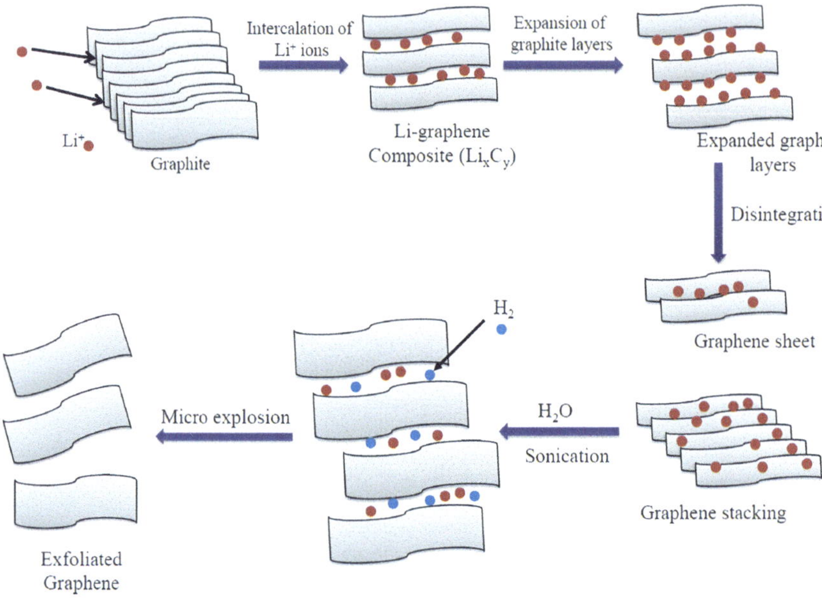

Fig. 1.5 Schematic diagram for proposed cathodic electrochemical exfoliation mechanism

lithium ion (Li^+)), sodium chloride (NaCl), dimethyl sulfoxide ($Me_2S = O$), and quaternary ammonium salt ($R_4N^+X^-$) (Li et al. 2020a, b).

In cathode electrode exfoliation, for example, graphite is used as the cathode and LiOH (lithium hydroxide) as an electrolyte, which is melted at 600 °C and electrolyzed by a current of 15 A for 30 min (Huang et al. 2012). Lithium ions (Li^+) move to the graphite cathode, are intercalated into the layers of graphite, and expand the graphite by forming a lithium graphite composite (Li_xC_y).

This composite is further ultrasonicated in distilled water, and hydrogen gas is generated, leading to the micro-explosion and exfoliation of graphite to graphene with fewer layers and defects, as presented schematically in Fig. 1.5 (Huang et al. 2012). However, the efficiency of cathode exfoliation is low as compared to the anode exfoliation. Besides these, the high cost and environmental issues associated with lithium ion solutions restrict their practical use.

1.3.3 Double Electrode Exfoliation Mechanism

When anodic or cathodic exfoliation is carried out by exploring a single working electrode for exfoliation without entirely using the counter electrode, it leads to chemical and energy losses. As a result, this limits the effectiveness of the exfoliation process. As an alternative, simultaneous dual-electrode exfoliation can significantly increase the pace of production and energy use efficiency. The anode's and cathode's layered materials may or may not be the same. The intercalation of anode and cathode electrode materials occurs by anions and cations respectively while performing the electrochemical process (Lu et al. 2023).

In this method, the cathodic and anodic potentials have merged. Initially, intercalation, followed by exfoliation, is carried out by alternating changes in applied potential. The double-electrode exfoliation system encompasses graphite as a cathode and anode in a suitable electrolyte and their exfoliation simultaneously by altering positive and negative bias. On the basis of power supply, the double-electrode exfoliation can be categorized as an alternating current (AC) and direct current (DC) exfoliation (Li et al. 2020a, b).

The high-quality graphene production by the double-electrode electrochemical exfoliation method was first reported in 2017, in which an aqueous solution of organic sulfate (tetra-n-butyl ammonium bisulfate, TBA.HSO_4) was employed as an electrolyte under AC current (Yang et al. 2017). The intercalation of anode and cathode graphite can be achieved by the action of applied AC. At the graphite-water interface, several oxygen-containing free radicals (hydroxyl radical OH· and oxygen radical O) are triggered by applying a high over potential (approximately +10 V) to the graphite anode.

The sulfate ions (SO_4^{2-}) migrated into the graphite layer by opening their edges, which was accompanied by the attack of OH· and O· free radicals at their intrinsic defect or boundaries of the graphite. The introduced sulfate anion gets reduced to SO_2 gas bubbles upon changing the anodic bias from positive to negative. These gas bubbles create a tremendous force inside the graphite to facilitate its exfoliation into graphene. Meanwhile, tetra-n-butyl ammonium cations (TBA^+) entered the graphite layers of the cathode side. On changing the working potential from negative to positive, sulfate ions get introduced inside the graphite layers. The gases (SO_2, H_2, O_2, and CO_2) are produced as a consequence of the electrochemical redox reaction and ultimately promote the expansion and exfoliation of graphite to graphene, as represented in Fig. 1.6 (Yang et al. 2017).

1.3.4 Electrochemical Exfoliation via Alternative Intercalation

It is important to note that anodic and cathodic exfoliations each have benefits and drawbacks. Consider graphene synthesis as an example of anodic exfoliation. This method produces a high yield (~70%) of graphene thin-sheet flakes with a large lateral size (~5 μm) at a very good production rate (Parvez et al. 2014). Oxygen-encompassing radicals (O and HO) generated by electrochemical water splitting attack the graphite sheets at high voltage (Rao et al. 2015) to form GO with a significant quantity of oxygen functional groups. However, cathodic exfoliation of graphite can ensure the graphene's quality while taking advantage of the reduction potential (Abdelkader et al. 2014a, b). An advanced method for combining the benefits of

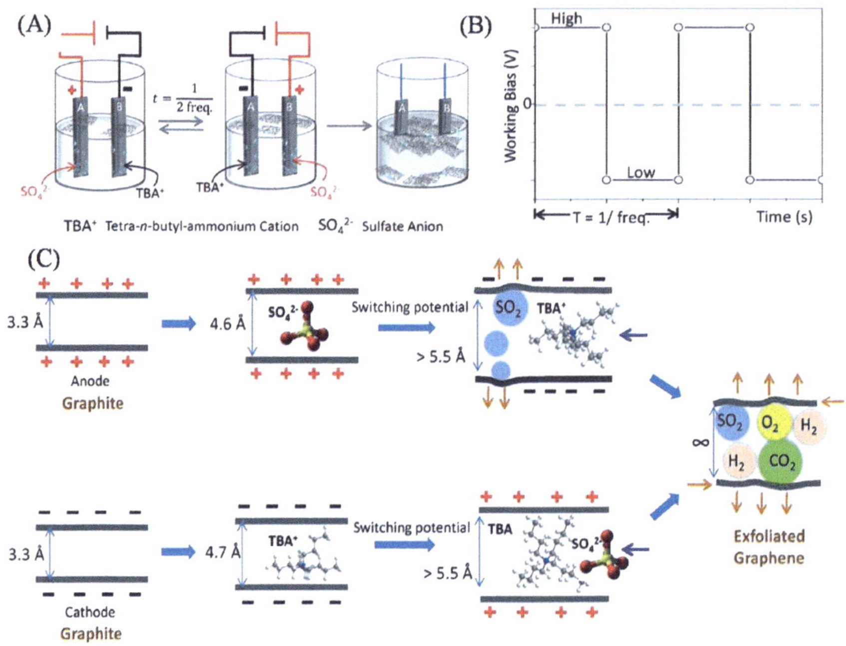

Fig. 1.6 A schematic representation of double electrode exfoliation mechanism involving **A** The graphite double electrode in TBA sulfate electrolyte, **B** Alternate changing of working bias, and **C** Exfoliation mechanism under the influence of working bias. Represented with permission from Yang et al. 2017

both anodic and cathodic exfoliation is alternative intercalation. In this method, the layered materials initially undergo anodic exfoliation by applying a positive potential first, and then cathodic exfoliation by introducing a negative potential. This process would significantly improve the quality and exfoliation efficiency of 2D materials.

1.3.5 Electrochemical Exfoliation via Multistep Intercalation

Multistep intercalation is a process where ions penetrate layered materials in multiple steps. Typically, the first step entails the intercalation of tiny ions to gradually widen the gap between adjacent layers. In order to create isolated 2D nano-flakes or nanosheets, bigger cations or anions were added to the materials in a subsequent stage. This increased the interlayer gap and broke the van der Waals forces. Multi-step intercalation technique with successive well-ordered ion absorption processes has substantially superior efficiency than one-step anodic or cathodic exfoliation (Lu et al. 2023).

1.4 Factors Affecting Electrochemical Exfoliation

There are following factors which affect the quality of produced graphene as a result of electrochemical exfoliation. These are listed below:

1.4.1 Types and Nature of Electrode

The nature and type of electrode play a significant role in the electrochemical exfoliation of graphene. Graphite is an allotrope of carbon that possesses a high melting point and is a good conductor. It is mostly employed as a precursor material for electrochemical exfoliation. It encompasses graphite as an electrode in the following forms: graphite rod, graphite powder, graphite foil, graphite flakes, highly oriented/ordered pyrolytic graphite (HOPG), and waste metal–carbon-based battery graphite (Danial et al. 2021).

The mechanical durability of the graphite electrodes plays significant roles in synthesizing GO since it affects GO amount, rate of exfoliation, and effectiveness. A small amount of GO formed because graphite foil and fakes exfoliated more quickly than HOPG and graphite rods. Additionally, compared to natural graphite, expanded graphite was exfoliated more effectively, but less amount of GO was produced. When graphite rods of high density are used for GO synthesis then functional groups are introduced mostly at their edges and basal plane devoid of it. Highly flexible, good strength possesses graphite paper or graphite foil which are able to tolerate the expansion of volume, was employed for pre-intercalation in 98% concentrated sulfuric acid and subsequently oxidation in 50% sulfuric acid.

The highest quality of GO was also prepared via electrochemical exfoliation of the pencil core which was explored as an electrode. It has been illustrated in many articles, that a single electrode like graphite rod/foil was utilized for synthesis of graphene sheets. But its quantity, degree of oxidation, and scalable formation are limited. Chen et al. observed the same problems and emphasized that when GO was synthesized from graphite flakes rather than graphite foil under identical experimental parameters, gives 2–3 layers of GO with small size. Therefore, the mechanically supported electrochemical method was adapted to synthesize GO from graphite flakes and resolved such issues. The electrochemically yielded GO displays less structural defect with good degree of oxidation compared to chemically synthesized GO. In case of graphite rod which is obtained from dry cell batteries showed GO production with oxidation degree expected to be less than 24.1% while GO produced via Hummers method displays approximately 43.5% of oxidation level (Liu et al. 2022). The synthesis of GO with variable characteristics using different types of graphite electrodes are tabulated in Table 1.1.

Table 1.1 Effect of graphite electrode over GO synthesis and their characteristics

Types of electrodes	Yield of GO	Lateral sizes	Thickness	I_D/I_G	C/O	Rate of synthesis	Refs.
Graphite rod, graphite fakes, graphite foil	11.97–18.84%	200–400 nm	0.6–2.1 nm	0.598	4.27–7.3	–	Zhang et al. (2018a, b)
Pencil cores	–	1–5 μm	3–9 nm	0.71	–	–	Liu et al. (2013)
Flexible graphite foil	~ 96 wt %	1–10 μm	1.0 nm	–	1.7	12 g/h	Pei et al. (2018)
Graphite rod extracted from Zn-C dry cell batteries	–	–	–	–	–	0.0305 g/h	Neog et al. (2022)

1.4.2 Types of Electrolytes

The quality, number of layers, and yield of graphene produced as a consequence of electrochemical exfoliation can also depend on the selection of electrolytes, like inorganic electrolytes (acidic, basic or neutral), organic electrolytes (tetra butyl ammonium sulphate), and surfactant. Different types of electrolytes may generate different types of graphene, like suspensions of graphene with surfactant, laminated graphene or complete exfoliated graphene (Danial et al. 2021). The different electrolytes produced different quantities of graphene oxide based on their capability of oxidation extent. Sulfuric acid is most commonly used electrolyte for electrochemical exfoliation of graphite to graphene oxide (GO) and augmented the functionality of GO. This is because of matching of interlayer distance of graphite with size of sulphate ions (SO_4^{-2}), fast rate of exfoliation, and formation of gases (SO_2 and O_2) further favoured the exfoliation process. Table 1.2 shows the effect of different electrolytes on the GO characteristics.

1.4.3 Concentration of Electrolytes

The concentration of ions present in the electrolyte affects the mechanism involved in the intercalation and expansion of graphite into graphene via electrochemical exfoliation. Higher intercalation of electrodes occurred in the presence of a higher

Table 1.2 The role of electrolytes on the GO quality and characteristics

Effect of electrolytes	Yield of GO	Lateral sizes	Thickness	C/O	I_D/I_G	Refs.
H_2SO_4	~5–8 wt %	1 – 40 μm	<3 nm	–	–	Su et al. (2011)
H_2SO_4, $HClO_4$	–	10 μm	2.0 nm	3.0 – 5.7	–	Tian et al. (2017a; b)
$HClO_4$	6.63 – 34.2%	–	–	1.43–11.55	0.16–0.75	Gurzeda et al. (2016)
H_2SO_4, $HClO_4$, HNO_3	49–50%	120–220 nm	2.7 – 5.6 nm	1.54–1.76	1.005–1.072	Sahoo et al. (2016)
H_2SO_4, H_3PO_4	–	1–5 μm	3–9 nm	–	0.71	Liu et al. (2013)
LiCl, NaCl, KCl, RbCl, CsCl	–	1–3 μm	5 nm	57.8	0.06–0.12	Dalal et al. (2021)
$(NH_4)_2SO_4$, H_2SO_4, Na_2SO_4, $MgSO_4$	>80%	0.5–3.0 μm	1.08 nm	–	0.85	Parvez et al. (2016)
$(NH_4)_2SO_4$, $NH_3{\cdot}H_2O$	34 wt %	–	1.9 – 3.5 nm	7.15	1.33	Lou et al. (2016)
> 95% H_2SO_4, 0.1 M $(NH_4)_2SO_4$	71%	–	<1.5 nm	4.6	1.48	Cao et al. (2017)
NaOH, PTA	87.3%	4.2 μm	2.0 nm	4.02	0.98	Wang et al. (2018)
0.2 M of sodium citrate (pH 3.7), HNO_3	–	–	1.1 nm	4.0	–	Abdelkader et al. (2014a; b)
Oxone (potassium monopersulfate, $KHSO_5{\cdot}0.5KHSO_4{\cdot}0.5K_2SO_4$)	~60.1%	1–5 μm	2.5 nm		~1.24	Tian et al. (2017a; b)

concentration of electrolyte at their active site. The graphite oxide produced electrochemically in dilute acid (less than 10 M) is amorphous with less oxidation extent on their basal plane but more carboxylic group at their edges owing to water hydrolysis and CO_2 formation. On the other hand, more crystalline and hydroxyl groups are introduced in graphite oxide in a higher concentration of acid electrolyte as a consequence of intercalation and the addition of oxygen onto the basal plane of the graphite sheet (Lowe et al. 2019). One of the reports illustrated that the partial electrochemical exfoliation of graphite flakes is observed experimentally in the acid electrolyte containing a mixture of sulfuric acid (95%) and nitric acid (65%) of variable volume fraction in different ratios. The extent of exfoliation rises by increasing the volume fraction of sulfuric acid (Gurzeda and Krawczyk 2019). In the case of surfactant as an electrolyte, critical micellar concentration (CMC) is needed for the production of micelles, which in turn interact with the electrode and determine the stability of graphene suspension in the selected electrolyte (Danial et al. 2021). Table 1.3 summarized the effect of electrolyte concentration.

1.4.4 Applied Working Bias/potential

The various working potentials or biases applied to the electrode in a suitable electrolyte can greatly influence the electrochemical exfoliation mechanism of graphene (Danial et al. 2021), as tabulated in Table 1.4.

Electrochemical exfoliation of a graphite rod in the mixture of sulfuric acid and nitric acid (H_2SO_4 and HNO_3) of 0.5 M in a ratio of 3:1 produces graphene oxide (GO) with few-layered graphene and multi-layered graphene. Mass production of GO is observed by increasing the anodic bias ranging from 2.5 to 6 V in the mixture of H_2SO_4 and HNO_3 (Coros et al. 2016).

1.4.5 Applied Electrical Current

More structural defects are created in the graphene oxide with increasing electrical current density in the electrochemical system, along with a higher number of carboxylic groups at their edges and hydroxyl/epoxy groups at the basal plane, but in a smaller number. Moreover, higher electrical current intensity gives small GO particles in a bulk in the aqueous electrolyte. The intensity of electrical current influences water hydrolysis, intercalation of anions and evolution of gas produced as a result of electrochemical processes (Sun et al. 2022).

Table 1.3 The effect concentration of electrolytes on to the GO characteristics

Concentration of electrolytes	Yield of GO	Lateral sizes	Thickness	C/O	I_D/I_G	Refs.
H_2SO_4 (10−100 wt %)	–	1−10 μm	1.0 nm	1.7–5.7	–	Pei et al. (2018)
H_2SO_4 (2, 5, 7.1, 10, 11.6, 14, and 16 M)	~22.5%	1 − 5 μm	1.7−2 nm	–	~1.2 −1.6	Lowe et al. (2019)
H_2SO_4:HNO_3 (3:1) (0.5, 1.0, 2.0, and 3.0 M)	7−9%	–	9.16 nm	–	1.22−1.24	Coros et al. (2016)
$HClO_4$ (4.6, 7.0, 9.2, and 11.6 M)	19.1−37.3%	–	–	–	–	Tian et al. (2017a, b)
$HClO_4$ (0.5, 1.0, 1.5, and 2.0 M)	40−44%	107−140 nm	3−6 layers	1.59−1.67	0.96−1.04	Sahoo and Mallik (2015)
$(NH_4)_2SO_4$ (0.5 and 1.0 mol/L)	29.88−32.22%	–	10−11 μm	3.26−3.31	1.66−1.69	Chen et al. (2021a; b)

Table 1.4 Preparation of graphene via electrochemical exfoliation of graphite in different electrolytes with or without surfactant at variable applied potential

S. No	Working Electrode (WE)	Applied Potential/ Current	Electrolyte	Product (P)	Thickness (nm)	Size (nm)	Number of layers	Refs.
1	Graphite	10 V	H_3PO_4,	Graphene	–	8.5, 9.1, 9.5	Multi	Htwe et al. (2018)
			H_2SO_4,			8.3, 8.7, 12.4		
			$(NH_4)_2SO_4$			8.4, 9.2, 9.6		
2	Graphite	2–10 V (5 min)	Na_2SO_4	MoS_2@graphene	–	–	Single and multi	Ali et al. (2019)
3	Graphite	10 mA	SDBS	Graphene	0.6	–	2	Wang and Zhang (2019)
			NaCl		1.1		3	
			DTAB		1.6		4	
			PVP		7.0		20	
			NaOH		9.0		30	
4	Graphite	−1 V and +2 V	SDS	Graphene flakes	–	500	Single	Alany-alioglu et al. (2011)
5	Graphite foil	10 V	$NaBF_4$ (aq.)	F@graphene	1.6	12,000	≤ 3	Zhou et al. (2018)
6	Graphite foil	15 V	$Na_2S_2O_3/H_2SO_4$ (5:1)	S-doped graphene	0.79 and 0.85	–	Single	Lee et al. (2019)
7	Graphite foil	−5 V	$LiClO_4$ in PC → TBA → aryl diazonium salt	Functionalized graphene flakes	–	–	<5	Zhong and Swager (2012)

(continued)

Table 1.4 (continued)

S. No	Working Electrode (WE)	Applied Potential/ Current	Electrolyte	Product (P)	Thickness (nm)	Size (nm)	Number of layers	Refs.
8	Graphite foil	+10 V	$(NH_4)_2SO_4$ (aq) + TEMPO	Graphene	2.08	–	2	Zhao et al. (2019b)
9	Graphite foil	5 V	$(NH_4)_2SO_4 + NH_3{\cdot}H_2O$	Graphene based bio-mimetic skin	–	–	Multi	Tang et al. (2019)
10	Graphite foil	+10 V	Na_2SO_4	Graphene	1–1.2	1850	1, 2, and multi	Sharif et al. (2019)
			$(NH_4)_2SO_4$			1580		
			NH_4NO_3			5210		
			$(NH_4)_2SO_4 + (NH_4)_2HPO_4$			2360		
11	Graphite foil	15 V	Sodium-p-styrenesulph-onate (aq)	Polystyrene sulphonatefunctionalized graphene	–	>1	–	Zhao and Cai et al. (2019a)
12	Graphite powder	+20 V (50 s) and −20 V (20 s)	H_2SO_4	Graphene	3	–	–	Lim et al. (2020)
13	Natural graphite flakes	+10 V	H_2SO_4	Graphene	0.86 and 1.5	10,000	1–4	Parvez et al. (2013)
14	Paraffin-coated graphite foil	3	NaOH	Graphene Flakes	0.64 and 1.27	–	<3	Wang et al. (2017)

1.4.6 Exfoliation Temperature and Amount of Hydrogen Peroxide

The exfoliation temperature and amount of hydrogen peroxide both influence the exfoliation mechanism and regulate the defects, stability, and quality of graphene prepared by electrochemical exfoliation. The oxygen-containing functional groups or defects created at the edges of graphene decreased as the electrochemical exfoliation temperature increased from ambient to 95 °C. As the amount of hydrogen peroxide increases and the exfoliation temperature increases, the thermal stability of graphene rises (Hossain and Wang 2016). However, high graphene oxide (GO) production is reported at ambient temperatures. It claimed that high-temperature resonance occurred due to the negative thermal expansion coefficient of graphene. On increasing temperature, which provides energy to the ions as heat and facilitates the number of ions to intercalate into the graphite layers, the van der Waals weakens and consequently increases the interlayer spacing (Hsieh and Hsueh 2016). The exfoliation yield is enhanced up to 4.5 times on account of the high temperature compared to the ambient temperature (Tripathi et al. 2015). However, the amount of GO decreased while heating owing to bubble formation (Hossain and Wang 2016). Thus, the optimization of temperature is needed to improve the GO yield (Fang et al. 2019).

1.5 Influence of Electrolytes onto the 2D Materials

Htwe and co-workers examined the role of three different nature of electrolytes such as phosphoric acid (H_3PO_4), sulphuric acid (H_2SO_4), and ammonium sulphate ($(NH_4)_2SO_4$) on to the electrochemical exfoliation of graphene with varied sonication time. They have observed that the graphene produced in sulphuric acid electrolyte after sonication (45 min) was more stable with small crystallite size possesses lesser defect and excellent electrical conductivity (Htwe et al. 2018). On the other hand, molybdenum disulfide decorated graphene (MoS_2@graphene) composite synthesized electrochemically (Ali et al. 2019), where graphite rod with MoS_2 crystal served as the anode and platinum wire act as the cathode. They have observed that the MoS_2@graphene composite exhibited low electrical resistance and good electrochemical stability by evaluating specific capacitance 227 F/g in sodium sulphate (Na_2SO_4) salt electrolyte of 1 M concentration at initial applied voltage 2 V for 5 min followed by final voltage 10 V for 5 min.

Table 1.4 summarized the various electrochemical routes to exfoliate graphite into graphene by employing a different electrolyte, potentials bias and surfactant which weakened the van der Waals interaction between the layers of graphite via introducing ions or molecules in between the layers. The different geometries of graphite, for example, graphite flakes, foils, rods, powders, or plates, can be employed as working electrodes (WE) in liquid electrolytes and platinum rod was used as the counter electrode (CE).

The formation of graphene with controlled size and thickness was achieved by employing variable working potential on the anodic graphite in the presence of a surfactant, for example, sodium dodecyl sulfate (SDS), as illustrated in Fig. 1.7 (Danial et al. 2021; Alanyalioglu et al. 2011).

Initially, positively charged graphite triggered the anionic surfactant and intercalated the graphite layers via ionic interaction. Thus, the interlayer spacing increases as a consequence of the weakening of the van der Waal force of attraction between the layers. Micelles of surfactant are formed in the solution when the concentration of SDS surfactant exceeds CMC (critical micellar concentration). Then, these micelles interact with graphite layers via the negatively charged head of the SDS surfactant with the positively charged graphite. The expansion of interlayer spacing between the graphite layers is further enhanced as the micelles grow. During electrochemical exfoliation, anodic graphite was changed to a cathodic electrode by switching the working bias of the electrode to negative. The surfactant-intercalated graphite is again electrochemically exfoliated into the stable graphene-SDS suspensions owing to the

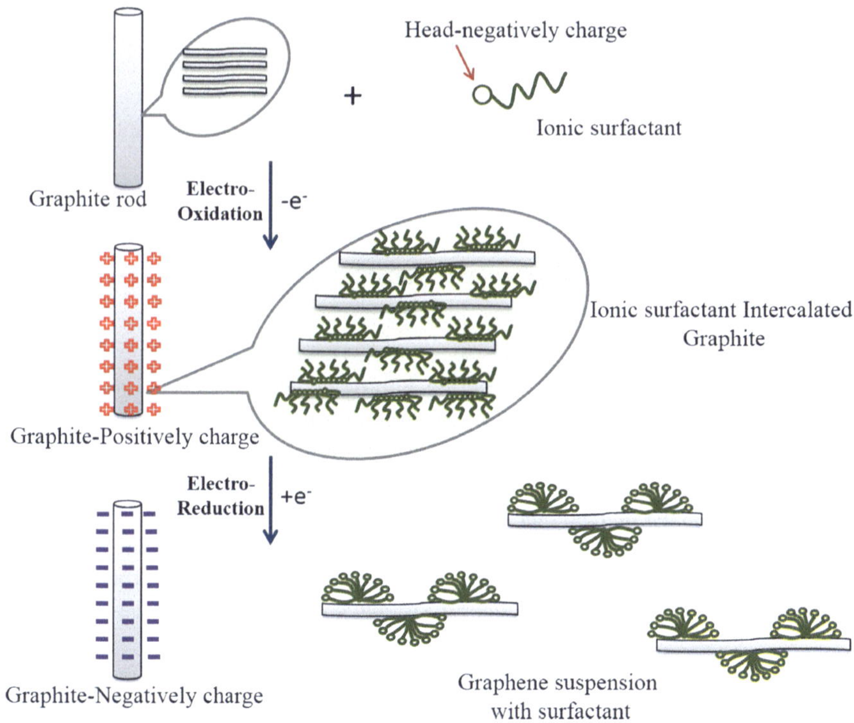

Fig. 1.7 A schematic diagram displaying the general mechanism of electrochemical exfoliation of graphene

electrostatic repulsion of negatively charged graphite cathode and anionic surfactant (Alanyalioglu et al. 2011).

1.6 Influence of Intercalation on the 2D Materials

As shown in Fig. 1.8, in cathodic exfoliation, the positively charged ions (e.g., Li^+) in the electrolyte would migrate towards the cathodic (negative bias) graphite electrodes. On the other hand, in anodic exfoliation, negatively charged ions (e.g., SO_4^{2-}) might be attracted to anodic electrodes (positive bias). In the case of graphite, its sheets or layers are intercalated with ions, electrolyte molecules, or co-intercalating species present in electrolytes. It has been anticipated that the electrochemical reactions would be appropriate for driving forces to disrupt the van der Waals forces among graphene sheets, causing the structural expansion and exfoliation of graphite (Yu et al. 2015).

The graphene nanomaterials with different defect densities, numbers of layers, lateral sizes, and oxygen contents have been obtained by regulating operational parameters like applied currents and electrical potentials, the composition of electrolytes, and processing time.

Furthermore, during electrochemical exfoliation, concurrent chemical reactions with functionalizing agents may take place to accomplish in situ chemical doping or functionalization of graphene nanomaterials and acquire numerous graphene-based nanocomposite materials.

1.7 The Application of Electrochemically Exfoliated 2D Materials

Numerous energy conversion and storage technologies, such as water splitting, supercapacitors, metal-air batteries, lithium-ion, sodium-ion, and lithium-sulfur batteries, are now exhibiting significant potential for use in future energy supplies (Wang et al. 2020a; Ji et al. 2020; Shen et al. 2018; Wang and Liu 2020b). The development of these energy systems depends on the preparation of electrode materials. 2D materials have sparked a lot of passion because of their characteristic qualities. Herein, we discuss the more recent uses of 2D materials that have been electrochemically exfoliated and utilized in common energy storage and conversion devices.

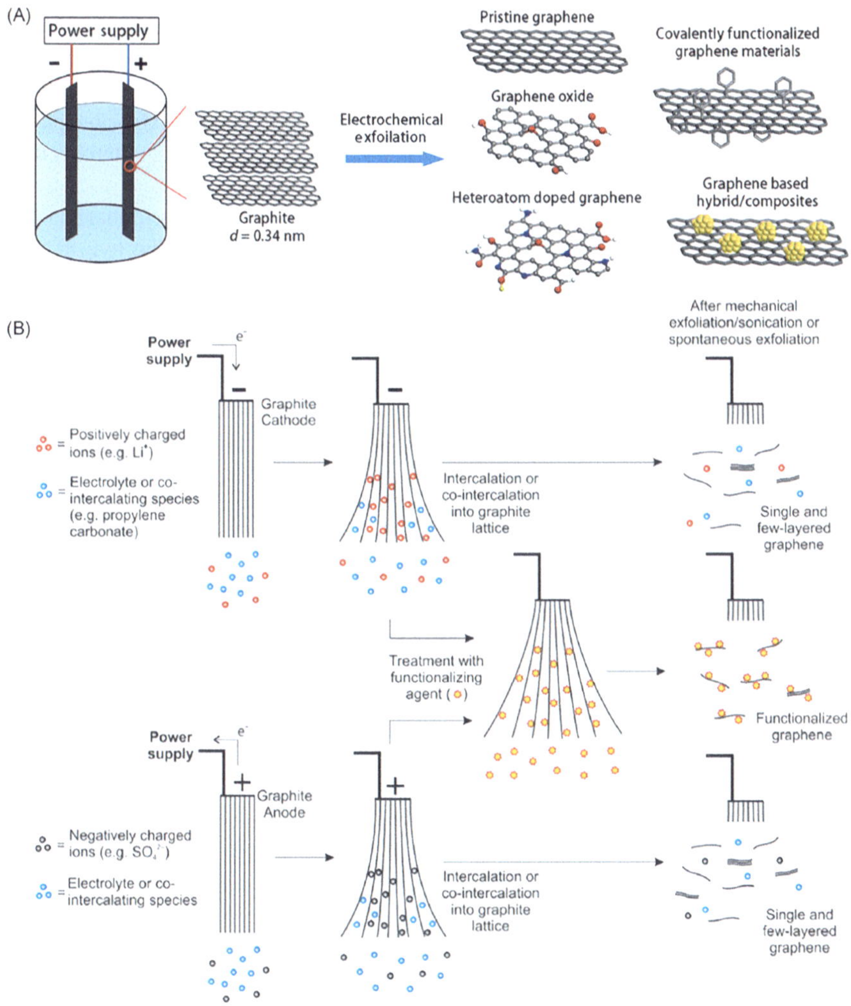

Fig. 1.8 Schematic diagrams presenting **A** Electrochemical exfoliation of graphite to create doped graphene, and **B** sketch of electrochemical anodic and cathodic graphite exfoliation. Represented with permission from (Yu et al. 2015)

1.7.1 2D Materials for Energy Storage

By adjusting the applied voltage, Ji et al. utilized the cation insertion approach to produced black phosphorene (BP) with tunable-layer (Huang et al. 2017). The cathode in this experiment was BP crystal, and the anode was Pt. DMF was used to dissolve tetrabutylammonium hexafluorophosphate, $Bu_4N^+PF_6^-$ (TBAP) to prepare the electrolyte. The interplanar distance measured by HRTEM was found to be 0.333 nm, which is equivalent to the (021) plane of exfoliated BP. Two layers with thicknesses ranging from 0.76 to 0.79 nm were revealed by AFM images (Batmunkh et al. 2016). With a 100 mA g^{-1}current density, the produced phosphorene can be used as anode materials for sodium-ion batteries, providing a capacity of 1968 mAh g^{-1}. It demonstrated the promising potential of the electrochemically exfoliated phosphorene by maintaining a capacity of 603.3 mAh g^{-1} after 100 cycles when cycling at 1500 mAh g^{-1}.

Shuai et al. prepared a phosphorene–graphene sandwich like hybrid material by using a concurrently delamination technique (Shuai et al. 2019). The hybrid was chemically activated to introduce P–O–C and P–C bonds, resulting in an excellent specific capacity (2.311 Ah g^{-1}) at 100 mA g^{-1}current density, enabling good sodium storage execution. After 100 cycles, the capacity seized at 83.9%. The advance activity was ascribed to the graphene layers' flexible space, which was coated on the black phosphorene surface and reduced the phosphorene's volumetric expansion. Furthermore, the P–C and P–O–C bonds immobilized phosphorene and acted as bridges for electron transport. Sun et al. reported a scalable and high-production exfoliation process that produced 1–5 layers of graphene with a 76% yield in ternary eutectic melts including urea, ammonium nitrate, and acetamide (Zhang et al. 2018a, b). A large area supercapacitance of 0.120 Fcm^{-2} and 97.2% high retention were achieved by graphene based solid-state flexible supercapacitors after 10,000 cycles.

The 2D MXene Ti_3C_2 nanosheets were prepared by anodic exfoliation of Max-phase Ti_3AlC_2 using an etching technique that was free from fluoride (Yang et al. 2018). A high yield of carbide nano-flakes (>90%) extracted by the removal of Al and the intercalation of tetrabutyl ammonium cation (TBA^+). The AFM result revealed that the thickness of MXene equivalent to mono- or bilayer, with a size of up to 18.6 μm. In all-solid-state supercapacitor, these exfoliated MXene nanosheets demonstrated excellent performance, displaying high volumetric and areal capacitances of 439 Fcm^{-3} and 220 $mFcm^{-2}$ at a scan rate of 0.01 V/s, respectively. At a current density of 0.005 A/cm^2, long-standing stability was also achieved with 94.2% retention of capacitance up to 10,000 discharge–charge cycles.

Feng and co-workers prepared graphene with a high yield of exfoliation of 80% was obtained by exfoliating graphite using alternating current (Yang et al. 2017). The SEM and AFM data reveals the lateral diameters of exfoliated graphene (EG) were ranged from 1 to 5 μm. The thin EG could be uniformly coated on the surface of $LiFePO_4$ by combining the dispersion of EG with commercial $LiFePO_4$ via sonication. This showed a capacity of 0.157 Ah/g at the 20th cycle. The capacity increased to 0.167 Ah/g after activation at the 200th cycle, and it remained steady until the 500th cycle.

Liu and colleagues formed micro-supercapacitors with a 160 Fcm^{-3} volumetric capacitance and 98.1% capacitance even after 250,000 cycles by preparing graphene nanosheets doped with chlorine using an easy-to-use yet environmentally friendly electrochemical exfoliation method (Liu et al. 2022).

1.7.2 Energy Conversion Using 2D Materials

Electrochemical delamination is a significant process that yields 2D materials for energy conversion, besides to energy storage. Protons and functional groups interacted on the surface of the double transition metal MXene ($Mo_2TiC_2T_x$), as demonstrated by Wang and colleagues, which allowed for the production of numerous Mo vacancies in the outside layers of the nanosheets produced by employing electrochemical delamination (Zhang et al. 2018a, b). Pt foil was employed as the counter electrode throughout the exfoliation procedure, and this allowed the molybdenum (Mo) vacancies to immobilize individual Pt atoms. As compared to pure $Mo_2TiC_2T_x$ nanosheets, the as-produced $Mo_2TiC_2T_x$–PtSA provided increased catalytic performance toward the hydrogen (H_2) evolution reaction. The mass activity was approximately 40 times larger than the commercial Pt/C catalyst, and enormously stumpy over potentials of 0.30 and 0.77 V were attained to give the current density of 0.010 and 0.100 A cm^{-2}. The positively charged Pt single atoms' strong covalent bonds with the ultrathin MXene nanosheets may be the cause of the exceptional performance.

Molybdenum disulfide (MoS_2) quantum dots were synthesized by Shaijumon and co-workers using a one-step electrochemical method (Gopalakrishnan et al. 2015). The applied DC voltage might be changed to effectively control the size distribution. With an initial potential of approximately 210 mV, these MoS_2 quantum dots demonstrated strong catalytic performance for the hydrogen evolution process (HER). Zhang and co-workers synthesized graphene-MoS_2 composite materials by using a concurrent electrochemical exfoliation technique (Li et al. 2018). For HER, a 2D composite with an ideal composition of 38.3% graphene needs just 0.125 V to reach a current density of 0.01A cm^{-2}. The measured Tafel value of 41 mVdec^{-1}reveals a Volmer Heyrovsky HER mechanism.

1.8 Summary and Future Prospect

Based on the literature survey, it is clear that among the various existing procedures for graphene synthesis through a top-down approach, for example, electrochemical exfoliation is a well-accepted technique. Moreover, it is feasible, fast, simple, cost-effective, and one of the most efficient techniques for nanomaterial synthesis, especially nanosheets of graphene of high quality and high yield. On the other hand, the proper choice of electrode materials, bias voltage, electrolytic solution or dispersion medium, exfoliation time, and temperature are also important factors for governing the size, number of layers, quality, and yield of the products as formed by carrying out electrochemical exfoliation of graphite, as discussed in the chapter. However, the formation of good-quality single-layer graphene is still challenging from GO, i.e., through the controllable oxidation of graphite. Usually, single-layer graphene always provides better results than multi-layer in the case of forming composites with other layered materials for a wide range of applications, such as sensing, energy storage, catalysts, conducting films, anti-corrosion, anti-inflammatory, hybrid membranes, etc. Depending upon the compatibility and characteristics of the materials, various excellent exfoliation approaches have been established, such as liquid-phase exfoliation, electrochemical bubbling, electrochemical delamination, pulse electrochemical exfoliation, electrochemical intercalation, etc. Electrochemically exfoliated graphene or other 2D materials like black phosphorene, MXene, MoS_2, siligene, and dichalcogenides produced possess high purity and excellent quality, which makes them applicable to industrial scale-up. This chapter has not only discussed the different mechanisms involved in the electrochemical exfoliation of graphite to graphene but also offered an insight into the challenges faced by bottom-up and top-down methods over the advantages of the electrochemical exfoliation method, which are fascinating to fabricate 2D materials for use in a variety of fields, particularly energy conversion and storage, such as metal-air batteries, hydrogen evolution, CO_2 reduction, metal-ion batteries, nitrogen reduction reaction, oxygen evolution, and so forth, depending on their quantity and quality. Despite the many benefits of the electrochemical exfoliation technique, there are also issues with this technology. The following crucial points must to be taken into account for upcoming research. First and foremost, when performing anionic exfoliation, it is important to pay close attention to how a high working voltage may oxidize the raw materials or exfoliated nanosheets. Additionally, a high voltage has the potential to cause gaseous species and electrolyte breakdown, which would peel the layered materials that have not been exfoliated away from the working electrode. The efficiency of exfoliation would also be impacted by electrolyte ingestion. Secondly, it's still unclear how electrochemical exfoliation works. With the help of an electric field, anions and cations would typically intercalate in layered materials and break the interlayer van der Waal force of attraction. The exfoliation process may potentially be aided by solvent molecules with a significantly bigger radius. The gaseous species created by the solvent's breakdown, which would likewise cause the interlayers to expand, is another significant component. Further work is needed to develop in situ methodologies for the systematic elucidation of electrochemical exfoliation mechanisms and process optimization. Moreover, the precursors for the electrochemical delamination process are of excellent quality. Bulk crystals were typically utilized as the initial materials, and in order to complete the electron capture process, they needed to be conductive or semi conductive. The issue of high contact resistance and easy decompaction in the electrolyte, which is particularly problematic when gaseous bubbles are present, is addressed by powder crystals. Consequently, it is important to investigate suitable precursors or delamination techniques in response to various 2D material types. Furthermore, essential to electrochemical exfoliation is choosing the right electrolyte. To enable simple intercalation, the solvated anions and cations must first fit the interlayer space of the layered starting materials. It is imperative that the solvent exhibits appropriate viscosity and matching surface tension in proximity to the generated 2D materials to inhibit the restacking of exfoliated nanosheets. Thirdly, in order to support the delamination/exfoliation process and enable fast ion consumption at high potentials, the electrolyte needs to have an excellent ionic conductivity. Recent developments have revealed the enormous promise of electrochemical exfoliation in production of 2D materials. It is expected that the controlled synthesis of single layer or a few layered 2D nanosheets with high-quality and high-yield are achieve by optimized electrochemistry processes, which will aid future energy storage and conversion applications. In this way, this chapter may unlock the opportunities in the field of 2D material development and their wide range applications.

Acknowledgements The authors acknowledge the Bhabha Atomic Research Centre, Mumbai and Defence Materials and Stores Research & Development Establishment (DMSRDE), Kanpur for providing facility to accomplishing this work.

References

Abdelkader, A.M., Kinloch, I.A., Dryfe, R.A.W.: Continuous electrochemical exfoliation of micrometer-sized graphene using synergistic ion intercalations and organic solvents. CS Appl. Mater. Interfaces **6**, 1632–1639 (2014a)

Abdelkader, A.M., Kinloch, I.A., Dryfe, R.A.W.: High-yield electro-oxidative preparation of graphene oxide. Chem. Commun.commun. **50**(61), 8402–8404 (2014b)

Achee, T.C., Sun, W., Hope, J.T., Quitzau, S.G., Sweeney, C.B., Shah, S.A., Habib, T., Green, M.J.: High yield scalable graphene nano sheet production from compressed graphite using electrochemical exfoliation. Sci. Rep. **8**, 14525 (2018)

Alanyalioglu, M., Segura, J.J., Oro-Sol, J., Casan-Pastor, N.: The synthesis of graphene sheets with controlled thickness and order using surfactant-assisted electrochemical processes. Carbon **50**, 142–152 (2011)

Ali, G.A.M., Thalji, M.R., Soh, W.C., Algarni, H., Chong, K.F.: One step electrochemical synthesis of MoS_2/graphene composite for supercapacitor application. J. Solid State Electrochem.electrochem. **24**, 25–34 (2019)

Ambrosi, A., Pumera, M.: Exfoliation of layered materials using, electrochemistry. Chem. Soc. Rev. **47**, 7213–7224 (2018)

Ambrosi, A., Chua, C.K., Bonanni, A., Pumera, M.: Electrochemistry of graphene and related materials. Chem. Rev. **114**, 7150–7188 (2014)

Batmunkh, M., Erdene, M.B., Shapter, J.G.: Phosphorene and phosphorene-based materials-prospects for future applications. Adv. Mater. **28**, 8586–8861 (2016)

Cao, J., He, P., Mohammed, M.A., Zhao, X., Young, R.J., Derby, B., Kinloch, I.A., Dryfe, R.A.W.: Two-step electrochemical intercalation and oxidation of graphite for the mass production of graphene oxide. J. Am. Chem. Soc. **139**(48), 17446–17456 (2017)

Chen, J., Perez-Page, M., Ji, Z., Zhang, Z., Guo, Z., Holmes, S.: One step electrochemical exfoliation of natural graphite flakes into graphene oxide for polybenzimidazole composite membranes giving enhanced performance in high temperature fuel cells. J. Power. Sources **491**, 229550 (2021a)

Chen, Z., Sui, Y., Li, J., Kang, H., Wang, S., Zhao, S., Gao, X., Peng, S., Jin, Z., Liu, X., Zhang, Y., Yu, G.: Conversion of the stacking orientation of bilayer graphene through high-pressure treatment. Carbon **172**, 480–487 (2021b)

Coros, M., Pogacean, F., Rosu, M.C., Socaci, C., Borodi, G., Magerusan, L., Biris, A.R., Pruneanu, S.: Simple and cost-effective synthesis of graphene by electrochemical exfoliation of graphite rods. RSC Adv. **6**(4), 2651–2661 (2016)

Dalal, M.H., Lee, C.Y., Wallace, G.G.: Cathodic exfoliation of graphite into graphene nanoplatelets in aqueous solution of alkali metal salts. J. Mater. Sci. **56**(4), 3612–3622 (2021)
Dali, S.G., Paredes, J.I., Munuera, J.M., Rodil, S.V., Alonso, A.M., Tascon, J.M.D.: An aqueous cathodic delamination route towards high quality graphene flakes for oil sorption and electrochemical charge storage applications. Chem. Eng. J. **372**, 1226–1239 (2019)
Danial, W.H., Norhisham, N.A., Noorden, A.F.A., Majid, Z.A., Matsumura, K., Iqbal, A.: A short review on electrochemical exfoliation of graphene and graphene quantum dots. Carbon Lett. **31**(3), 371–388 (2021)
Fang, S., Lin, Y., Hu, Y.H.: Recent advances in green, safe, and fast production of graphene oxide via electrochemical approaches. ACS Sustain. Chem. Eng. **7**(15), 12671–12681 (2019)
Gao, L., Ren, W., Xu, H., Jin, L., Wang, Z., Ma, T., et al.: Repeated growth and bubbling transfer of graphene with millimetre-size single-crystal grains using platinum. Nat. Commun.commun. **3**, 699 (2012)
Gao, Y., Liu, Z., Sun, D.M., Huang, L., Ma, L.P., Yin, L.C., et al.: Large-area synthesis of high-quality and uniform monolayer WS2 on reusable Au foils. Nat. Commun.commun. **6**, 8569 (2015)
Geim, A.K., Novoselov, K.S.: The rise of graphene. Nat. Mater. **6**, 183–191 (2007)
Gopalakrishnan, D., Damien, D., Li, B., Gullappalli, H., Pillai, V.K., Ajayan, P.M., Shaijumon, M.M.: Electrochemical synthesis of luminescent MoS_2 quantum dots. Chem. Commun.commun. **51**, 6293–6296 (2015)
Gui, G., Li, J., Zhong, J.: Band structure engineering of graphene by strain: first-principles calculations. Phys. Rev. B **78**, 075435 (2008)
Gurzeda, B., Florczak, P., Kempinski, M., Peplinska, B., Krawczyk, P., Jurga, S.: Synthesis of graphite oxide by electrochemical oxidation in aqueous perchloric acid. Carbon N Y **100**, 540–545 (2016)
Gurzeda, B., Krawczyk, P.: Eletrochemical formation of Graphite Oxide from the Mixture of sulfuric and nitric acids. Electrochim. Acta **310**, 96-103 (2019)
Hossain, S.T., Wang, R.: Electrochemical exfoliation of graphite: effect of temperature and hydrogen peroxide addition. Electrochim. Acta. Acta **216**, 253–260 (2016)
Hsieh, C.T., Hsueh, J.H.: Electrochemical exfoliation of graphene sheets from a natural graphite flask in the presence of sulfate ions at different temperatures. RSC Adv. **6**(69), 64826–64831 (2016)
Htwe, Y.Z.N., Chow, W.S., Suda, Y., Thant, A.A., Mariatti, M.: Effect of electrolytes and sonication times on the formation of graphene using an electrochemical exfoliation process. Appl. Surf. Sci. **469**, 951–961 (2018)
Huang, H., Xia, Y., Tao, X., Du, J., Fang, J., Gan, Y., Zhang, W.: Highly efficient electrolytic exfoliation of graphite into graphene sheets based on Li ions intercalation-expansion-microexplosion mechanism. J. Mater. Chem. **22**, 10452–10456 (2012)
Huang, Z., Hou, H., Zhang, Y., Wang, C., Qiu, X., Ji, X.: Layer-tunable phosphorene modulated by the cation insertion rate as a sodium-storage anode. Adv. Mater. **29**(1–7), 1702372 (2017)
Jaiswal, R., Saha, U., Goswami, T.H., Srivastava, A., Prasad, N.E.: Pillar effect' of chemically bonded fullerene in enhancing supercapacitance performances of partially reduced fullerenol graphene oxide hybrid electrode material. Electrochim. Acta. Acta **283**, 269–290 (2018)
Jaiswal, R., Saha, U., Goswami, T.H., Maji, P.K.: Effect of partially reduced fullerenol-graphene hybrid nanofiller on photophysical and super capacitance properties of fluorescence conducting polymer nanocomposites. Polym. Compos.. Compos. **44**, 3401–3415 (2023)
Ji, H., Wang, M., Liu, S., Sun, H., Liu, J., Qian, T., Yan, C.: In-situ observation as activity descriptor enables rational design of oxygen reduction catalyst for zinc-air battery. Energy Storage Mater **27**, 226–231 (2020)
Kepic, D.P., Markovic, Z.M., Jovanovic, S.P., Perusko, D., Budimir, M., Holclajtner-Antunovic, I.D., Pavlovic, V.B., Markovic, T.B.: Preparation of PEDOT: PSS thin films doped with graphene and graphene quantum dot. Synth. Met. **198**, 150–154 (2014)

Kepic, D.P., Ristic, I.S., Marinovic-Cincovic, M.T., Perusko, D.B., Spitalsky, Z., Pavlovic, V.B., Budimir, M.D., Siffalovic, P., Dramicanin, M.D., Micusik, M., Kleinova, A., Janigova, I., Markovic, Z.M., Todorovic Markovic, B.M.: Simple route for the preparation of graphene/poly(styrene-b-butadiene-b-styrene) nanocomposite films with enhanced electrical conductivity and hydrophobicity. Polym. Int.. Int. **67**, 1118–1127 (2018)

Lee, J., Noh, S., Pham, N.D., Shim, J.H.: Top-down synthesis of S-doped graphene nanosheets by electrochemical exfoliation of graphite: metal-free bifunctional catalysts for oxygen reduction and evolution reactions. Electrochim. Acta. Acta **313**, 1–9 (2019)

Li, X., Yu, J., Wageh, S., Ghamdi, A.A.A., Xie, J.: Graphene in photocatalysis: a review. Small **12**, 6640–6696 (2016)

Li, F., Xue, M., Zhang, X., Chen, L., Knowles, G.P., MacFarlane, D.R., Zhang, J.: Advanced composite 2D energy materials by simultaneous anodic and cathodic exfoliation. Adv. Energy Mater. **8**, 1702794 (2018)

Li, L., Zhang, D., Deng, J., Fang, J., Gou, Y.: Review-preparation and application of graphene-based hybrid materials through electrochemical exfoliation. J. Electro. Chem. Soc. **167**, 086511 (2020a)

Li, Z., Young, R.J., Backes, C., Zhao, W., Zhang, X., Zhukov, A., Tillotson, E., Conlan, A.P., Ding, F., Haigh, S.J., Coleman, J.N., Novoselov, K.S.: Mechanism of liquid phase exfoliation for the production of graphene. ACS Nano **14**(9), 10976–10985 (2020b)

Lim, S., Han, J.H., Kang, H.W., Lee, J.U., Lee, W.: Preparation of electrochemically exfoliated graphene sheets using DC switching voltages. Carbon Lett. **30**, 409–416 (2020)

Lin, L., Peng, H., Liu, Z.: Synthesis challenges for graphene industry. Nat. Mater. **18**, 520–524 (2019)

Liu, J., Yang, H., Zhen, S.G., Poh, C.K., Chaurasia, A., Luo, J., Wu, X., Yeow, E.K.L., Sahoo, N.G., Lin, J., Shen, Z.: A green approach to the synthesis of high-quality graphene oxide flakes via electrochemical exfoliation of pencil core. RSC Adv. **3**(29), 11745–11750 (2013)

Liu, B., Zhang, Q., Zhang, L., Xu, C., Pan, Z., Zhou, Q., Zhou, W., Wang, J., Gu, L., Liu, H.: Electrochemically exfoliated chlorine-doped graphene for flexible all-solid-state micro-supercapacitors with high volumetric energy density. Adv. Mater. **34**, 2106309 (2022)

Lou, F., Buan, M.E.M., Muthuswamy, N., Walmsley, J.C., Ronning, M., Chen, D.: One-step electrochemical synthesis of tunable nitrogen-doped graphene. J Mater Chem. A **4**(4), 1233–1243 (2016)

Lowe, S.E., Shi, G., Zhang, Y., Qin, J., Jiang, L., Jiang, S., AlMamun, M., Liu, P., Zhong, Y.L., Zhao, H.: The role of electrolyte acid concentration in the electrochemical exfoliation of graphite: mechanism and synthesis of electrochemical graphene oxide. Nano Mater Sci. **1**(3), 215–223 (2019)

Lu, X., Cai,M., Wu, X., Zhang, Y., Li, S., Liao, S., Lu, X.: Controllable synthesis of 2D materials by electrochemical exfoliation for energy storage and conversion application. Small **19**, 2206702 (2023)

Markovic, Z.M., Trajkovic, L.M.H., Markovic, B.M.T., Kepić, D.P., Arsikin, K.M., Jovanović, S.P., Pantovic, A.C., Dramićanin, M.D., Trajkovic, V.S.: In vitro comparison of the photothermal anticancer activity of graphene nanoparticles and carbon nanotubes. Biomaterials **32**, 1121–1129 (2011)

Markovic, Z.M., Kepic, D.P., Matijasevic, D.M., Pavlovic, V.B., Jovanovic, S.P., Stankovic, N.K., Milivojevic, D.D., Spitalsky, Z., Antunovic, I.D.H., Bogdanovic, D.V.B., Niksic, M.P., Markovic, B.M.T.: Ambient light induced antibacterial action of curcumin/graphene nanomesh hybrids. RSC Adv. **7**, 36081–36092 (2017)

Neog, A., Biswas, R., Bharali, N.: A novel electrochemical synthesis of graphene oxide: converting waste product to utility. EPL **137**, 16001 (2022)

Parvez, K., Li, R., Puniredd, S.R., Hernandez, Y., Hinkel, F., Wang, S., Feng, X., Mullen, K.: Electrochemically exfoliated graphene as solution processable, highly conductive electrodes for organic electronics. ACS Nano **7**, 3598–3606 (2013)

Parvez, K., Wu, Z.S., Li, R., Liu, X., Graf, R., Feng, X., Mullen, K.: Exfoliation of graphite into graphene in aqueous solutions of inorganic salts. J. Am. Chem. Soc. **136**, 6083–6091 (2014)

Parvez, K., Rincon, R.A., Weber, N.E., Cha, K.C., Venkataraman, S.S.: One-step electrochemical synthesis of nitrogen and sulfur co-doped, high-quality graphene oxide. Chem. Commun. **52**(33), 5714−5717 (2016)

Pei, S., Cheng, H.M.: The reduction of graphene oxide. Carbon **50**, 3210–3228 (2012)

Pei, S., Wei, Q., Huang, K., Cheng, H.M., Ren, W.: Green synthesis of graphene oxide by seconds timescale water electrolytic oxidation. Nat. Commun.commun. **9**(1), 1–9 (2018)

Pogacean, F., Coros, M., Magerusan, L., Mirel, V., Turza, A., Katona, G., et al.: Exfoliation of graphite rods via pulses of current for graphene synthesis: sensitive detection of 8-hydroxy-2′-deoxyguanosine. Talanta **196**, 182–190 (2019)

Prekodravac, J.R, Kepic, D.P., Colmenares, J.C., Giannakoudakis, D.A., Jovanovic, S.P.: A comprehensive review on selected graphene synthesis methods: from electrochemical exfoliation through rapid thermal annealing towards biomass pyrolysis. J. Mater. Chem. C **9**, 6722-6748 (2021)

Rao, K.S., Sentilnathan, J., Cho, H.W., Wu, J.J., Yoshimura, M.: Soft processing of graphene nanosheets by glycine-bisulfate ionic-complex-assisted electrochemical exfoliation of graphite for reduction catalysis. Adv. Funct. Mater. **25**, 298–305 (2015)

Saha, U., Jaiswal, R., Goswami, T.H., Maji, P.K.: Role and Effect of electrolytes selection on supercapacitance behaviour of aminated graphenes. Electrochim. Acta. Acta **435**, 141400 (2022)

Sahoo, S.K., Mallik, A.: Synthesis and characterization of conductive few layered graphene nanosheets using an anionic electrochemical intercalation and exfoliation technique. J Mater Chem C **3**(41), 10870–10878 (2015)

Sahoo, S.K., Ratha, S., Rout, C.S., Mallik, A.: Physicochemical properties and supercapacitor behavior of electrochemically synthesized few layered graphene nanosheets. J. Solid State Electrochem.electrochem. **20**(12), 3415–3428 (2016)

Sahu, S., Rout, G.C.: Band gap opening in graphene: a short theoretical study. Int. Nano Lett **7**, 81–89 (2017)

Sharif, F., Zeraati, A.S., Ganjeh-Anzabi, P., Yasri, N., Perez-Page, M., Holmes, S.M., et al.: Synthesis of a high-temperature stable electrochemically exfoliated graphene. Carbon **157**, 681–692 (2019)

Shen, X., Qian, T., Chen, P., Liu, J., Wang, M., Yan, C.: Bioinspired polysulfiphobic artificial interphase layer on lithium metal anodes for lithium sulfur batteries. ACS Appl. Mater. Interfaces **10**, 30058–30064 (2018)

Shuai, H., Ge, P., Hong, W., Li, S., Hu, J., Hou, H., Zou, G., Ji, X.: Electrochemically exfoliated phosphorene-graphene hybrid for sodium-ion batteries. Small Methods **3**, 1800328 (2019)

Sood, A.K., Lund, I., Puri, Y.R., Efstathiadis, H., Haldar, P., Dhar, N.K., Lewis, J., Dubey, M., Zakar, E., Wijewarnasuriya, P., Polla, D.L., Fritze, M.: Graphene-new trends and developments. In: Tech, pp. 59–89 (2015)

Su, C.Y., Lu, A.Y., Xu, Y., Chen, F.R., Khlobystov, A.N., Li, L.J.: High-quality thin graphene films from fast electrochemical exfoliation. ACS Nano **5**(3), 2332–2339 (2011)

Sun, H., Xu, G., Lian, W., Kastiukas, G., Zhang, J., Zhang, X., Liu, W., Xing, F., Ren, J.: Electrochemical synthesis and Property characterisation of graphene oxide using water as electrolyte. Chem. Phys. Lett. **786**, 139206 (2022)

Tang, W., Sun, D., Liu, S., Li, B., Sun, W., Fu, J., et al.: One step electrochemical fabricating of the biomimetic graphene skins with super hydrophobicity and super oleophilicity for highly efficient oil-water separation. Sep. Purif. Technol. **236**, 116293 (2019)

Tang, Z., Neumann, C., Winter, A., Turchanin, A.: Electrochemical delamination assisted transfer of molecular nanosheets. Nanoscale **12**, 8656–8663 (2020)

Tian, S., Yang, S., Huang, T., Sun, J., Wang, H., Pu, X., Tian, L., He, P., Ding, G., Xie, X.: One-step fast electrochemical fabrication water-dispersible graphene. Carbon **111**, 617 (2017a)

Tian, Z., Yu, P., Lowe, S.E., Pandolfo, A.G., Gengenbach, T.R., Nairn, K.M., Song, J., Wang, X., Zhong, Y.L., Li, D.: Facile electrochemical approach for the production of graphite oxide with tunable chemistry. Carbon N Y **112**, 185−191 (2017)

Tripathi, P., Patel, P.C.R., Dixit, A., Singh, A.P., Kumar, P., Shaz, M.A., Srivastava, R., Gupta, G., Dhawan, S.K., Gupta, B.K., Srivastava, O.N.: High yield synthesis of electrolyte heating assisted electrochemically exfoliated graphene for electromagnetic interference shielding applications. RSC Adv. **5**(25), 19074–19081 (2015)

Wang, X., Zhang, L.: Green and facile production of high quality graphene from graphite by the combination of hydroxyl radicals and electrical exfoliation in different electrolyte systems. RSC Adv. **9**, 3693–3703 (2019)

Wang, H., Wei, C., Zhu, K., Zhang, Y., Gong, C., Guo, J., et al.: Preparation of graphene sheets by electrochemical exfoliation of graphite in confined space and their application in transparent conductive films. ACS Appl. Mater. Interfaces **9**, 34456–34466 (2017)

Wang, H.S., Tian, S.Y., Yang, S.W., Wang, G., You, X.F., Xu, L.X., Li, Q.T., He, P., Ding, G.Q., Liu, Z., Xie, X.M.: Anode coverage for enhanced electrochemical oxidation: a green and efficient strategy towards water-dispersible graphene. Green Chem. **20**(6), 1306–1315 (2018)

Wang, M., Ji, H., Liu, S., Sun, H., Liu, J., Yan, C., Qian, T.: Single-atom scale metal vacancy engineering in heteroatom-doped carbon for rechargeable zinc-air battery with reduced overpotential. Chem. Eng. J. **393**(1–7), 124702 (2020a)

Wang, M., Liu, S., Ji, H., Liu, J., Yan, C., Qian, T.: Unveiling the essential nature of lewis basicity in thermodynamically and dynamically promoted nitrogen fixation. Adv. Funct. Mater. **30**(1–7), 2001244 (2020b)

Xia, Z., Maccaferri, G., Zanardi, C., Christian, M., Ortolani, L., Morandi, V., et al.: Dispersion stability and surface morphology study of electrochemically exfoliated bilayer graphene oxide. J. Phys. Chem. C **123**, 15122–15130 (2019)

Yang, S., Zhang, P., Wang, F., Ricciardulli, A.G., Lohe, M.R., Blom, P.W.M., Feng, X.: Fluoride-free synthesis of two-dimensional titanium carbide (MXene) using a binary aqueous system. Angew. Chem. Int. Ed.. Chem. Int. Ed. **57**, 15491 (2018)

Yang, S., Ricciardulli, A.G., Liu, S., Dong, R., Lohe, M.R., Becker, A., Squillaci, M.A., Samori, P., Mullen, K., Feng, X.: Ultrafast delamination of gaphite into high-quality graphene using alternating currents. Angew Chem. Int. Edi. **56**, 6669–6675 (2017)

Yu, P., Lowe, S.E., Simon, G.P.: Zhong YL. Electrochemical exfoliation of graphite and production of functional graphene. Curr. Opin. Colloid Interface Sci. **20**, 329–338 (2015)

Zhang, Y., Xu, Y., Zhu, J., Li, L., Du, S.X.: Electrochemically exfoliated high-yield graphene in ambient temperature molten salts and its application for flexible solid-state supercapacitors. Carbon **127**, 392–403 (2018a)

Zhang, J., Zhao, Y., Guo, X., Chen, C., Dong, C.L., Liu, R.S., Han, C.P., Li, Y., Gogotsi, Y., Wang, G.: Single platinum atoms immobilized on an MXene as an efficient catalyst for the hydrogen evolution reaction. Nat. Catal.catal. **1**, 985 (2018b)

Zhao, Z., Cai, W., Xu, Z., Mu, X., Ren, X., Zou, B., et al.: Multirole p-styrene sulfonate assisted electrochemical preparation of functionalized graphene nanosheets for improving fire safety and mechanical property of polystyrene composites. Compos. B Eng. **181**, 107544 (2019a)

Zhao, X., Li, H., Han, F., Dai, M., Sun, Y., Song, Z., et al.: Electrochemical exfoliation of graphene as an anode material for ultralong cycle lithium ion batteries. J. Phys. Chem. Solids **139**, 109301 (2019b)

Zhong, Y.L., Swager, T.M.: Enhanced electrochemical expansion of graphite for in situ electrochemical functionalization. J. Am. Chem. Soc. **134**, 17896–17899 (2012)

Zhou, F., Huang, H., Xiao, C., Zheng, S., Shi, X., Qin, J., et al.: Electrochemically scalable production of fluorine-modified graphene for flexible and high-energy ionogel-based microsupercapacitors. J. Am. Chem. Soc. **140**, 8198–8205 (2018)

Chapter 2
Advantages of Electrochemical Exfoliation Method Over Conventional Methods

Melkamu Biyana Regasa and Tebello Nyokong

Abstract Graphene and its derivatives received considerable interest for various technological uses due to their exceptional characteristics such as mechanical strength, physical properties, and excellent electrical conductivities. The preparation of these important materials for a particular purpose is very important. Graphene synthesis can be performed by top-down (mechanical, and chemical exfoliation, and chemical synthesis) and bottom-up (chemical vapour deposition, pyrolysis, epitaxial growth, and arc discharge methods) approaches. The top-down approach is the division of large precursors into small particles while the bottom-up is the buildup of small carbon based materials to form graphene structure. The later method gives the product with good quality than the first one but needs well-developed operational facilities and systems and thus it is relatively expensive. However, electrochemical exfoliation as a top-down route is simple, fast, easy controllable and scalable to give the desired products under mild conditions. Thus, this chapter presents the advantage of electrochemical exfoliation over the conventional approaches (bottom-up and top-down). Based on the literature reports, the benefits and drawbacks of each graphene preparation method will be discussed and forwarded.

Keywords Graphene · Graphene oxide · Reduced graphene oxide · Bottom-up · Top-down · Electrochemical exfoliation

---

M. B. Regasa (✉)
Department of Chemistry, College of Natural and Computational Sciences, Wollega University, Nekemte, Ethiopia
e-mail: malkako2011@gmail.com

T. Nyokong
Insititute of Nanotechnology Innovation (INI), Rhodes University, Makhanda, South Africa
e-mail: t.nyokong@ru.ac.za

R. Khan et al. (eds.), *Electrochemical Exfoliation of Graphene and its Derivatives*, Engineering Materials, https://doi.org/10.1007/978-981-97-2128-3_2

2.1 Introduction

Graphene is the carbon material with the smallest cross section in contrast to its length. It is made up of flattened (2D) layers of carbon atoms arranged densely in honeycomb crystal lattices that were discovered in 2004 (Novoselov et al. 2004). In other ways, graphene is made up of sp^2-hybridized pure carbon atoms that are interconnected by van der Waals forces and are covalently bonded into a continuous hexagonal lattice in the same plane (Katariya 2022). The strong sigma (σ) bonds in the chemical structure maintain the rigid backbone of the hexagonal structure, while the out-of-plane pi (π) bonds control the interaction between different graphene layers (Choi et al. 2010). Graphene materials have a very large surface area, high mechanical strength, good biocompatibility, and unique electronic and optoelectronic properties which make them ideal for the fabrication of the preparation of electronic apparatus and structures. The derivatives of graphene include graphite (multiple layers of graphene), glassy carbon, fullerene, graphene quantum dot, graphene oxide, reduced graphene oxide, and carbon nanotube. All these functional ingredients are derived from 2D graphene. These advanced materials are the basic building blocks in nanotechnology due to their extraordinary physical and chemical properties. However, the possible applications of these materials directly depend on their production in large quantities with distinctive properties and multi-functionalities (Sun et al. 2020). The preparation of graphene-based functional materials can be performed using different preparation methods. Generally, the production of graphene materials can be performed by two main approaches namely top-down (demolition) and bottom-up (building) methods. The destruction approach (top-down) is the synthesis of graphene by breaking down the structure of large carbonaceous materials into single-layer followed by separation of the interlayers to obtain graphene sheets. The other approach, the building approach (bottom-up) is the assembly of graphene from smaller carbon units (Ikram et al. 2020). Currently, there are various state-of-the-art synthesis methods employed for the development of graphene materials either in the destruction or building approach. These methods include micromechanical cleavage of graphite, epitaxial growth by chemical vapor deposition on a metal substrate, epitaxial growth by thermal deposition from silicon carbide (SiC) surface, colloidal suspension from graphite oxide, electrochemical exfoliation, and many others. Yet, it is challenging to develop a synthetic route that is cost-effective, highly reliable, and scalable as well as high product yield with the best quality (Zhong et al. 2015). Among the fabrication and preparation methods of graphene materials mentioned here, electrochemical is considered as a promising method that provides high purity, great performance, low cost, and thin product with the potential for huge production capacity. Graphene is the strongest material that attracted more attention from technological applications demanding lightweight but strong materials. Hence, is an advanced carbon functional material with inherent unique properties that make it suitable for a wide range of technological uses (Yan et al. 2020). The extraordinary properties of ultrathin graphene and its derivatives (Fig. 2.1) make them suitable for various applications including electronic devices, field emission displays, field

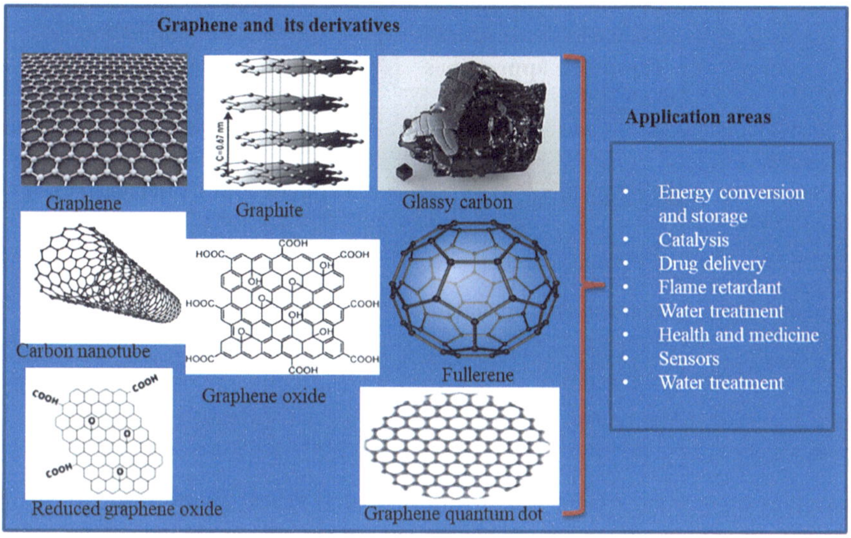

Fig. 2.1 Graphene materials and their selected applications

effect transistors, energy conversion and storage, transparent electrodes, photocatalysis, antibacterial agents, nanomedicines or dental materials, flame retardancy, sensor fabrication, water treatment, and liquid-phase chemistry (Zhu et al. 2010a, b).

2.2 Synthesis Methods for Graphene and Its Derivatives

Graphene and its derivatives are regarded as a basis for engineering, scientific innovation, and microelectronic applications in the twenty-first century due to their peculiar physicochemical properties. Therefore, the synthesis of graphene and its other forms from different starting materials including graphite is critical. The best-known starting material for graphene preparation is graphite—a stacked layer of many graphene sheets bonded together by the weak van der Waals force. Thus, the synthesis of a graphene sheet can be realized by breaking these weak forces or bonds using either mechanical or chemical energy. In this regard, exfoliation and cleavage use mechanical or chemical energy to break these weak bonds and separate individual graphene layers. Compared to graphite, graphene is the thinnest sp^2 carbon network that is very important due to its extraordinary carrier mobility and probable applications in electronics (Guo and Dong 2011). Therefore, the wide application of graphene and its derivatives mainly depends on its successful preparation. Hence, graphene must be synthesized at a cost competitive to that of current materials. This means the effective preparation of these materials can be evaluated in terms of cost and the quality of the product. It is of great challenge to develop synthesis processes with the following attributes: cost-effective, highly reliable, and scalable production

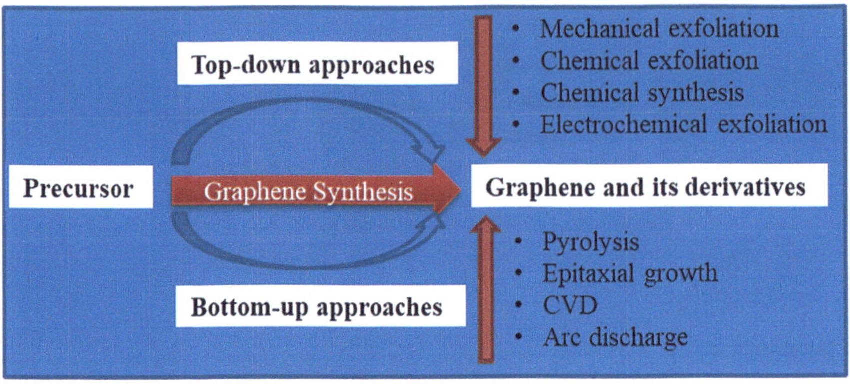

Fig. 2.2 General approaches for the synthesis of graphene and its derivatives

as well as high product yield and quality (Lee et al. 2019; Zhong et al. 2015). A single-layer or few layers of graphene can be synthesized on different substrates by different methods. The quantity of product obtained, purity of the product, structural integrity, and the way to process or pattern the obtained material are the major issues that need to be addressed during graphene production (Yan et al. 2020). Nowadays, graphene can be prepared by two general approaches namely top-down and bottom-up approaches (Fig. 2.2). Hence, this section focuses on the discussion of the different synthesis methods.

The top-down methods include mechanical exfoliation (adhesive tape and atomic force microscopy (AFM) tips) (Sun et al. 2020; Yu et al. 2016), chemical exfoliation (Ni et al. 2008), electrochemical exfoliation (Tung et al. 2009), and chemical synthesis (sonication and reduced graphene oxide) (Yang et al. 2016; Guker et al. 2016). All these methods convert graphite materials into single-layer, bi-layers, and few-layers graphene through continuous peeling-off processes. In these preparation approaches, larger precursors such as graphite and other carbon-based precursors are destroyed to form nano-sized graphene. Generally, some top-down methods are highly scalable and produce high-quality products, but have difficulties in forming products with consistent properties, low yield, and rely heavily on the finite graphite precursor. On the other hand, the bottom-up method is the assembly of graphene from smaller carbonaceous elements onto a substrate by using methods like pyrolysis (Baig et al. 2018) epitaxial growth (Illakkiya et al. 2016), chemical vapor deposition (CVD) (thermal and plasma) (Yang et al. 2013), arc discharge (Allen et al. 2009), and other methods (Subrahmanyam et al. 2009). These methods construct graphene materials from atomic-sized precursors. Among these bottom-up methods, CVD and epitaxial growth methods are promising for fundamental study and electronic application of graphene because they can produce tiny amounts of large-size and defect-free graphene nanosheets (Table 2.1).

By and large, the bottom-up methods produce graphene products with an almost defect-free and large surface area but they are limited by the high production cost

Table 2.1 Selected precursors used for the synthesis of graphene and its derivatives

Precursor	Synthesis method	References
HOPG/Graphite	Mechanical exfoliation	Novoselov et al. (2004)
	Electrochemical exfoliation	Abdelkader et al. (2015)
	Chemical exfoliation	Chen et al. (2008)
	Oxidation–reduction	Rao and Sood (2013)
	Arc discharge	Kane et al. (2020)
Hydrocarbon	CVD	Yang et al. (2016)
SiC	Epitaxial growth	Borovikov and Zangwill (2009)
Petroleum asphalt	Arc discharge	Wu et al. (2009)
Methane	CVD	Zhu and Xu (2020)
Biomass	Pyrolysis	Chen et al. (2017)
Camphor	CVD	Fang et al. (2015)
Natural oil	Pyrolysis	Wu et al. (2009)

and the need for a sophisticated operational setup. The type of feedstock would greatly affect the commercial viability of graphene materials preparation as its mass production requires a continuous supply of carbon precursors (Zhong et al. 2015). For example, graphite is used as the universal feedstock in the synthesis of graphene using the top-down approach due to its abundance in nature and its low price. Also different raw materials can be used as a precursor to prepare various graphene forms. These precursors include graphite oxide, polymers, biomass from plants and animals, solid carbon dioxide, SiC, alcohols, fossil fuels, and carbon nanotubes. Nevertheless, the selection of the starting materials is determined by the purpose of application and the available fabrication method.

2.3 Top-Down Approaches

Graphene materials synthesis by the top-down method uses the direct method and the indirect method. Under the direct method, graphene is directly prepared from graphite by using mechanical exfoliation, direct sonication, and electrochemical and super acid dissolution methods. The peeling of graphite layers can be performed by chemical and mechanical exfoliation methods. The indirect methods involve two steps such as the the oxidation of graphite to form graphene oxide and its chemical or thermal treatment to yield reduced graphene oxide (Wang et al. 2019; Yi and Shen 2015).

2.3.1 Mechanical Exfoliation

The physical shearing or rubbing something to remove the external layer is knowm as mechanical exfoliation. In this method, mechanical energy is applied to break down the van der Waals forces holding graphite layers together and separate the individual sheets. Mechanical exfoliation, also known as the adhesive tape technique, is one effective way of producing single-layer and few-layers graphene through a continual peeling process. This method uses directional force, either normal or lateral, to peel off the graphene sheet from the bulk graphite by overcoming the weak electrostatic forces of attraction between the layers (Chen et al. 2012). The process requires a starting graphite feedstocks such as highly oriented pyrolytic graphite (HOPG), single-crystal graphite, or natural graphite (Gao et al. 2018), and an applied force of around 300 nN μm^{-2} (Lu et al. 1999). The typical mechanical exfoliation techniques include micromechanical cleavage, continuous mechanical cleavage, shear exfoliation, and explosive exfoliation. While all these methods utilize mechanical force, the equipment by which this force is exerted varies from mills and blenders to tape. These methods are also attempted at relatively low processing temperatures apart from the explosive technique, which requires high energy for detonation. Thus, mechanical exfoliation is the pulling out of graphene flakes on a substrate using an external force for mechanical cleaving by dividing down the layers from graphitic materials. This graphene synthesis approach can be attained using several agents including scotch tape, electric field, and ultrasonication (Lang 1975). In line with this, thin multilayered graphite with about 200 nm thickness was prepared by mechanical exfoliation using an atomic force microscopy (AFM) tip (Gao et al. 2018). Later on, a single graphene layer with a thickness of about less than 10 nm was fabricated by the application of the adhesive tape substrate and HOPG precursor (Novoselov et al. 2004). The process involves the continuous peeling off of layers of the highly oriented precursor. The exfoliation of graphene sheets using the adhesive tape method, where graphite is placed on a substrate and repeatedly peeled using an adhesive tape until a monolayer sheet is obtained (Fig. 2.3). The product was characterized using Raman spectroscopy and the graphene sheets have shown the existence of structural defects (Yi et al. 2015) that greatly upset the electrical conductivity due to the disruption of the electronic structure.

The graphene produced by this method is relatively good but it is challenged by the difficulty of scalability and reproducibility. This process is slow and imprecise; hence the material produced is most often used to study the properties of graphene rather than using it commercially (Stankovich et al. 2006). Furthermore, graphene produced by this method is one of the highest quality and purity as evaluated with Raman spectroscopy, atomic force microscopy (AFM), and optical microscope with a reduced production (Yi et al. 2015). However, one disadvantage of using this method is its yield directly affecting its scalability (Edwards and Coleman 2013) even though it is the basis for different exfoliation techniques resulting in quite an affordable production cost (Yi and Shen 2015). From the top-down methods, mechanical exfoliation offers original graphene by an easy and inexpensive repetitive

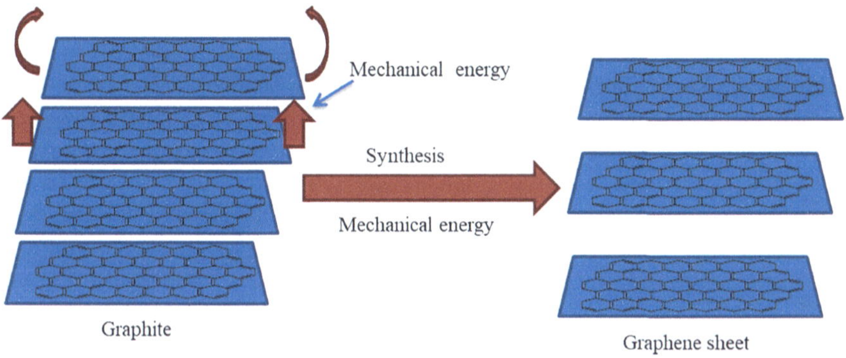

Fig. 2.3 Schematic of preparation of graphene from graphite using mechanical exfoliation

stick-and-peel process from highly ordered pyrolytic graphite HOPG (Pirzado et al. 2019). This is because it avoids the high temperature, long processing time, and metal catalyst in other methods like in epitaxial growth (Soldano et al. 2010) and structural damage in chemical exfoliation (Li et al. 2009). Nevertheless, mechanical exfoliation is slow, labor-intensive, and expensive. It leaves glue residues on samples that require high-temperature post-treatments (Xu et al. 2008). Therefore, further modification of the method is important to improve both the quality and quantity of the yield.

2.3.2 *Chemical Exfoliation*

Chemical exfoliation or liquid-phase exfoliation is a method that uses strong oxidants to oxidize the aromatic network to create graphene oxide based on the Hummers method (Moser et al. 2007). This method involves two general steps:-preparation of graphene-intercalated compounds from graphite using alkali metals (Zaaba et al. 2017) and heating or sonication of graphene-intercalated compounds through exfoliating graphene sheets (Rao and Sood 2013). This process leads to the formation of graphene-intercalated compounds with alkali metals (Betancur et al. 2018), surfactants (Viculis et al. 2005), and other electrolytes that are used to increase the interlayer spacing by reducing the van der Waals forces of attraction between graphene layers. In this method, solution-dispersed graphite is exfoliated by injecting large alkali ions between the graphite layers to minimize the van der Waals forces between the graphite layers (Fig. 2.4). It is one of the most effective techniques for fabricating graphene due to its simplicity and quickness (Vallés et al. 2008). The reduction of graphite oxide by different chemicals is one of the conventional procedures to prepare graphene in large quantities (Chen et al. 2008). Thus, graphite oxide is usually synthesized through the oxidation of graphite using oxidants (Fig. 2.5) including concentrated sulfuric acid, nitric acid, and potassium permanganate based on the Brodie method (Hummers and Offeman 1958), Staudenmaier method (Brodie 1860), and Hummers method

(Staudenmaier 1898), respectively. Finally, graphene layers are prepared from the graphene oxide formed by either thermal, chemical treatment or electrochemical reduction as a simple one-step, solvothermal reduction method to produce reduced graphene oxide dispersion in the organic solvent (Marcano et al. 2010). However, electrochemical reduction shows the best results in terms of production and quality.

A simple, green, and low-cost technique was developed by using a surfactant to exfoliate precursor graphite into high-quality layered graphene with the free basal

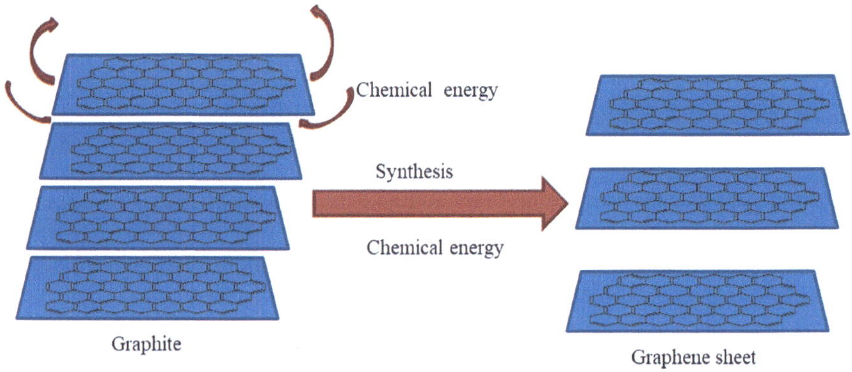

Fig. 2.4 Schematic of preparation of graphene from graphite using chemical exfoliation

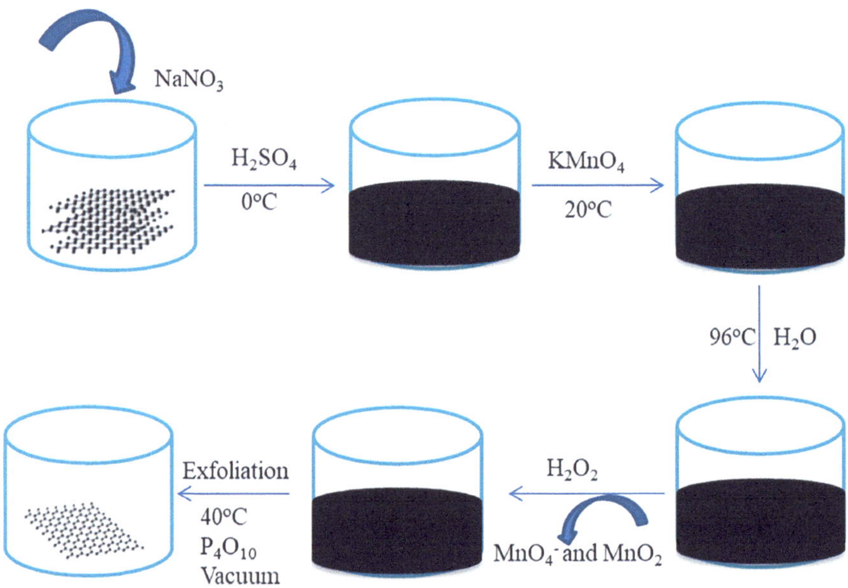

Fig. 2.5 Schematic diagram of chemical synthesis of graphene oxide

plane defect. The peeling off of graphene sheets are realized through centrifugation, and treatment of the product with thermal and annealing processes to get the product with minimum defect ratio and better electrical conductivity (Badri et al. 2017). The effort to find the development of cost-effective preparation methods for high-quality graphene materials is important for its commercialization. Hence, a novel binary-component system comprised of peroxyacetic acid and sulfuric acid for the chemical exfoliation of graphite to few-layer graphene was reported (Liu et al. 2019). The graphene produced by this method exhibited a high quality with few structure defects and superior electrical conductivity indicating that the method is promising for the mass production of high-quality graphene sheets. The Hummers' method and its modified ones are low-cost and solution-processed fabrication, but the electronic properties of the graphene oxides produced by these methods are poor when compared with pristine graphene (Su et al. 2011). However, these methods have the benefits of using cheap precursors and the simplicity of sythesis routes. Therefore, chemical exfoliation seems imperative to produce graphene and unique as it can produce a huge amount of graphene at a low temperature. Furthermore, the important feature of this method is its prospective for the possible mass production of graphene and functionalized graphene materials.

2.3.3 Chemical Synthesis

Graphene synthesis based using the chemical oxidation–reduction methods is the common method used for the production of suffient amount of graphene materials. This method involves the formation of graphite oxide, its dispersion in a solution, and its reduction with strong reductants (Park and Ruoff 2009). Graphene nanoplates were prepared using chemical exfoliation and potassium alkali metal (Betancur et al. 2018) through the reaction of potassium with graphite at about 200 °C under an inert atmosphere to form a potassium graphite (KC_8) intercalated compound. Then, the dispersion of KC_8 in aqueous ethanol obtained graphene nanoplates.

$$KC_8 + CH_3CH_2OH \rightarrow 8C + KOCH_2CH_3 + \frac{1}{2}H_2$$

Alkali metals were replaced by ternary potassium salt tetrahydrofuran to synthesize the single-layer graphene (Viculis et al. 2005). However, the product was reassembled to give the graphite that adversely affects the final product in terms of quality and quantity. The application of surfactants into the solvents used, foms immiscible liquids (heptane and water), ferric chloride, supercritical carbon dioxide, and nitromethane which were introduced as other intercalation chemicals (Kamali and Fray 2013; Safavi et al. 2012).

The Brodie method was modified to get the Hummers' method which has become a primary method in producing reduced graphene oxide. Furthermore, the Hummers' method was modified by many researchers which led to the ability to scale up the production method (Fig. 2.5). The strong oxidizing agents (H_2SO_4 and $KMnO_4$) employed in the Hummers' method paved the way for the preparation of additional graphene oxide sheets and bulks of carbon structure with oxygenated-capped edges by oxygen hydroxyl groups (Fathy and Abdel 2014). In more general terms, this oxidative exfoliation-reduction is considered the most viable route as it offers bulk production of reduced graphene oxide sheets from graphite feedstock at a relatively low cost. From this process, approximately 80% of monolayer-reduced graphene oxide sheets could be generated (Guo et al. 2009; MaAllister et al. 2007). However, the resulting product may not be pure due to the deformation of hexagonal carbon chains which in turn can affect its overall quality. Graphene material produced by this method has been utilized as a precursor in adsorbent development, drug delivery, and composite reinforcement (Wei et al. 2018; Hussien et al. 2018). Toxic chemicals such as hydrazine, borohydrides, and aluminum hydride have been used as the reducing agents in the oxidative exfoliation-reduction method, thereby limiting the environment indicator preference score in the process evaluation.

In the last two decades, many researchers have been trying to develop green synthesis of graphene. Among these achievements, the chemical production of graphene from less toxic chemicals, and the natural raw materials are regarded as a green synthesis method. With this regard, graphene was prepared from reducing sugar (Zhu et al. 2010a, b), table sugar and gelatin (Ruiz-Hitzky et al. 2011), carbon black (Muramatsu et al. 2014) as well as rice husk and potassium hydroxide (Singh et al. 2017). This method is a very simple, single-step, and cost-effective chemical approach to prepare graphene from natural precursors.

2.3.4 Electrochemical Exfoliation

Electrochemistry plays a great role in our day-to-day life because of its many common applications. For example, electrosynthesis is widely used for the production of various industrial products including graphene materials. In the electrochemical exfoliation method, the application of electric potential that causes the intercalation of a cation or anion from the electrolyte becomes among the promissing graphene production methods. This method has emerged as a likely bulk method for the production of graphene materials through the application of voltage that drives ionic species to intercalate into graphite where they form gaseous species that expand and exfoliate individual graphene pieces (Liu and Aziz 2022). Exfoliation of graphite by an electrochemical technique has become a simple yet high-yielding method in recent times for the preparation of bulk graphene (Su et al. 2011). Briefly, this synthesis method involves the intercalation of a cation or anion from the electrolyte with the help of applied potential. This method involves the use of various forms of graphite such as graphite foils, plates, rods, and powders as electrodes in an aqueous or non-aqueous electrolyte and electric current to bring about the expansion of electrodes. The electrodes could be of two types based on the power applied, that is, cathodic (negative) and anodic (positive) electrodes (Liu et al. 2019). The exfoliation mechanisms depend on the type of potential used either anodic or cathodic (Liu et al. 2019) (Fig. 2.6). The method was based on the programmed potential modulation strategies to balance the ion intercalation/deintercalation, surface tailoring, and bubbling dispersion processes. This approach has the potential to solve the problems associated with excessive surface tailoring and oxidation giving small size, high defect density, and poor yield during the frequently reported electrolysis of graphite at a high constant potential. In anodic exfoliation, the accumulation of positive charge at the anode facilitates the intercalation of bulky anions through the attraction forces and increasing the interlayer spacing between graphite layers. During the cathodic graphite layers exfoliation, the graphene layers were intercalated, opened, expanded, and exfoliated with the help of the cations attracted to the graphite working electrode by the negative bias. Therefore, electrochemical preparation of graphite has been considered to be an effective method for the mass production of superior graphene due to its easy, simple, and eco-friendly synthetic features (Park et al. 2021). For this method, various types of graphite materials such as graphite flakes, graphite rods, and HOPG can be used as precursors (Aghamohammadi et al. 2020). Water-dispersible graphene with high quality, and yield, high surface area, and low oxidation degree with remarkable conductivity can be synthesized. This kind of graphene preparation method is very important both economically and technologically. The electrolytes utilized in the electrochemical method could be divided into two categories: aqueous solutions and ionic-liquid-based electrolytes. For electrochemical processes conducted in an aqueous solution, it is hard to obtain and disperse graphene directly. Thus, organic solvents with low surface energy were usually employed to disperse the graphene prepared in an aqueous electrolyte. Accordingly, the direct electrochemical means for the preparation of ionic-liquid functionalized graphene using an ionic-liquid /water-electrolyte and the exfoliated yield was dispersed in dimethylformamide successfully (Liu et al. 2008). Although ionic liquids can promote green chemistry applications by reducing the utilization of toxic chemicals (Wang et al. 2010) it is not cost effective when applied to the large-scale production of graphene. Most importantly, electrochemical synthesis of graphite is a fast and simple method to prepare graphene materials with a variety of geometries, such as graphite powders, foils, rods, flakes, or plates, which can be used as working electrodes in liquid electrolytes.

Cathodic exfoliation occurs in organic solvents such as dimethyl sulfoxide and propylene carbonate comprising alkylammonium/lithium salts as electrolytes. In this method, a negative potential is applied to the graphite electrode in organic solvents using lithium ions as the intercalating metal. The other form of the electrochemical method is anodic exfoliation which is performed in an ionic liquid or water mixture of aqueous acids. The delamination of the graphite layers will proceed through the intercalation of anions amongst graphite sheets using electrolysis in an aqueous solution (Dang et al. 2020). 2D graphene materials are able to be prepared by using anodic-oxidation cationic-intercalation and cathodic-exfoliation using liquid-electrolyte to expand the structure of the materials with the applied potential. Thus, the electrochemical technique is also employed as a quick and controllable tool for lithium/

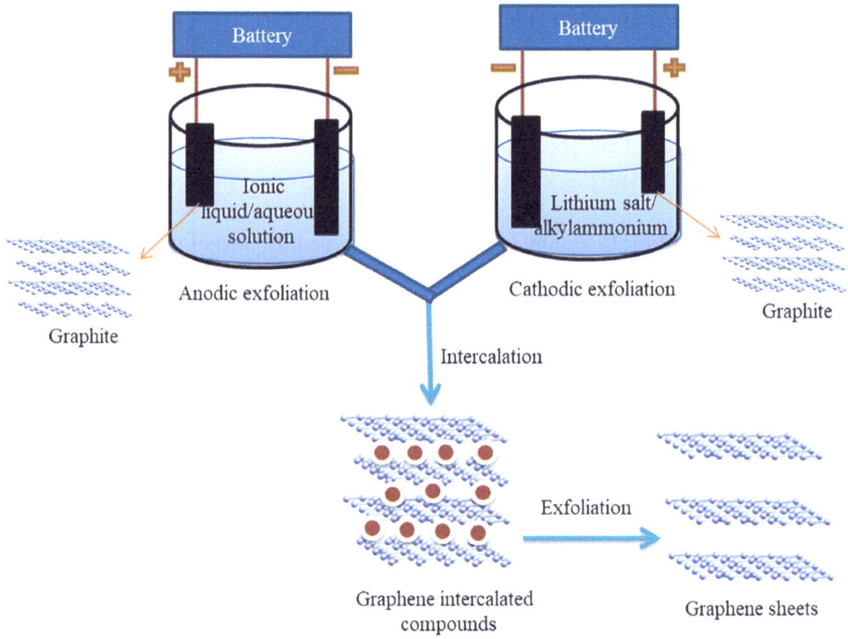

Fig. 2.6 Electrochemical exfoliation of graphene from graphite precursor

non-lithium intercalations and is considered as an effective technique for obtaining a single or multi-layered 2D nanosheets (Luheng et al. 2009). Pure graphite as electrodes and polysodium-4-styrenesulfonate solubilized in deionized water were used to incarnate the electrolyte (Luheng et al. 2009). Graphite rods were accommodated in an electrochemical cell which was filled with the electrolyte. A persistent current of 5 V was applied. A few minutes of electrolysis resulted in the accumulation of black material at the anode. The exfoliation process was carried on for four hours to isolate the product from the cell. It was then centrifuged at 1000 rpm and then the product was gradually poured out. The dispersion obtained was found to be extremely stable.

Recently, electrochemical exfoliation has drawn pronounced consideration because of its simplicity, cost-effectiveness, and greenness method for bulky production of high-quality graphene slips (Abdelkader et al. 2015). Consequently, the exfoliation of graphite using electrochemical technique leads to the production of a high-quality graphene with high crystallinity, low oxidation degree, uniform size distribution, and few layers (Shams et al. 2015).

In another research work, powdered dry graphene was prepared by washing the dispersion with deionized water and alcohol followed by vacuum drying. Then the yield was calculated by weighing the dry powder and sediment (Parvez et al. 2013).

Solution-processable thin-layer graphene is an intriguing nanomaterial with tremendous potential for electronic applications. To realize this idea, the exfoliation of graphite by using electrochemical method was proved to furnish graphene materials of extraordinary quality. Interestingly, the electrochemically exfoliated graphene contains a high yield (>80%) of one- to three-layer graphene flakes with a high carbon-to-oxygen (C/O) ratio of 12.3 and low sheet resistance (4.8 kΩ/0 for a single graphene sheet). Due to the solution processability of graphene, a vacuum filtration method in association with the dry transfer is introduced to produce graphene sheets with large area and superior conductivity on various substrates. Moreover, it was demonstrated that the patterned graphene can serve as a high-performance source/drain electrode for organic field-effect transistors (Achee et al. 2018). The requirement for the graphite monoliths to prepare graphite electrodes and the degradation of such electrodes are the main factors that limited the practical applications of this method. To eliminate these problems, graphite powders were contained and compressed within a permeable and expandable containment system to continuously intercalate, expand and exfoliate graphite to obtaing graphene with better quality (Achee et al. 2018; Roscher et al. 2019). The technique is encouraging for large-scale production of graphene, however, the efficiency of the system depends on the shape of the reactor used, graphite compression, and electrolyte transport through the system. Similarly, Roscher and his co-authors (2019) reported a highly efficient, single-step, cathodic exfoliation process of graphite to produce single- to few-layer graphene with a yield of over 70% from natural graphite flakes. In this work, a boron-doped diamond electrode was used to introduce an electrochemical one-pot reactor that enables the use of natural graphite flakes.

Furthermore, an attempt was made to solve the chemical distraction of the the surfaces of graphite during electrochemical exfoliation in different conditions through the combined approach of combining visual microscopy and Raman spectroscopy to map it in real-time (Xia et al. (2021). The mechanism of graphene formation was performed in acidic (0.5 M H_2SO_4) and neutral (0.5 M Li_2SO_4) electrolytes and compared with that performed in non-destructive intercalation of graphite in an organic electrolyte (1 M $NaClO_4$ in acetonitrile). The final graphene showed a high yield with a low defect.

Exfoliation of graphite using electrochemical methods would be interesting with the proper selection of intercalating salts and electrochemical potentials to control flake size, and defects of the final product. Thus, this method is emerging as an essential approach to obtaining several graphene materials. However, the efficiency of this method entirely depends on many factors such as graphite electrodes, electrolytes, and various additives in electrolytes, equipment setups, and power supplies to various operational parameters.

Advantages of electrochemical exfoliation means over conventional methods include:

- Controlling the characteristics (lateral size, thickness, electronic and chemical properties) of the resulting graphene sheets and their derivatives by changing the electrolyte type.
- Preparation of water-dispersible graphene materials which is highly important for various industrial applications.

- It is an easy, simple, fast, and eco-friendly method to prepare high-quality and high-quantity of graphene. The possible production of high-quality and quantity graphene by this method is due to the use of electrochemical activation and intercalation of electrolyte ions without the need for hazardous chemicals for cleaning.
- It is easy to operate, cheap, and less time-consuming method.
- Control and potential for scaling up

To this end, the one-step exfoliation of graphite materials and their functionalization for a particular application in many industries is very important. This huge demand needs the development of a suitable method for large-scale production based on the application of proper potentials. However, the extensive production of graphene and its derivatives has more challenges in terms of electrochemical setup, mixing of electrolytes, preparation of graphite electrodes, etc. that should be investigated in the future (Abdelkader et al. 2015).

2.4 Bottom-Up Approaches

The synthesis of graphene materials using the bottom-up approach can be performed via different common techniques like chemical vapor deposition (CVD), arc discharge, and epitaxial growth on SiC. Compared to the bottom-up method, the top-down method is advantageous because it can produce good-quality of graphene products but the method is not economically feasible. However, the bottom-up approach has the advantage of obtaining large quantity graphene with large surface areas (Bhuyan et al. 2016).

2.4.1 Epitaxial Growth

The type of crystal growth from gas precursors or the vapor phase from molecular beams on a substrate is known as an epitaxial growth. In this method, graphene is produced on the substrate (a semiconductor in nature) through the organization of the carbon atoms on the surface of the support materials. Thus, the graphene fabricated using this process is named epitaxial graphene which can be of homo-epitaxial layer (if the film was deposited on a substrate of the same material) and hetero-epitaxial layer (if the film and substrate are different materials) (Copel and Reuter 1989). Generally, it is the growth of a layer of the monocrystalline layer graphene with the same orientation as the substrate on a single mineral substrate as if the original crystal had been extended outward (Jilani et al. 2018). The epitaxial graphene could be grown on various substrates such as SiC, hexagonal boron nitride, ruthenium (Ru), iron (Fe), copper (Cu), and nickel (Ni) (Mishra et al. 2016). Among these substrates, SiC is used as a likely substrate for the development of epitaxial graphene. The crystal SiC as the starting material is heated at 1200–1600 °C under vacuum conditions, then Si atoms evaporate and carbon atoms remained on the surface, leading to the formation of graphene (Borovikov and Zangwill 2009). Thus, substrate-based graphene can be prepared by an annealing process, which involves gradual cooling after heating to deposit carbon atoms on the Si face as an epitaxial sheet to produce graphene (Riedi et al. 2010). However, the graphene sheet produced by this method lacks homogeneity. Berger and his co-authors (2004), from the Georgia Institute of Technology, produced a graphene thin layer that is similar in area to the original SiC on the (0001) crystal plane by heating a large area of single crystal SiC at high temperatures. However, Si may still evaporate during the deposition of carbon atoms (Norimatsu and Kusunoki 2014). To overcome this problem, a vacuum condition is substituted by introducing an argon gas atmosphere at near ambient pressure, which successfully hinders the transport of Si atoms away from the SiC surface. This reduces the overall sublimation rate and increases the graphitization temperature by several hundred degrees (Manukyan et al. 2013). The regulation of the heating system temperature helps to control the number of graphene layers. Conversely, the homogeneity of graphene produced through this energy-intensive method is rarely observed. This means the quality of graphene could further degrade by the development of a carbon-polar face rather than the Si-polar face that facilitates the homogeneous growth of graphene (Gupta, et al. 2014). Later on, a novel exothermic reaction between SiC and polytetrafluoroethylene has been reported to synthesize graphene (Kong et al. 2020). However, environmental concern was raised against this method due to the release of tetrafluoroethylene (C_2F_4) during synthesis apart from the production of superior graphene (<1 wt% of oxygen) and simple operation. Based on the current synthesis condition, epitaxial growth of graphene would be expensive due to the energy-intensive process and limited size of commercial SiC substrate (Guo and Dong 2011). Furthermore, the epitaxial growth can produce different polar faces such as Si-face or C-face which hinder the graphene product quality. Si-face graphene formation is preferred as it ensures the homogeneous growth of graphene. Using this method, the number of layers of graphene that depend on heating temperature can be easily modulated (Gupta et al. 2014). This synthesis method is still under investigation as documents on the growth mechanisms and interaction between graphene and the substrate are scarce (Choi et al. 2010). Moreover, the epitaxial growth method has the potential for the preparation of high-quality graphene. The potential application of epitaxial growth in the industry is generally limited by the strict operating conditions and the use of special equipment such as high temperature, ultrahigh vacuum, inert conditions, and utilization of single-crystal substrate (Yu et al. 2016). Therefore, it is important to modify the method to prepare large-area graphene with the desired quality and quantity.

2.4.2 Pyrolysis

Heating organic materials at or above 500 °C to destroy the strong bio-polymers is known as pyrolysis. Thus, the pyrolysis of biomass opened a new opportunity for high-value materials preparation. Additionally, the use of natural materials as a precursor is very important in minimizing the cost of producing graphene materials. Accordingly, the design and pyrolysis of biomass precursors has been merged as an alternative approach for the generation of graphene-like materials (Juang et al. 2009). The pyrolysis method is the separation of graphene layers by the application of heat energy. Graphene synthesis by pyrolysis involves heating sodium-ethanol in an equal molar ratio in a sealed vessel to form an intermediate consisting of fused graphene layers realized by heating the solvothermal product of sodium and ethanol without oxygen. Finally, graphene materials are separated by sonication in ethanol. The pyrolysis method has some benefits such as cheap graphene production with high purity at low temperatures using low-cost starting materials. Later on, sodium was substituted by sodium borohydride as a mild catalyst to improve the method and the product obtained. Furthermore, the addition of the surfactant greatly improved the quality of graphene produced by this method. Few-layer graphene can be fabricated on the metal surface through the arrangement of carbon atoms up on the thermal decomposition of SiC (Parvez et al. 2013). The formation of few-layer of graphene occurs from carbon atoms freed from Si at high temperatures. This technique has received a significant improvement through the continuous production of graphene films in millimeter scale at a temperature of 750 °C on a thin film of nickel coated on SiC substrate (Zhu and Xu 2020). The advantage of this method is the continuous production of graphene films over the entire SiC-coated surface and a high-purity graphene mono-layer (Kumar et al. 2014). However, this method cannot be used in the synthesis of graphene on a large scale. A similar approach is applied in the thermal decomposition of ethylene above 700 °C.

Spray pyrolysis involves the spraying of graphene dispersion onto a heated substrate to obtain a thin film of graphene. It is a low-cost and easy method for the fabrication of high-quality products for various applications (Chen et al. 2017). The graphene dispersion is first pumped into a nebulizer and a carrier gas, which is usually compressed air. The venture effect created by the compressed air flowing through the jet enables the solution to be nebulized. The baffle placed at the tip of the jet produces fine aerosol droplets while blocking and returning the larger ones into the reservoir, yielding a stream of uniform and fine droplets. These uniform and fine droplets are then sprayed onto a heated glass substrate to form graphene films. The advantage of the spray pyrolysis method is that thin and homogeneous graphene film can be produced. Additionally, no high temperatures are required as the glass substrate only needs to be heated up to 200 °C. As the reduction occurs thermally, reducing agents are not required. Moreover, vacuum condition is not required, and hence the cost to generate a vacuum environment is avoided. The only utility cost to account for in this method is the compressed air (1 bar), increasing the production cost slightly (Wang et al. 2019).

2.4.3 Chemical Vapor Deposition

Another bottom-up approach for the preparation of graphene and its derivatives is chemical vapor deposition (CVD). It involves the deposition of gaseous reactants onto the surface of a substrate to form graphene sheets (Fig. 2.7). First, a gas-phase carbon precursor is adsorbed onto the surface of a transition metal catalyst, which undergoes catalytic decomposition to carbon atoms, diffused and deposited onto the metal surface at high temperatures. The dissolved carbon atoms are then precipitated out and segregated onto the metal surface. Up on cooling, surface nucleation initiates, followed by graphene growth (Jia et al. 2020). The resulting graphene quality is mainly determined by the giving out parameters such as catalysts, precursors, gas flow rate, temperature, pressure, and time. Transition metals with an incomplete shell (e.g. Ni and Co) exhibit a certain degree of carbon solubility and thus produce the few-layer crystalline graphene with a larger area. On the other hand, transition metals with filled shells (e.g. Cu and Zn) possess a low affinity to carbon, and hence both dissolution and subsequent segregation steps are not possible. In this case, carbon atoms derived from hydrocarbon precursors could directly diffuse onto the metal surface and build up thermodynamically stable graphene. This process easily leads to the formation of large-area single-layer graphene using binary Ni–Cu alloys as the catalytic substrates (Illakkiya et al. 2016). Thus, CVD leads to very high-quality, large-area, and single-crystal graphene and is thus frequently the synthesis method of choice. Recently, the cold-wall CVD method has attracted interest over the conventional approach featuring a less complicated reactor, short deposition times, and rapid heating and cooling rates, which is instrumental in achieving fast growth of graphene. In this process, both the substrate and the chamber walls remain at room temperature, negating heating the entire chamber. On the other hand, the cold-wall method can suppress the gas phase reactions associated with high-temperature graphene growth ensuing the production of super-clean graphene (Boas et al. 2019). Microwave-plasma-enhanced CVD is carried out at reduced temperatures of 760 °C, which can achieve rapid processing of graphene in a single step (Akimoto and Ueno 2019). Also, current-enhanced CVD utilizing nickel catalysts at a low temperature of 464 °C can produce the high-crystallinity multilayer graphene. The dimensions of the synthesized graphene are not constrained to the size of the initial bulk graphite, allowing the polycrystalline mono-layer graphene to be produced as large as the primary substrate. The feature of the graphene produced is sensitive for electronic applications, as the graphene grows in a self-limiting way to a single or few layers only. The graphene produced is easily moved from the metal surface to an arbitrary substrate by supporting the graphene with a polymer and etching away from the metal. However, the transfer may contaminate and damage the graphene sheet.

The synthesis cost of graphene using the CVD method is moderate; as it is cheaper than the cost of epitaxial growth but more expensive than the cost of reducing graphene oxide (Tian et al. 2016). At high-temperature decomposable gases (CH_4 and C_2H_2) generate carbon atoms that can be adsorbed onto the surface of the substrate (such as Cu, Pt, Ni, Ru, and Ir) by thermal annealing to form a single graphene

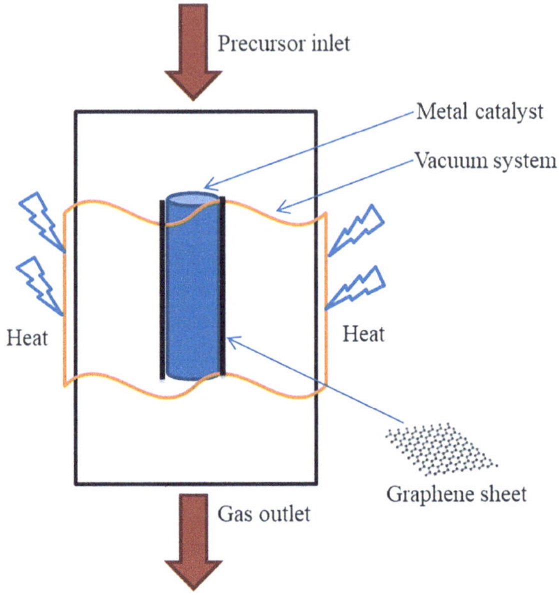

Fig. 2.7 Schematic diagram of graphene synthesis using the CVD approach

sheet which is applicable in the field of sensors (Ravani et al. 2013). Thus, CVD is considered the best technique for preparation of the monolayer or few-layer graphene which can be realized by applying thermal or plasma for the deposition processes. Furthermore, the CVD technique is talented with extracting carbon atoms from a carbon-rich sources by the reduction process. However, the major drawback of this technique is the difficulty in finding a appropriate substrate to grow graphene layers on as well as the complication of removing the graphene layers from the substrate without altering or destructing the graphene's atomic structure. Graphene with good quality was synthesized by CVD which makes this method interesting but the amount of graphene obtained by this method depends on the size of the support materials used (Tan et al. 2019). The high-quality product from this method is very important in electronic and optical applications. Apart from its advantage such as scalability, this technique is costly in comparison to oxidative exfoliation-reduction as the process is inhibited by the size of the substrate, type of precursor, and energy cost. Additionally, this method is restricted by the challenge of transferring the graphene product developed on the supporting material. Structural defects and impurities may be introduced into the graphene products during the transfer of the materials from the substrate and thus needs due consideration (Tian et al. 2016). Lastly, CVD source materials are highly toxic or flammable and the method needs high temperatures which demands the improvement or careful design and operation of the system.

2.4.4 *Arc Discharge*

An arc discharge (or electric arc) is a method in which a direct-current arc voltage is applied across two high purity graphite electrodes immersed in an inert gas such as He. The two electrodes are held at short distance apart under inert condition to evaporate some of the carbon atoms from the anode, and re-condensed as a hard cylindrical deposit on the cathodic rod. It is a bottom-up technique to synthesize bulk graphene nanoplatelets is an arc discharge method. Arc-discharge plasma technology uses different sizes of graphite rods as the cathode and anode, to synthesize carbon nano-materials in a relatively confined space in a special atmosphere. It was confirmed that this technology can synthesize multilayer graphene successfully (Li et al. 2010). The most important feature of this route is its suitability to successfully synthesize graphene flakes with and without the enhancement of a magnetic field (Fang et al. 2015). Furthermore, it has been demonstrated that plasmas of glow discharge are suitable for different experimental conditions to prepare graphene materials (Wu et al. 2009). It becomes common that the arc-discharge method is used for producing large-scale graphene sheets with fewer defects (functional groups). A few layers of graphene pieces are obtained in a mixture of hydrogen and helium (Lu et al. 1999) by helium arc discharge (Li et al. 2010), and N-doped multi-layered graphene sheets were prepared by arc discharge with a mixture of ammonia and helium as the buffer gas. A few layered graphene sheets were synthesized using the arc method created between two graphite electrodes subjected to a continuous electric current by division of graphite in helium used as a buffer gas (Kane et al. 2020). In another work, a few layered and good quality graphene was produced using the highly efficient and large-scale arc-discharge process in a buffer of carbon dioxide and helium. The arc-discharge method has its unique advantages: no metal catalyst is used; the process is time-saving and facile; high-quality graphene could be synthesized in decagram-scale at a low cost. Graphene production with the arc discharge method is a promising technique, as it is environmentally friendly and yields high-purity graphene flakes compared to chemical assisted graphite exfoliation. Plasmas of the arc discharge can also provide a convenient way to add functional groups in situ during synthesis. The high temperature between the electrodes (3000–4000 °C) that causes sublimation of the carbon precursors may limit the application of this method for future commercialization.

2.5 Comparison of Electrochemical Exfoliation with Other Methods

The broad use of graphene and its derivatives, high-efficiency and low-cost industrial-scale methods ought to be developed to produce superior 2D materials. Up to now, various top-down methodologies (e.g. scotch-tape isolation, chemical exfoliation, and liquid-phase exfoliation) and bottom-up methods (e.g. epitaxial growth, pyrolysis, and chemical vapor deposition) have been used to prepare 2D graphene materials. The advantages and disadvantages of different graphene preparation techniques are summarized in Table 2.2. However, the top-down methods are limited by being time-consuming, hazardous chemical usage, and the generation of defects. Similarly, the bottom-up methods are capable of preparing large-area ultrathin 2D materials but they are complex and costly and require high temperatures and high vacuum, and they need an additional transfer step to transfer the 2D products from the metal surface to the target substrates (Yang et al. 2019). Thus far, there is a need to address, the synthesis of structurally controlled graphene materials for realizing its real-world applications. A key target of graphene synthesis is to produce almost original graphene with low oxygen content, fewer layers, and a large lateral size. However, graphite exfoliation in liquid electrolytes in ambient conditions commonly introduces oxygen in the carbon lattice of graphene and breaks exfoliated graphene into small pieces. Recent investigations have shown remarkable improvements about the top-down approach, regarding time-consuming, hazardous chemical nature, and more defects generation. Comparatively, epitaxial growth, and CVD, bottom-up approaches have the considerable capability of fabricating ultrathin 2D materials containing large surface areas (Adetayo and Runsewe 2019). Nevertheless, aforesaid bottom-up methods are so complex that they are expensive and need high temperatures and pressure. Furthermore, there is a need to transfer the 2D materials fabricated products from the metal surface to the targeted substrate, making it difficult for controlling the synthesis process, and may induce defects and impurities into the products.

Electrochemical technologies are usually carried out under mild conditions, in comparison with, other synthesis technologies, as they have proven convenient and controllable conditions (Park et al. 2021). Electrochemical exfoliation methods have demonstrated some controllability in the structural properties of graphene materials in addition to its simplicity (Wei et al. 2018). These advantages need to be fully investigated with careful consideration and control of the operation parameters, such as electrolytes and applied potential biases. Thus, the electrochemistry-based technique is deliberated as a promising approach that provides high yield, great performance, low cost, and excellent up-scalability. Notably, playing with electrochemical parameters not only allows tunable properties but also enhances the content variety from graphene to a wide spectrum of 2D semiconductors (Ikram et al. 2020). Generally speaking, the production of graphene and its derivatives can be performed using different techniques. Each method employed in the synthesis of graphene has a major problem either the use of expensive equipment, low yield of graphene production, or the use or release of toxic chemicals.

Table 2.2 Advantages and disadvantages of different graphene production methods

Method	Advantages	Disadvantages	References
Mechanical exfoliation	Need relatively low processing temperatures, unmodified and large-size graphene sheets	Slow and imprecise, not suitable for large-scale production	Gao et al. (2018), Adetayo and Runsewe (2019), Pirzado et al. (2019)
Chemical exfoliation	Produces graphene in large quantities	Needs complex chemical processes and bulky defects	Zaaba et al. (2017), Badri et al. (2017)
Chemical synthesis	Yield large quantities of graphene	Low-quality yield and environmentally not safe	Betancur et al. (2018), Muramatsu et al. (2014)
Electrochemical exfoliation	Simple, cheap, and suitable for mass production, gives highly stable, good quality and conducting graphene, avoid the use of harsh chemicals	Needs expensive ionic liquids, degradation of graphite electrode	Park et al. (2021), Aghamohammadi et al. (2020), Dang et al. (2020)
Epitaxial growth	Good control over the number of graphene layers formed and a very large area of pure graphene	Expensive and produces poor-quality graphene, Very small-scale production	Jilani et al. (2018), Kong et al. (2020)
Chemical vapor deposition	It produces very-high-quality, large-area, and single-crystal graphene	The graphene transfer to another substrate may contaminate and damage the graphene sheet and relatively expensive	Jia et al. (2020), Tan et al. (2019)
Pyrolysis	Produces thin and homogeneous graphene film, no reducing agents required	Moderately expensive	Zhu and Xu (2020), Chen et al. (2017)
Arc discharge	The green method yields a large quantity of graphene with high purity	Expensive due to the need for a vacuum	Fang et al. (2015), Kane et al. (2020)

2.6 Conclusion

Advanced carbon nanomaterials such as graphene and its byproducts have gained substantial consideration in the nanotechnology and nanoelectronics industries. These wider applications are due to their extraordinary physicochemical properties including high specific surface area, excellent mechanical, chemical and thermal stability, good electrical conductivity and transparency. Therefore, the preparation of these relatively remerging materials plays a great role in synthesizing graphene materials with the desired standard and quantity for particular application. The industrial applications of graphene directly depend on the effective and efficient synthesis. Generally, common approaches such as top-down (mechanical, chemical, and electrochemical exfoliation, and chemical synthesis), and bottom-up (pyrolysis, epitaxial growth, arch discharge, and chemical vapor deposition) are used to synthesize graphene-based materials. However, each preparation method has its own advantage and disadvantage, therefore, the selection of the synthesis methods entirely depends on the available precursor and the final application. From the various techniques such the oxidative exfoliation-reduction, electrochemical method as a top-down route shows substantial advantages to prepare graphene-based materials in terms of cost, time, quality of the final product and environmental friendliness.

Acknowledgements This work was supported by the Department of Science and Innovation (DSI) and National Research Foundation (NRF), South Africa through the DSI/NRF South African Research Chairs Initiative for Professor of Medicinal Chemistry and Nanotechnology (UID 62620), DSI/Mintek Nanotechnology Innovation Centre, and Rhodes University, South Africa.

References

Abdelkader, A.M., Cooper, A.J., Dryfe, R.A.W., Kinloch, I.A.: How to get between the sheets: a review of recent works on the electrochemical exfoliation of graphene materials from bulk graphite. Nanoscale **7**, 6944–6956 (2015)

Achee, T.C., Sun, W., Hope, J.T., Quitzau, S.G., Sweeney, C.B., Shah, S.A., Habib, T., Green, M.J.: High-yield scalable graphene nanosheet production from compressed graphite using electrochemical exfoliation. Sci. Rep. **8**, 14525 (2018)

Adetayo, A., Runsewe, D.: Synthesis and fabrication of graphene and graphene oxide: a review. Open J. Compos. Mater. **9**, 207–229 (2019)

Aghamohammadi, H., Eslami-Farsani, R., Torabian, M., Amousa, N.: Recent advances in one-pot functionalization of graphene using electrochemical exfoliation of graphite: a review study. Synth. Met. **269**, 116549 (2020)

Akimoto, T., Ueno, K.: High crystallinity multilayer graphene deposited by a low-temperature CVD using Ni catalyst with applying current. In: Electron Devices Technology and Manufacturing Conference (EDTM), pp. 351–353 (2019)

Allen, M.J., Tung, V.C., Kaner, R.B.: Honeycomb carbon: a review of graphene. Chem. Rev. **110**, 132–145 (2009)

Badri, M.A.S., Salleh, M.M., MdNoor, N.F., Rahman, M.Y.A., Umar, A.A.: Green synthesis of few-layered graphene from aqueous processed graphite exfoliation for graphene thin film preparation. Mater. Chem. Phys. **193**, 212–219 (2017)

Baig, Z., Mamat, O., Mustapha, M., Mumtaz, A., Munir, K., Sarfraz, M.: Investigation of tip sonication effects on the structural quality of graphene nanoplatelets (GNPs) for superior solvent dispersion. Ultrason. Sonochem.. Sonochem. **45**, 133–149 (2018)

Berger, C., Song, Z., Li, T., Li, X., Ogbazghi, A.Y., Feng, R., Dai, Z., Marchenkov, A.N., Conrad, E.H.: First, P.N., deHeer, W.A.: Ultrathin epitaxial graphite: 2D electron gas properties and a route toward graphene-based nanoelectronics. J. Phys. Chem. B. **108**(52), 19912–19916 (2004)

Betancur, A.F., Ornelas-Soto, N., Garay-Tapia, A.M., Perez, F.R., Salazar, A., García, A.G.: A general strategy for direct synthesis of reduced graphene oxide by chemical exfoliation of graphite. Mater. Chem. Phys. **218**, 51–61 (2018)

Bhuyan, M.S.A., Uddin, M.N., Islam, M.M., Bipasha, F.A., Hossain, S.S.: Synthesis of graphene. Int. Nano Lett. **6**, 65–83 (2016)

Boas, C.R.S.V., Focassio, B., Marinho, E., Larrude, D.G., Salvadori, M.C., Leao, C.R., Dos Santos, D.J.: Characterization of nitrogen doped graphene bilayers synthesized by fast, low temperature microwave plasma-enhanced chemical vapour deposition. Sci. Rep. **9**, 13715 (2019)

Borovikov, V., Zangwill, A.: Step-edge instability during epitaxial growth of graphene from SiC(0001). Phys. Rev. B **80**(12), 121406(R) (2009)

Brodie, B.C.: Sur le poids atomique du graphite. Ann. Chim. Phys. **59**, 466–472 (1860)

Chen, H., Müller, M.B., Gilmore, K.J., Wallace, G.G., Li, D.: Mechanically strong, electrically conductive, and biocompatible graphene paper. Adv. Mater. **20**(18), 3557–3561 (2008)

Chen, J., Duan, M., Chen, G.: Continuous mechanical exfoliation of graphene sheets via a three-roll mill. J. Mater. Chem. **22**(37), 19625 (2012)

Chen, M., Haddon, R.C., Yan, R., Bekyarova, E.: Advances in transferring chemical vapour deposition graphene: a review. Mater. Horiz. **4**, 1054–1063 (2017)

Choi, W., Lahiri, I., Seelaboyina, R., Kang, Y.S.: Synthesis of graphene and its applications: a review. Crit. Rev. Solid State **35**(1), 52–71 (2010)

Copel, M., Reuter, M.C.: Surfactants in epitaxial growth. Phys. Phys. Lett. **63**, 632–635 (1989)

Dang, M.N., Nguyen, T.H., Nguyen, T.V., Thu, T.V., Le, H., Akabori, M., Ito, N., Nguyen, H.Y., Lu, T., Nguyen, T.H., Nguyen, V.T., Phan, N.H.: One-pot synthesis of manganese oxide/graphene composites via a plasma-enhanced electrochemical exfoliation process for supercapacitors. Nanotechnology **31**, 345401 (2020)

Edwards, R.S., Coleman, K.S.J.N.: Graphene synthesis: relationship to applications. **5**(1), 38–51 (2013)

Fang, X., Donahue, J., Shashurin, A., Keidar, M.: Plasma-based graphene functionalization in glow discharge. J. Appl. Phys. **4**, 1 (2015)

Fathy, M., Abdel, T.: Incorporation of multi-walled carbon nanotubes in microspheres used as anion exchange resin via suspension polymerization. Appl. Nanosci. **4**, 543–549 (2014)

Gao, E., Lin, S.Z., Qin, Z., Buehler, M.J., Feng, X.Q., Xu, Z.: Mechanical exfoliation of two-dimensional materials. J. Mech. Phys. Solids **115**, 248–262 (2018)

Guker, O., Guler, S.H., Selen, V., Albayrak, M.G., Evin, E.: Production of graphene layer by liquid-phase exfoliation with low sonication power and sonication time from synthesized expanded graphite. Fuller Nanotubes Carbon Nanostruct. **24**, 123–127 (2016)

Guo, S., Dong, S.: Graphene nanosheets: synthesis, molecular engineering, thin film, hybrids, and energy and analytical applications. Chem. Soc. Rev. **40**(5), 2644–2672 (2011)

Guo, H.L., Wang, X.F., Qian, Q.Y., Wang, F.B., Xia, X.H.: A green approach to the synthesis of graphene nanosheets. ACS Nano **3**(9), 2653–2659 (2009)

Gupta, B., Notarianni, M., Mishra, N., Shalei, M., Lacopi, F., Motta, N.: Evolution of epitaxial graphene layers on 3C SiC (111) as a function of annealing temperature in UHV. Carbon **68**, 563–572 (2014)

Hummers, W.S., Offeman, R.E.: Preparation of graphitic oxide. J. Am. Chem. Soc. **80**(6), 1339 (1958)

Hussien, N.A., Isklan, N., Turk, M.: Aptamer-functionalized magnetic graphene oxide nanocarrier for targeted drug delivery of pacilitaxel. Mater. Chem. Phys. **211**, 479–488 (2018)

Ikram, M., Raza, A., Ali, S., Ali, S.: Electrochemical exfoliation of 2D advanced carbon derivatives. In: Ikram, M., Maqsood, A. (eds.) 21st Century Advanced Carbon Materials for Engineering Applications—A Comprehensive Handbook. Intech Open, London (2020). https://doi.org/10.5772/intechopen.94892

Illakkiya, J.T., Rajalakshmi, P.U., Oommen, R.: Nebulized spray pyrolysis: a new method for synthesis of graphene film and their characteristics. Surf. Coat. Technol. **307**, 65–72 (2016)

Jia, K., Ci, H., Zhang, J., Sun, Z., Ma, Z., Zhu, Y., Liu, S., Liu, J., Sun, L., Liu, X., Sun, J., Yin, W., Peng, H., Lin, L., Liu, Z.: Superclean growth of graphene using a cold-wall chemical vapor deposition approach. Angew. Chem. Int. Ed.. Chem. Int. Ed. **59**(39), 17214–17218 (2020)

Jilani, A., Othman, M.H.D., Ansari, M.O., Hussain, S.Z., Ismail, A.F., Khan, I.U., Inamuddin.: Graphene and its derivatives: synthesis, modifications, and applications in wastewater treatment. Environ. Chem. Lett. **16**, 1301–1323 (2018)

Juang, Z.Y., Wu, C.Y., Lo, C.W., Chen, W.Y., Huang, C.F., Hwang, J.C., Chen, F.R., Leou, K.C., Tsai, C.H.: Synthesis of graphene on silicon carbide substrates at low temperature. **47**(8), 2026–2031 (2009)

Kamali, A.R., Fray, D.J.: Molten salt corrosion of graphite as a possible way to make carbon nanostructures. Carbon **56**, 121–131 (2013)

Kane, A., Barry, A.H., Chérif, S.M., Farhat, S.: Few-layered graphene synthesis by the arc-discharge method. In: Proceeding des JIC2019—Proceeding of the JIC2019, vol. 02, pp. 30–36 (2020)

Katariya, S.M.: Review on: synthesis approaches and applications of graphene and its derivatives. Int. J. Res. Tre. Innov. (IJRTI) **7**(4), 1–6 (2022)

Kong, X., Zhu, Y., Lei, H., Wang, C., Zhao, Y., Huo, E., Lin, X., Zhang, Q., Qian, M., Mateo, W., Zou, R., Fang, Z., Ruan, R.: Synthesis of graphene-like carbon from biomass pyrolysis and its applications. Chem. Eng. J. **399**, 125808 (2020)

Kumar, R., Singh, R.K., Kumar, P., Dubey, P.K., Tiwari, R.S., Srivastava, O.N.: Clean and efficient synthesis of graphene nanosheets and rectangular aligned-carbon nanotube bundles using green botanical hydrocarbon precursor: Sesame oil. Sci. Adv. Mater. **6**(1), 76–83 (2014)

Lang, B.A.: LEED study of the deposition of carbon on platinum crystal surfaces. Surf. Sci. **53**(1), 317–329 (1975)

Lee, X.J., Hiew, B.Y.Z., Lai, K.C., Lee, L.Y., Gan, S., Thangalazhy-Gopakumar, S., Rigby, S.: Review on graphene and its derivatives: synthesis methods and potential industrial implementation. J. Tai. Inst. Chem. Eng. **98**, 163–180 (2019)

Li, X.S., Cai, W.W., An, J.H., Kim, S., Nah, J., Yang, D.X., Piner, R., Velamakanni, A., Jung, I., Tutuc, E., Banerjee, S.K., Colombo, L., Ruoff, R.S.: Large-area synthesis of high-quality and uniform graphene films on copper foils. Science **324**, 1312–1314 (2009)

Li, N., Wang, Z., Zhao, K., Shi, Z., Gu, Z., Xu, S.: Large-scale synthesis of N-doped multi-layered graphene sheets by the simple arc-discharge method. Carbon **48**, 255–259 (2010)

Liu, N., Luo, F., Wu, H., Liu, Y., Zhang, C., Chen, J.: One-step ionic-liquid-assisted electrochemical synthesis of ionic-liquid-functionalized graphene sheets directly from graphite. Adv. Func. Mater. **18**(10), 1518–1525 (2008)

Liu, M., Zhang, X., Wu, W., Liu, T., Liu, Y., Guo, B., Zhang, R.: One-step chemical exfoliation of graphite to ~100% few-layer graphene with high quality and large size at ambient temperature. Chem. Eng. J. **355**, 181–185 (2019)

Liu, W.W., Aziz, A.: Review on the effects of electrochemical exfoliation parameters on the yield of graphene Oxide. **7**(38), 33719–33731 (2022)

Lu, X., Yu, M., Huang, H., Ruoff, R.S.: Tailoring graphite with the goal of achieving single sheets. Nanotechnology **10**, 269–272 (1999)

Luheng, W., Tianhuai, D., Peng, W.J.C.: Influence of carbon black concentration on piezoresistivity for carbon-black-filled silicone rubber composite. **47**(14), 3151–3157 (2009)

MaAllister, M.J., Li, J.L., Adamson, D.H., Schniepp, H.C., Abdala, A.A., Liu, J., Herrera-Alonso, M., Milius, D.L., Car, R., Prudhomme, R.K., Aksay, I.A.: Single sheet functionalized graphene by oxidation and thermal expansion of graphite. Chem. Mater. **19**(18), 4396–404 (2007)

Manukyan, K.V., Rouvimov, S., Wolf, E.E., Mukasyan, A.S.: Combustion synthesis of graphene materials. Carbon **62**, 302–3011 (2013)

Marcano, D.C., Kosynkin, D.V., Berlin, J.M., Sinitskii, A., Sun, Z., Slesarev, A., Alemany, L.B., Lu, W., Tour, J.M.: Improved synthesis of graphene oxide. ACS Nano **4**(8), 4806–4814 (2010)

Mishra, N., Boecki, J., Motta, N., Lacopi, F.: Graphene growth on silicon carbide: a review. Phys. Status Solidi A **213**, 2269–2289 (2016)

Moser, J., Barreiro, A., Bachtold, A.: Current-induced cleaning of graphene. Appl. Phys. Lett. **91**, 163513 (2007)
Muramatsu, H., Kim, Y.A., Yang, K.S., Cruz-Silva, R., Toda, I., Yamada, T., Terrones, M., Endo, M., Hayashi, T., Saitoh, H.: Rice husk-derived graphene with nano-sized domains and clean edges. Small **10**, 2766–2770 (2014)
Ni, Z., Wang, Y., Yu, T., Shen, Z.: Raman spectroscopy and imaging of graphene. Nano Res. **1**(4), 273–291 (2008)
Norimatsu, W., Kusunoki, M.: Epitaxial graphene on SiC {0001}: advances and perspectives. Phys. Chem. Chem. Phys. **16**(8), 3501–3511 (2014)
Novoselov, K.S., Geim, A.K., Morozov, S.V., Jiang, D., Zhang, Y., Dubonos, S.V., Grigorieva, I.V., Firsov, A.A.: Electric field effect in atomically thin carbon films. Science **306**(5696), 666–669 (2004)
Park, S., Ruoff, R.S.: Chemical methods for the production of graphenes. Nat. Nanotechnol. **4**, 217–224 (2009)
Park, S.W., Jang, B., Kim, H., Lee, J., Park, J.Y., Kang, S.O., Choa, Y.H.: Highly water-dispersible graphene nanosheets from electrochemical exfoliation of graphite. Front. Chem. **9**, 699231 (2021)
Parvez, K., Li, R., Puniredd, S.R., Hernandez, Y., Hinkel, F., Wang, S., Feng, X., Mullen, K.: Electrochemically exfoliated graphene as solution-processable, highly conductive electrodes for organic electronics. ACS Nano **7**(4), 3598–3606 (2013)
Pirzado, A., Le Normand, F., Romero, T., Paszkiewicz, S., Papaefthimiou, V., Ihiawakrim, D., Janowska, I.: Few-layer graphene from mechanical exfoliation of graphite-based materials: structure-dependent characteristics. Chem. Eng. **3**, 37–47 (2019)
Rao, C.N.R., Sood, A.K.: Graphene: Synthesis, Properties, and Phenomena. Wiley, Weinheim, Germany (2013)
Ravani, F., Papagelis, K., Dracopoulos, V., Parthenios, J., Dassios, K.G., Siokou, A., Galiotis, C.: Graphene production by dissociation of camphor molecules on a nickel substrate. Thin Solid Films **527**, 31–37 (2013)
Riedi, C., Coletti, C., Starke, U.: Structural and electronic properties of epitaxial graphene on SiC (0001): a review of growth, characterization, transfer doping and hydrogen intercalation. J. Phys. D Appl. Phys. **43**(37), 374009 (2010)
Roscher, S., Hoffmann, R., Prescher, M., Knittel, P., Ambacher, O.: High voltage electrochemical exfoliation of graphite for high-yield graphene production. RSC Adv. **9**, 29305–29311 (2019)
Ruiz-Hitzky, E., Darder, M., Fernandes, F.M., Zatile, E., Palomares, F.J., Aranda, P.: Supported graphene from natural resources: easy preparation and applications. Adv. Mater. **23**, 5250–5255 (2011)
Safavi, A., Tohidi, M., Mahyari, F.A., Shahbaazi, H.: One-pot synthesis of large scale graphene nanosheets from graphite-liquid crystal composite via thermal treatment. J. Mater. Chem. **22**, 3825–3831 (2012)
Shams, S.S., Sheng, L., Hu, R., Zhang, R., Zhu, J.: Synthesis of graphene from biomass: a green chemistry approach. Mater. Lett. **161**, 476–479 (2015)
Singh, P., Bahadur, J., Pal, K.: One-step one chemical synthesis process of graphene from rice husk for energy storage applications. Graphene **6**, 61–71 (2017)
Soldano, C., Mahmood, A., Dujardin, E.: Production, properties and potential of graphene. Carbon **48**(8), 2127–2150 (2010)
Stankovich, S., Piner, R.D., Nguyen, S.T., Ruoff, R.S.: Synthesis and exfoliation of isocyanate-treated graphene oxide nanoplatelets. Carbon **44**, 3342–3347 (2006)
Staudenmaier, L.: Verfahren zur Darstellung der Graphitsaure. Eur. J. Inorg. Chem. **31**(2), 1481–1487 (1898)
Su, C.Y., Lu, A.Y., Xu, Y., Chen, F.R., Khobystov, A.N., Li, L.J.: High-quality thin graphene films from fast electrochemical exfoliation. ACS Nano **5**(3), 2332–2339 (2011)
Subrahmanyam, K., Panchakarla, L., Govindaraj, A., Rao, C.: Simple method of preparing graphene flakes by an arc-discharge method. J. Phys. Chem. C **113**(11), 4257–4259 (2009)

Sun, Z., Fang, S., Hu, Y.H.: 3D graphene materials: from understanding to design and synthesis control. Chem. Rev. **120**(18), 10336–10453 (2020)

Tan, H., Wang, D., Guo, Y.: A strategy to synthesize multilayer graphene in arc-discharge plasma in a semi-opened environment. Materials (Basel) **12**(14), 2279 (2019)

Tian, S., Yang, Y., Liu, Z., Wang, C., Pan, R., Gu, C., Li, J.: Temperature-dependent Raman investigation on suspended graphene: contribution from thermal expansion coefficient mismatch between graphene and substrate. Carbon **104**, 27–32 (2016)

Tung, V.C., Allen, M.J., Yang, Y., Kaner, R.B.: High-throughput solution processing of large-scale graphene. Nat. Nano Technol. **4**, 25–29 (2009)

Vallés, C., Drummond, C., Saadaoui, H., Furtado, C.A., He, M., Roubeau, O., Ortolani, L., Monthioux, M., Penicaud, A.: Solutions of negatively charged graphene sheets and ribbons. J. Am. Chem. Soc. **130**(47), 15802–15804 (2008)

Viculis, L.M., Mack, J.J., Mayer, O.M., Hahn, H.T., Kaner, R.B.: Intercalation and exfoliation routes to graphite nanoplatelets. J. Mater. Chem. **15**(9), 974–978 (2005)

Wang, X., Fulvio, P.F., Baker, G.A., Veith, G.M., Unocic, R.R., Mahurin, S.M., Chi, M., Dai, S.: Direct exfoliation of natural graphite into micrometer size few layers of graphene sheets using ionic liquids. Chem. Commun. **46**, 4487–4489 (2010)

Wang, J., Jin, X., Li, C., Wanga, W., Wu, H., Guo, S.: Graphene and graphene derivatives toughening polymers: toward high toughness and strength. Chem. Eng. J. **370**, 831–854 (2019)

Wei, M.P., Chai, H., Cao, Y.L., Jia, D.Z.: Sulfonated graphene oxide as an adsorbent for removal of Pb2+ and methylene blue. J. Colloid Interface Sci. **524**, 297–305 (2018)

Wu, Z.S., Ren, W.C., Gao, L.B., Zhao, J.P., Chen, Z.P., Liu, B.L., Tang, D.M., Yu, B., Jiang, C.B., Cheng, H.M.: Synthesis of graphene sheets with high electrical conductivity and good thermal stability by hydrogen arc discharge exfoliation. ACS Nano **3**, 411–417 (2009)

Xia, Z., Bellani, V., Sun, J., Palermo, V.: Electrochemical exfoliation of graphite in H_2SO_4, Li_2SO_4 and $NaClO_4$ solutions monitored in situ by Raman microscopy and spectroscopy. Faraday Discuss. **227**, 291–305 (2021)

Xu, Y.X., Bai, H., Lu, G.W., Li, C., Shi, G.Q.: Flexible Graphene films via the filtration of water-soluble noncovalent functionalized graphene sheets. J. Am. Chem. Soc. **130**, 5856–5857 (2008)

Yan, Y., Nashath, F.Z., Chen, S., Manickam, S., Lim, S.S., Zhao, H., Lester, E., Wu, T., Pang, C.H.: Synthesis of graphene: Potential carbon precursors and approaches. Nanotechnol. Rev. **9**, 1284–1314 (2020)

Yang, W., Chen, G., Shi, Z., Liu, C.C., Zhang, L., Xie, G., Cheng, M., Wang, M., Wang, D., Yang, R., Shi, D., Watanabe, K., Taniguchi, T., Yao, Y., Zhang, Y., Zhang, G.: Epitaxial growth of single-domain graphene on hexagonal boron nitride. Nat. Mater. **12**, 792–797 (2013)

Yang, S., Lohe, M.R., Müllen, K., Feng, X.: New-generation graphene from electrochemical approaches: production and applications. Adv. Mater. **28**, 6213–6221 (2016)

Yang, Z., Tian, J., Yin, Z., Cui, C., Qian, W., Wei, F.: Carbon nanotube- and graphene-based nanomaterials and applications in high-voltage supercapacitor: a review. Carbon **141**, 467–480 (2019)

Yi, M., Shen, Z.: A review on mechanical exfoliation for the scalable production of graphene. J Mater Chem A **3**(22), 11700–11715 (2015)

Yu, H., Zhang, B., Bulin, C., Li, R., Xing, R.: High-efficient synthesis of graphene oxide based on improved Hummers method. Sci. Rep. **6**, 36143 (2016)

Zaaba, N.I., Foo, K.L., Hashim, U., Tan, S.J., Liu, W.W., Voon, C.H.: Synthesis of graphene oxide using modified hummers method: solvent influence. Procedia Eng. **184**, 469–477 (2017)

Zhong, Y.L., Tian, Z., Simon, G.P., Li, D.: Scalable production of graphene via wet chemistry: progress and challenges. Mater. Today **18**(2), 73–78 (2015)

Zhu, Z., Xu, Z.: The rational design of biomass-derived carbon materials towards next-generation energy storage: a review. Renew. Sustain. Energy Rev. **134**, 110308 (2020)

Zhu, C., Guo, S., Fang, Y., Dong, S.: Reducing sugar: new functional molecules for the green synthesis of graphene nanosheets. ACS Nano **4**, 2429–2437 (2010a)

Zhu, Y., Murali, S., Cai, W., Li, X., Suk, J.W., Potts, J.R., Ruoff, R.S.: Graphene and graphene oxide: synthesis, properties, and applications. Adv. Mater. **22**(35), 3906–3924 (2010b)

Chapter 3
Graphene and Its Derivatives: Various Routes of Synthesis

Sudeepa Devi, Vijayeta Pal, Neha Yadav, Rimpa Jaiswal, Debmalya Roy, and Upendra K. Tripathi

Abstract Graphene and its derivatives have attracted significant attention among the scientific community due to their exceptional properties, such as their large surface area, zero band gap, high electrical conductivity, excellent mechanical stability, high optical transmission, and high thermal conductivity, which sparks intensive interests in various fields. Graphene is the first 2-dimensional atomic allotrope of carbon which consists of a single layer of sp^2-bonded carbon atoms arranged in a honeycomb lattice structure. Over the past decades, graphene technologies and its derivatives have experienced rapid development, which now encompasses a large family group that includes reduced graphene oxide, graphene oxide, vertical graphene, graphene sponge, graphene foam, and various 3D-graphene architectures. Although a large number of researchers have been despite the extensive studies on graphene and its derivatives, only limited comprehensive research is available on the various routes of synthesis of graphene and its derivatives. In this chapter, we are addressing the knowledge gap by exploring and discussing brief history of graphene, the general introduction of graphene derivatives, i.e., graphene oxide (GO), reduced graphene oxide (rGO), graphene quantum dots (GQDs), few layer graphene (f-LG) and various routes of synthesis of graphene and its derivatives. It also supported in detail the various routes of synthesis of graphene and its derivatives, namely top-down and bottom-up

S. Devi (✉) · U. K. Tripathi
Janta Maha Vidyalaya (CSJM University), Ajitmal, Auraiya 206121, Uttar Pradesh, India
e-mail: sudeepasharma113@gmail.com

D. Roy
Directorate of Nanomaterials, DMSRDE, Kanpur 208013, Uttar Pradesh, India

V. Pal
Materials Science and Technology, Indian Institute of Technology (BHU), Varanasi 221005, Uttar Pradesh, India

N. Yadav
Department of Biological Science & Bioengineering, Indian Institute of Technology, Kanpur 208016, Uttar Pradesh, India

R. Jaiswal
High Pressure and Synchrotron Radiation Physics Division, Bhabha Atomic Research Centre, Mumbai 400085, India

R. Khan et al. (eds.), *Electrochemical Exfoliation of Graphene and its Derivatives*, Engineering Materials, https://doi.org/10.1007/978-981-97-2128-3_3

approaches: including electrochemical exfoliation, mechanical exfoliation, plasma synthesis, epitaxial growth on silicon carbide, chemical vapour deposition (CVD), arc discharge and some other synthesis techniques.

Keywords Graphene · Synthesis · GO · rGO · GQDs · Few layer Graphene(f-LG) · Top-down approach and bottom-up approach

3.1 Introduction

Graphene and its derivatives have become popular nanomaterials in material science and nanotechnology. The most stable carbon form is graphite, used in pencil leads and as an industrial lubricant. Graphite is a layered sheet structure of carbon atoms in which all carbon atoms are present in a hexagonal pattern, basically graphene a single layer of graphite (Ando 2009). Graphene is a remarkable material that has earned many accolades. It is the strongest material ever researched and the thinnest material known to exist in the universe (almost 1 million times thinner than a human hair) (Geim 2009). Graphene is the first single-layered 2D atomic allotrope of carbon, consisting of a honeycomb lattice structure (Shen 2012). It is an electroactive and transparent material containing a zero-band gap. Further, it also has some additional unique properties like enzyme loading, high enzyme operating stability, strong catalytic activity, a rapid electron transfer rate for enzyme immobilisation, and a large surface area (Choudhary et al. 2020). Graphene shows exceptional electrical, mechanical, and thermal properties due to the presence of long-range π-conjugation, which piqued the interest of numerous theoretical investigations. The large-scale synthesis of graphene and its derivatives, such as few-layer graphene (f-LG), graphene oxide (GO), graphene quantum dots (GQDs), and reduced graphene oxide (r-GO), has been the subject of various research initiatives. Few-layer graphene and graphene quantum dots are fascinating materials due to their excellent optical, mechanical, and electrical characteristics (Liao et al. 2018). Graphene and its derivatives, such as graphene oxide (GO), reduced graphene oxide (r-GO), few-layer graphene (f-LG), graphene nanoribbons, graphene nanosheets, etc., are the most important nanomaterials of the graphene family. Due to their extraordinary physical and chemical properties, graphene and graphene-based materials are often referred to as "nanoparticles" or "nanomaterials".

Recent trends in graphene awareness are quite impressive, and they point to graphene as a potential new material source for the growth of the science, engineering, health, and composite sectors. The demand for graphene and its derivatives is rising on a global scale due to its potential catalysis and wide range of applications such as fuel cells (Jafri et al. 2010), batteries (Wu et al. 2010), adsorption (Chandra and Kim 2011; Gao et al. 2011), sensors (Shao et al. 2010) drug delivery, and biosensors. Graphene can be produced by using a variety of techniques, including reduction of graphene oxide (Pei and Cheng 2011), unzipping of carbon nanotubes, mechanical exfoliation of graphite (Novoselov et al. 2000), electrochemical exfoliation, and chemical vapor deposition (CVD) (Ambrosi and Pumera 2014). The synthesis of graphene and its derivatives has been a significant challenge, and many researchers have been working to find an easier and more efficient way to produce high-quality monolayer graphene. In the past decades, there have been significant advances in graphene synthesis using chemical vapor deposition (CVD), which has allowed for daily manipulation of monolayer graphene in research laboratories and potential use in industrial processes. The reproducible production of high-quality monolayer graphene sheets on a large scale has become possible due to the significant progress made by CVD techniques. Schedin et al. (2007) conducted the first examinations on exfoliated graphene and demonstrated its ability to detect individual gas particles using Hall effect measurements. This showcased graphene's potential for gas sensing applications. Liquid exfoliation produces graphene from graphite flakes, or highly ordered pyrolytic graphite (HOPG). Liquid exfoliation involves dispersing graphite flakes or HOPG in a solvent and subjecting the suspension to sonication. This disrupts the $\pi-\pi$ inter-sheet interactions between graphene layers and promotes their separation into individual graphene sheets or small stacks. The choice of solvent is crucial for efficient exfoliation. The solvent should have a surface tension similar to that of graphene to minimise re-aggregation of exfoliated sheets, as well as a high boiling point to enable sonication at elevated temperatures. Some common solvents used for liquid phase exfoliation includes: N-methyl pyrrolidone (NMP), dimethylformamide (DMF), and dimethylacetamide (DMAc). However, it is still challenging to produce high-quality graphene using a simple procedure. Recent years have seen a lot of interest in graphene produced by the electrochemical exfoliation of graphite because it provides an eco-friendly, simple, safe, and scalable procedure for making high-quality graphene (Gondosiswanto et al. 2015) (Fig. 3.1).

The electrochemical exfoliation production of graphene has been proposed as a substitute method of the classical liquid-exfoliation technique. This method can produce graphene of similar quality and quantity in just a few minutes. Recent research has shown that fast electrochemical production of graphene flakes can be achieved using anodic oxidation of graphite in room-temperature ionic liquids (RTILs) or in aqueous solutions. Electrochemically exfoliated-graphene can be synthesized in a variety of electrolytes, including acidic media (Liu et al. 2012, 2013), inorganic salts (Chen and Xue 2014; Parveen et al. 2015), and ionic liquids (Lu et al. 2009; Lu and Zhao 2013). Among all of these, it has been found that inorganic salts produce graphene with a higher lateral size and less oxygenated functional groups as compared to other electrolytes (Parvez et al. 2014). Graphene and its derivatives have shown immense potential in the field of biomedicine due to their tunable structure, unique physicochemical properties, and excellent biocompatibility. These materials have been extensively studied for their use in biomedical imaging, drug delivery, biosensing, and cancer therapy. The use of graphene and its derivatives has revolutionized the field of biomedicine and holds great promise for the development of new and improved biomedical technologies.

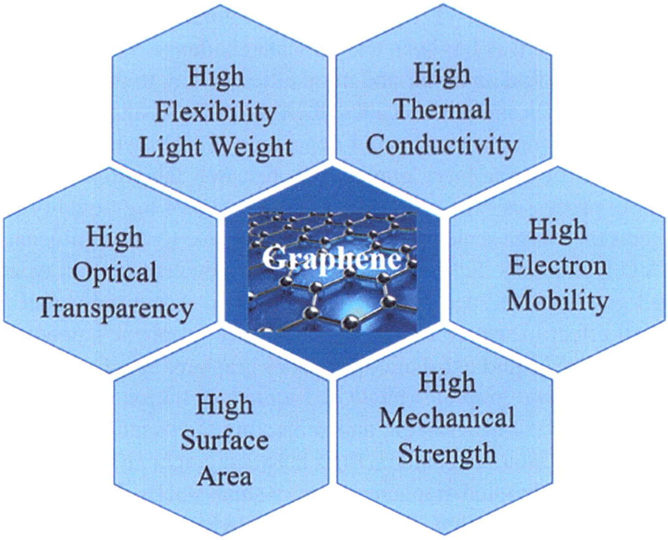

Fig. 3.1 Graphical representation of unique properties and possible applications of graphene

3.2 Brief History of Graphene

Over the past decades, graphene has attracted extensive worldwide attention and has become a prodigious nanomaterial in the fields of material science and nanotechnology. In 2004, "Graphene" was first time isolated by Novoselov et al. (2004). Andre Geim et al. won the Nobel Prize in physics in 2010 for their groundbreaking research on graphene. The word ***"Graphene"*** is derived from the prefix "Graph" which comes from graphite, and the suffix "ene" comes from the carbon–carbon double bond (Bianco et al. 2013). Graphene is the first two-dimensional atomic allotrope of carbon, which consists of a single layer of sp^2-bonded carbon atoms in the form of a packed hexagonal crystal that resembles a honeycomb lattice structure (Gaj et al. 2020; Novoselov et al. 2005). The newest carbon allotrope, "graphene," is the fundamental building block of other significant carbon allotropes, including 1D-carbon nanotubes (CNT), a cylindrical structure formed by rolling of graphene sheets; 3D-graphite, a stacked structure of multiple graphene sheets, and 0D-fullerene (C60) a Bucky ball sphere formed by wrapping of graphene sheets (Fig. 3.2). The terms monolayer, bilayer, and trilayer graphene, respectively, refer to single-layer, double-layer, and triple-layer graphitic sheets. Graphene with a thickness of 5–30 layers is commonly known as multilayer graphene or thick graphene. A single graphene sheet has a carbon–carbon bond distance of approximately 0.142 nm (1.42A°) and a layer thickness of 0.33 nm (3.3A°) (Novoselov et al. 2000; Shinohara 2015). The honeycomb lattice structure of graphene is responsible for its high electrical conductivity (Moser et al. 2007), superior stretchability, high thermal conductivity, transparency,

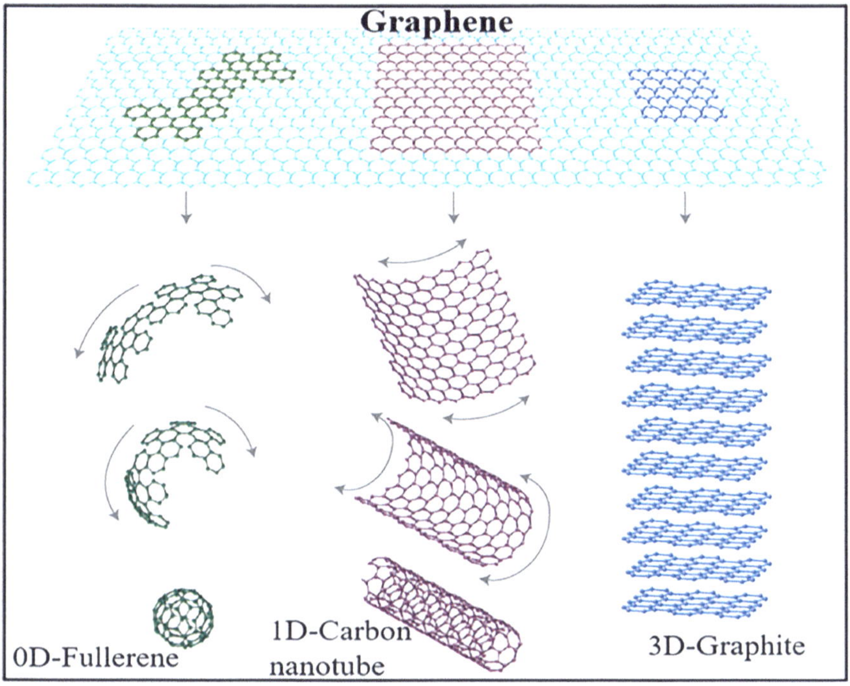

Fig. 3.2 2D hexagonal Graphene nanosheet as a building block of other carbon allotropes. An illustration of significant carbon allotropes arising from a 2D graphene nanosheet. Reprinted with Copyright permission from Geim et al. (2007). Copysight2007@Nature Materials

excellent mechanical stability, high specific surface area, and other exceptional properties (Huang et al. 2011). Huang et al. demonstrated that the tensile strength (young modulus) of graphene is 1.0 TPa and the maximum tensile stress (breaking strength) is ~42 Nm^{-1} (Lee et al. 2008), which is 200 times stronger than steel and makes it one of the most potent substances ever discovered. It is appropriate for practically every application, including electronics, biomedical, environmental, textile, etc. Furthermore, the high electronic transportation (high intrinsic mobility) of graphene at Dirac point is 200,000 cm^2v^{-1} s^{-1}(Bolotin et al. 2008; Morozov et al. 2008). Graphene also possesses a high specific surface area of 2630 m^2g^{-1} (Zhu et al. 2010; Dimiev and Eigler 2016), and another interesting strong property of graphene is its high thermal conductivity of 5000 m^{-1} K^{-1} (Balandin et al. 2008).

Graphene's new properties have been discovered by tuning its structure. These include superconductivity findings in twisted graphene superlattices, the opening of a bandgap in layered graphene and its derivatives, quantum tunnelling through vertical heterogeneous stacking, and the magnetic study of graphene in triangular flakes with zig-zag edges. High in-plane thermal conductivity and low out-of-plane thermal conductance are the two defining characteristics of the thermal properties of graphene. The specific heat of graphene, which is slightly higher than that of diamond and graphite below room temperature, is controlled by the phonons (Pop et al. 2012). The excellent thermal conductivity of graphene makes it a promising material for all electronic and photonic applications (Balandin 2011).

3.3 Graphene Derivatives

Different derivatives of graphene are being researched, including graphene oxide (GO), reduced graphene oxide (r-GO), few-layer graphene (f-LG), graphene quantum dots (GQDs), etc. (Wu et al. 2018). Among them, graphene oxide (GO) and reduced graphene oxide (r-GO) as a graphene-driven sheet are the closest relatives of graphene. These graphene derivatives have attracted extensive attention in the graphene family and are used in a wide range of applications due to their distinctive properties (Dideikin and Vul' 2019) (Fig. 3.3).

Graphene's Terminology: Based on the number of layers, graphene can be described in different categories and here all categories are described:

- Very-few-layer graphene (v-FLG): Graphene that consists of only 1–3 layers of carbon atoms It is sometimes also called monolayer or single layer graphene.
- Few-layer graphene (f-LG): f-LG refers to graphene that consists of 2–5 layers of carbon atoms. It is sometimes also called bi-layer or tri-layer graphene.

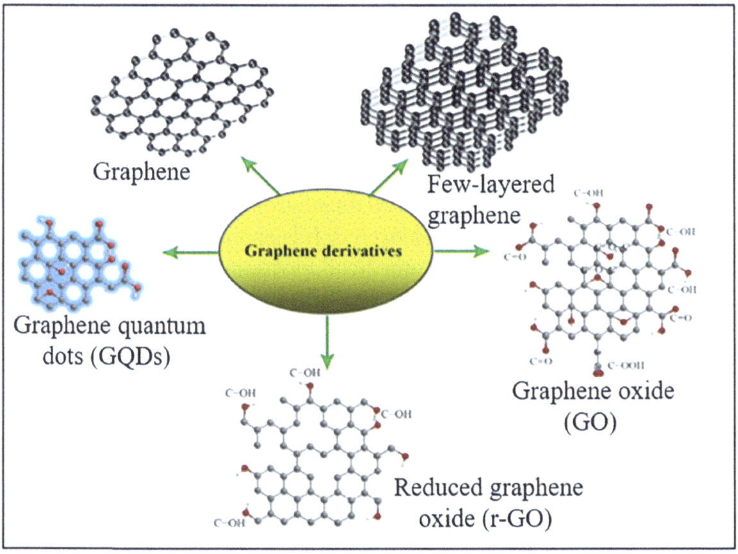

Fig. 3.3 Representation of chemical structures of graphene and its derivatives. Reprinted with permission from Ramezani and Dehghani (2019). Copyright 2019@Elsevier

- Multi-layer graphene (MLG): Graphene that consists of 2–10 layers of carbon atoms, abbreviated as multi-layer graphene.
- Graphene nanoplatelets (GNP): GNP refers to graphene that consists of more than 10 layers of carbon atoms. They are typically thicker (1 and 3 nm) and more irregularly shaped (lateral dimension ~100–100 μm) than other types of graphene and may have rough or curved edges.

3.3.1 Graphene Oxide (GO)

Graphene oxide (GO) is a single-layer planar 2D-dimensional nanostructure consisting of a honeycomb lattice structure, in which carbon atoms are arranged in sp^2 and sp^3 hybridization. It can be synthesized by using different fundamental techniques, i.e., Brodie, Staudenmaier, Hofmann, Hummers, and modified Hummers. Graphene oxides possess different active oxygen-containing functional groups, including carbonyl (–CO), epoxy (>O), carboxylic acid (–COOH), and hydroxyl (–OH). It contains amorphous as well as crystalline defect-regions, which reveal the presence of sp^3-hybridized carbon and oxygen functional groups (Luo et al. 2013; Wei and Qu 2012). The amorphous regions in graphene oxide have a disordered structure and contain sp^3 hybridized carbon atoms bonded to oxygen functional groups such as hydroxyl (–OH) and epoxy (–O–) groups. These groups create a negatively charged surface, which makes graphene oxide highly hydrophilic and water soluble. The crystalline regions in graphene oxide are more ordered and contain a mixture of sp^2 and sp^3-hybrized carbon atoms, which contribute to the acidity of graphene oxide. GO is hydrophilic in nature and easily dispersible in water, while graphene is hydrophobic in nature and difficult to do so. Researchers have experimentally demonstrated that graphene oxide (GO) possesses many unique and superior properties compared to graphene and its derivatives. Some of the most notable properties of GO include:

- High specific surface area: GO has a very large surface area per unit mass, which makes it an excellent material for adsorption applications.
- High water solubility: unlike graphene, which is hydrophobic, GO is highly water soluble due to the presence of oxygen-containing functional groups on its surface.
- Good biocompatibility: GO has been found to be biocompatible, which makes it a promising material for biomedical applications.
- High adsorption capacity: GO has a high adsorption capacity for a variety of substances, including heavy metals, organic pollutants, and dyes.
- Active-oxygen-containing functional groups: GO possesses a large number of oxygen-containing functional groups on its surface, which makes it a versatile material for various applications, including catalysis, energy storage, and drug delivery.

Recently, graphene has drawn a lot of interest as a potential pioneer in the large-scale production of graphene derivatives, leading to low synthesis costs as well as the possibility to customise the electrical properties as a semimetal, an insulator, or a semiconductor. These graphene-based materials have a wide range of applications in the fields of energy storage (Lightcap and Kamat 2013), separation, removal of environmental contaminants, purification, electronic devices (Eda and Chhowalla 2010; Wan et al. 2011), surfactant (Kim et al. 2010a, b), catalyst (Pyun 2011), and biomedical (Sun et al. 2008).

3.3.2 Reduced Graphene Oxide (r-GO)

Reduced graphene oxide (r-GO) is indeed an attractive member of the graphene family due to its unique properties and potential applications, which are distinct from both graphene oxide and pristine graphene. r-GO is synthesized by reducing graphene oxide, a material obtained by the oxidation of graphene. The reduction process involves removing the oxygen containing groups in the form of carbon dioxide (CO_2) from graphene oxide, which restores the sp^2 carbon–carbon bonding network and increases its electrical conductivity. Moreover, the synthesis of reduced graphene oxide (r-GO) using the reduction of graphene oxide by various reagents was also performed (MacHado and Serp 2012; Pumera 2010). Numerous methods have been reported for the partial reduction of graphene oxide, of which chemical and thermal reduction are important and easy for mass-scale production of r-GO. Chemical reduction uses a variety of hazardous reducing agents like hydrazine to produce GO with nitrogen heteroatoms (Stankovich et al. 2007b), hydrolic acid (Shin et al. 2009), vitamin C (Fernández-Merino et al. 2010), and $NaBH_4$ (Pei et al. 2010) to remove the oxygen groups from graphene oxide. The prevalence of carboxyl groups, numerous defects, and reduced oxygen moieties of r-GO make it differ from GO. On the other hand, thermal reduction, uses high-temperature annealing in a reducing atmosphere to restore the sp^2 bonding network. Biological reduction employs enzymes or bacteria to reduce graphene oxide to r-GO. r-GO has unique properties such as high electrical conductivity, large surface area, and excellent mechanical strength. These properties make it useful in a wide range of applications, including biosensing (Chae et al. 2016), tissue engineering, and cell culture (Jalili-Firoozinezhad et al. 2017), and for eliminating environmental pollutants like nitroaromatics (Dong et al. 2015) energy storage devices, sensors, catalysis, and electronic devices. r-GO has also been used as a filler material in composites to improve the mechanical and electrical properties of the composite material.

3.3.3 Graphene Quantum Dots (GQDs)

Graphene quantum dots (GQDs) are the newest member of the carbon family with a zero-dimensional (0D) structure that were first time synthesized by Ponomarenko and Geim in 2008 (Benítez-Martínez and Valcárcel 2015; Li et al. n.d.; Ponomarenko et al.

2008). Graphene quantum dots, or GQDs, are nanoscale materials that are composed of a single or a few layers of graphene sheets with a lateral dimension of less than 10 nm (Wang et al. 2015a, b). GQDs exhibit various shapes, such as circular, elliptical, triangular, and hexagonal. There are two main approaches for synthesizing GQDs: top-down and bottom-up strategies. Their properties are determined by synthetic methods, defects, and functional groups that influence their properties and have potential applications in various fields. Unlike carbon dots or CDs, GQDs possess a crystalline graphene lattice structure in their core, which gives them unique electronic and optical properties. Graphene quantum dots, or GQDs, are a unique class of nanomaterial that combine the electronic and optical properties of graphene with the quantum confinement effect of quantum dots (QDs) (Müller and Müllen 2007). In the case of graphene quantum dots (GQDs), the confinement of electrons within the small size of the GQD structure leads to changes in their electronic and optical properties. The different nature and density of sp^2 sites in the structure of QDs can significantly affect their optical characteristics. sp^2 sites refer to the carbon atoms in graphene that are bonded with three other carbon atoms in a trigonal planar configuration. The density and arrangement of these sites can affect the energy levels of electrons in the GQD and, hence, the optical properties. Smaller GQDs with a higher density of sp^2 sites tend to have higher energy bandgaps and emit light in the blue or ultraviolet range. On the other hand, larger GQDs with a lower density of sp^2 sites tend to have lower energy bandgaps and emit light in the red or near-infrared range. Therefore, they have a size-dependent bandgap due to their finite size, which makes them promising materials for a wide range of applications, including energy storage and conversion, electronics, optoelectronics (Tsai et al. 2016), and biomedicine (Li et al. 2017). Graphene quantum dots (GQDs) possess a large specific surface, structural/chemical stability (Luk et al. 2012), high electron mobility, strong fluorescence (Sapkota et al. 2017), and excellent photoluminescence (Feng et al. 2017; Park et al. 2020) properties, making them suitable for use in catalysis, sensors (Fan et al. 2015), solar cells (Tang et al. 2017), optoelectronics, energy conversion, and photovoltaic devices (Li et al. 2011). Their edge effects also make them attractive for use in electronic devices, as they can provide enhanced charge transport properties and act as active sites for chemical reactions. In addition, GQDs exhibit excellent biocompatibility (Wu et al. 2018), low-toxicity (Kim et al. 2016; Song et al. 2015), and high-water solubility (Li et al. 2010), making them a promising nanomaterial of the carbon family for biomedical applications such as bioimaging (Lin et al. 2016; Nair et al. 2017), drug delivery (Dong et al. 2018; Qiu et al. 2015; Wang et al. 2013a, b), and theranostic (Revia and Zhang 2016; Schroeder et al. 2016). Overall, the combination of graphene and quantum dots (QDs) properties in GQDs provides a unique set of characteristics which make them highly versatile and suitable for a wide range of applications.

3.3.4 *Few-Layer Graphene*

Few-layer graphene (f-LG) refers to graphene with 2–10 layers. As the number of layers increases, there are changes in the electronic properties of the material and an increase in thermal conductivity as compared to monolayer graphene (Zhou et al. 2018). f-LG also provides a large surface area and high quality (Zhou et al. 2018). Few-layer graphene has attracted attention in various applications, including energy storage and conversion, electronics, and biomedicine. Its unique properties, such as high electrical conductivity, mechanical strength, and flexibility, make it an attractive nanomaterial for a wide range of applications. With ongoing research and development, f-LG is poised to have a significant impact on various fields in the future. Few-layer graphene can be synthesized using different methods, including exfoliation (Wee et al. 2017), chemical vapour deposition (CVD) (Park et al. 2009), and Microfluidization (Cooper et al. 2013). Electrochemical exfoliation is a more environmentally friendly and economically cheaper method to synthesizing few-layer graphene (f-LG) at large scale by simply controlling potential operations at ambient temperature and pressure. f-LG also has promising applications in bioimaging due to its ability to manipulate interlayer interactions and control electronic states (Subrahmanyam et al. 2010). It can also be used for nanoparticle coating, which provides an interesting design for bioimaging (Bachmatiuk et al. 2013; Mendes et al. 2012; Schinwald et al. 2012). However, it is important to note that few-layer graphene possesses low solubility as well as cytotoxicity in biological environments. The surface binding of proteins and other bovine serum (BSA) on f-LG has been found to enhance its biocompatibility, making it a promising material for biomedical applications (Mao et al. 2016). The incorporation of other biocompatible molecules with f-LG can further improve its properties and expand its potential applications in the fields of biomedicine. By utilising f-LG as a platform for surface modifications, it is possible to create tailored materials with improved biocompatibility, biodegradability, and functionality, which are crucial characteristics for biomedical materials.

3.4 Various Routes of Synthesis of Graphene and Its Derivatives

Graphene has been a topic of intense research and interest for several years now, and its unique properties have made it a promising material for various fields of applications. There have been numerous studies on graphene synthesis, and several techniques for graphite exfoliation into graphene and graphene-based materials have been reported. In 2004, a ground breaking method for producing two-dimensional graphene was discovered by Novoselov et al. known as "scotch tape" or "peeling off" (Novoselov et al. 2000). Shortly after this discovery, Novoselov et al. demonstrated the growth of graphene on a single-crystal platinum substrate by thermal decomposition of carbon. This technique, named chemical vapour deposition (CVD), involves heating a carbon-containing gas such as methane or ethylene in the presence of a metal catalyst, which leads to the deposition of graphene onto the substrate. Since then, a lot of research has been conducted to find simple and cost-effective methods for large-scale industrial production of this material (Guo and Dong 2011; Novoselov et al. 2012). Different synthesis techniques play a crucial role in determining the quality and performance of graphene. Researchers have developed several techniques to produce graphene with various properties, including surface chemistry and morphology. Avouris and Dimitrakopoulos (2012), Terrones et al. (2010) have studied and documented the impact of these techniques on the production of high-quality graphene (Avouris and Dimitrakopoulos 2012; Terrones et al. 2010). The selection of an appropriate method for synthesizing graphene is crucial for its intended application. Graphene can be synthesized using the solid, liquid, or gas phase of graphite (Dato et al. 2008; Hernandez et al. 2008; Stankovich et al. 2007b). The production of graphene primarily relies on two main approaches: top-down (destruction) and bottom-up (construction) approaches. These approaches are the backbone of high-quality graphene production on a large scale. The top-down synthesizing approach involves breaking down a larger material into smaller pieces, ultimately producing nano-sized graphene sheets. This method is typically used to produce graphene from bulk graphite or graphite derivatives. The class of top-down approaches includes liquid phase exfoliation (Güler et al. 2016; Haar et al. 2016), unzipping of carbon nanotube (CNT) (Chua and Pumera 2014; Dimiev et al. 2018), mechanical exfoliation (Novoselov et al. 2000), oxidative exfoliation reduction (Chua and Pumera 2014; Dimiev et al. 2018), arc discharge (IEEE Staff 2016) and electrochemical exfoliation. Similarly, the bottom-up synthesizing approach involves building up graphene from smaller building blocks or molecules. This approach is typically used to produce graphene through chemical vapour deposition (CVD Jacobberger et al. 2015; Son and Ham 2017), epitaxial growth (Yazdi et al. 2016), substrate-free gas-phase synthesis (SFGP) (Koo and Paek 2021), total organic synthesis (Singh et al. 2011). The carbon community has documented electrochemical exfoliation of graphite and mechanical exfoliation as the top-down, low-cost, and simple approaches for graphene synthesis to account for the enormous surge in curiosity in graphene synthesis. Both approaches i.e., top-down and bottom-up have their advantages and disadvantages depending on the desired application and properties of the resulting graphene (Fig. 3.4).

3.4.1 Electrochemical Exfoliation

Electrochemical exfoliation is a promising method for producing high quality graphene and graphene's derivatives on an industrial scale, with the benefits of efficiency, cost-effectiveness, and environmental friendliness. Exfoliation is usually executed through an electrochemical cell utilising a graphite rod, highly oriented pyrolytic graphite (HOPG), or film as the anode or cathode in organic or aqueous electrolytes. Electrolytes used in this process can be categorized into ionic-liquid

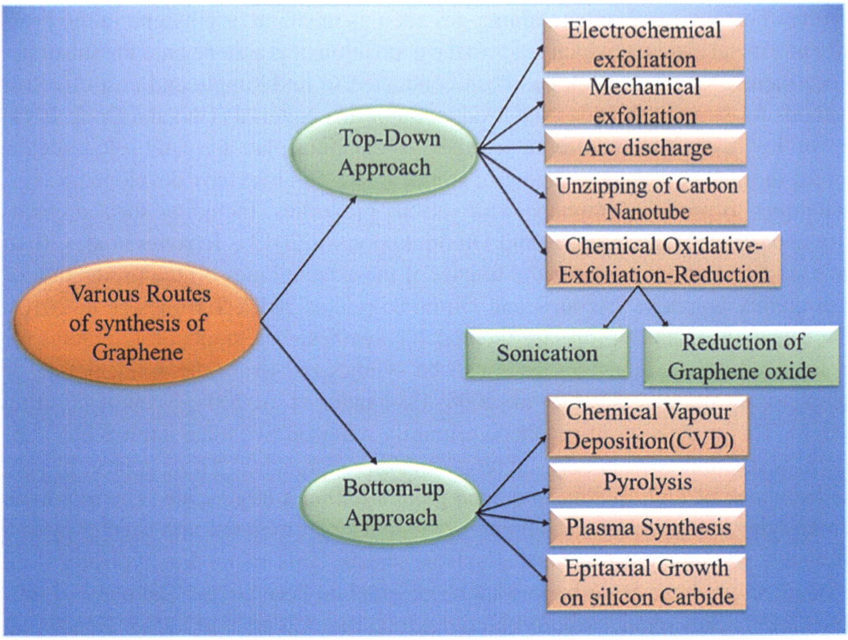

Fig. 3.4 The process flow chart of various routes of synthesis of graphene

(IL) and aqueous solution-based electrolytes. While aqueous solutions are commonly used, they make it difficult to obtain and disperse graphene directly. However, the use of IL-based electrolytes has shown promise in improving the quality and quantity of graphene produced through electrochemical exfoliation. It involves the intercalation of ions from an electrolyte under an applied potential, either anodic or cathodic. The mechanism of electrochemical exfoliation depends on the electrochemical intercalation of anions or cations such as SO_4^{2-}, Cl^-, NO_3^-, TBA^+, Li^+, etc. as either two types of potential, i.e., anodic or cathodic, into graphite layers, followed by decomposition reactions and exfoliation under applied voltages of 5–35 V. During anodic exfoliation, bulky anions are intercalated into the graphite layers, increasing their inter-layer spacing and causing exfoliation. In contrast, during cathodic exfoliation, cations are attracted to the negatively biased graphite working electrode, leading to intercalation and expansion of the graphene layers. This method produces a high-quality yield with 2–3 layers of graphene (with >60% yield) and larger flake sizes of up to ~30 μm (Cooper et al. 2013; Low et al. 2012). Liu et al. explored a novel electrochemical exfoliation method to prepare ionic-liquid (IL)-functionalized graphene by using an Ionic-liquid (IL)/water electrolyte, and the exfoliated graphene's precipitate was then dispersed in DMF, resulting in a more stable and homogeneous dispersion compared to other methods and eliminating the need for organic solvents. While ionic-liquid (IL) has potential for reducing the use of toxic chemicals as well as promoting green chemistry applications (Wang et al. 2010a, b), it is comparatively expensive for large-scale graphene production on an industrial level. Su et al. reported graphite powder-based continuous electrochemical exfoliation in which melamine was used as an additive electrolyte (Chen et al. 2015). Parvez et al. introduced a method for electrochemical stripping of graphite using inorganic salts. Graphite electrodes were submerged in an ammonium sulphate [$(NH_4)_2SO_4$, H_2O] electrolyte solution (Fig. 3.5a). Applying a positive potential of +10 V, water reduction takes place at the cathode, resulting in the formation of strong nucleophilic hydroxyl (–OH) ions. These energetic gas species facilitate the separation of the graphite layers, resulting in the production of layered graphene (Fig. 3.5b). These hydroxyl ions attack on the sp^2 carbons of the graphite anode, forming C–OH and vicinal OH groups (reaction 1) and sulphate ions intercalated between the graphite layers. Then epoxy rings (reaction 2) can be formed and further oxidised to carbonyl groups (reaction 3). During this step, water molecules co-intercalate with SO_4^{2-} ions. Carbon (C) and water react to form CO_2 (reaction 4), and the oxidation of water molecules generates the gas species CO_2 and O_2 (reaction 5) (Fig. 3.5c). These reactions provide insights into the mechanism underlying graphite anode degradation in aqueous electrolytes and are crucial for developing advanced materials for electrochemical applications (Parvez et al. 2014). Wu et al. synthesized the low-oxidative few-layer graphene due to the quick exfoliation caused by the release of strong gases, which resulted expanded graphitic layers (Wu et al. 2014).

Prashant Tripathi et al. reported a new electrochemical exfoliation route for graphene synthesis, using a pure alkaline electrolyte (KOH in water) instead of strong acids, which can damage graphene's structure. They used graphite as an anode electrode and Pt as a cathode electrode, but recognised that Pt is expensive. However, using KOH or H_2SO_4 in this process cannot remove metal ions from the graphitic sheet; their pure alkaline (KOH) electrolyte could prevent oxidation of the graphitic anode at the edges. This innovative approach presents a promising alternative for graphene synthesis without damaging its structure (Tripathi et al. 2013). In order to control the properties of graphene sheets produced through electrochemical exfoliation of graphite, researchers have experimented with changing the electrolyte used (Yang et al. 2016), which allows them to control the characteristics of graphene materials such as thickness, lateral size, chemical properties, and electronic properties. For industrial applications, it is economically and technologically beneficial to obtain water-dispersible graphene. However, achieving water dispersible graphene materials by introducing oxygen functional groups (–OH and –COOH groups) can result in the loss of the inherent structural, physical, and chemical properties of graphene materials (Wang et al. 2018).

Therefore, an efficient electrochemical approach is necessary to strike a balance between achieving high water dispersion and preventing excessive oxidation during the exfoliation of graphene. Hence, there is a need for an effective electrochemical approach that can strike a balance between achieving high water dispersion and preventing over-oxidation of graphene during the exfoliation process. In their study, Parvez et al. employed graphite as the anode electrode and platinum (Pt) as the cathode electrode. The electrodes were submerged in a sulfuric acid (H_2SO_4) electrolyte solution and subjected to a potential of +10 V for a duration of 10 min.

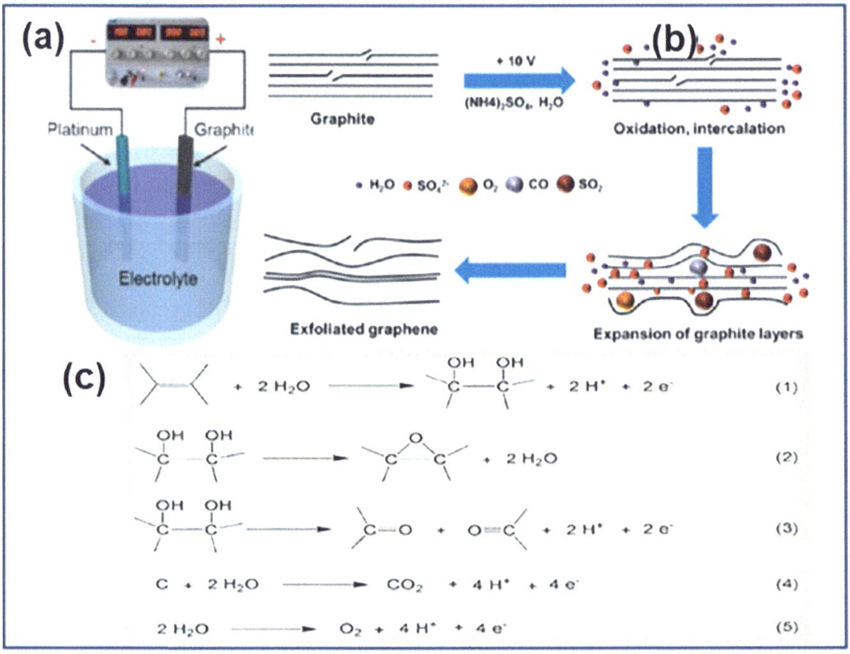

Fig. 3.5 Schematic Illustration of the electrochemical exfoliation unit setup (**a**), Schematic representation of mechanism of electrochemical exfoliation of graphite in aqueous solution of $(NH_4)_2SO_4$ salt (**b**), Proposed chemical reaction of the Electro-chemical oxidation process of graphite electrode (**c**). Reprinted with permission from Parvez et al. (2014). Copyright 2014@ Journal of American Chemical Society

The process yielded a product consisting of up to 1–3 graphene layers and achieving a 60% overall yield (Parvez et al. 2013).

3.4.2 *Mechanical Exfoliation*

Mechanical exfoliation, also known as the "peel-off" or "Scotch tape" method, is a long-standing technique used for decades to isolate monolayer graphene and few-layer graphene. It encompasses various top-down techniques like micromechanical cleavage, intercalation, sonication, ball milling, wet-ball milling, dry-ball milling, fluid dynamics, and liquid-phase exfoliation, enabling the production of graphene. The micromechanical cleavage produces high-quality monolayer graphene with lateral dimensions in the nm scale, finding wide applications. In 2004, Novoselov et al. pioneered the use of micromechanical cleavage (known as the Scotch tape method) to extract monolayer graphene flakes from highly oriented pyrolytic graphite

(HOPG), and in 2010, they earned the Nobel Prize in physics for their groundbreaking research on graphene (Novoselov et al. 2004, 2005). Dasari et al. demonstrated the micromechanical cleavage stripping of graphene sheets by using the Scotch tape technique (Fig. 3.6a). In this process, they placed graphite on a substrate and repeatedly peeled it with adhesive tape to obtain a monolayer graphene sheet. However, this method has some limitations, as the obtained product has a low yield and may also contain several structural defects (Dasari et al. 2017). Mechanical exfoliation mechanism of graphite into graphene can be classified into two primary routes based on the applied forces: shear/lateral force and normal force (Fig. 3.6b). Both of these mechanical routes are essential prerequisites for the production of graphene, as they enable the exfoliation process. A recent study has explored new routes of normal force synthesis peeling of graphite, particularly for graphene production utilizing modern machinery, specifically an ultra-sharp single-crystal diamond wedge (wedge-based mechanical exfoliation) (Jayasena and Subbiah 2011). Shear force-based approaches, particularly fluid dynamic and ball milling methods, have attracted significant attention. Ball milling is generally employed in grinding minerals, fertilizers, cement, ceramics and utilizes grinding media to generate mechanical force through impact and attrition, leading to the exfoliation of graphite and the formation of graphene flakes (Fig. 3.6c). This technique enables the production of graphene with nanoscale thickness (Yi and Shen 2015). The quality of graphene synthesized through ball milling is influenced by various process parameters, including the type of grinding media, speed, matrix, time, and initial size of the precursor material (Zhuang et al. 2017). However, the drawback of ball milling is its time-consuming nature. The reaction time for the exfoliation process can range from 24 to 48 h, which is not favourable for large-scale production. As a result, finding a solution to reduce reaction time and increase scalability is crucial. Overall, the potential of ball milling in the production of high-quality graphene permits further investigation, and improvements in its scalability would be highly desirable. In the process of exfoliation, an additional factor to consider is the fragmentation effect, illustrated in Fig. 3.6c. The force produced during exfoliation can cause the breakdown of large graphite particles or graphene flakes, resulting in smaller fragments. This fragmentation effect has both positive and negative consequences. On the one hand, it reduced the lateral size of graphene, which is undesirable to obtain high-surface-area graphene sheets. On the other hand, it aids in the exfoliation process by making smaller graphite flakes easier to separate. This is due to the reduced collective van der Waals interaction forces between the layers in smaller graphite flakes. Thus, the fragmentation effect plays a dual role in the exfoliation of graphene materials.

3.4.3 Arc Discharge

Arc discharge, a well-established method for synthesizing fullerene (Kyesmen et al. 2016) and carbon nanotubes (CNT) (Vekselman et al. 2017), has proven to be remarkably versatile for synthesizing graphene as well (Wu et al. 2016). This conventional

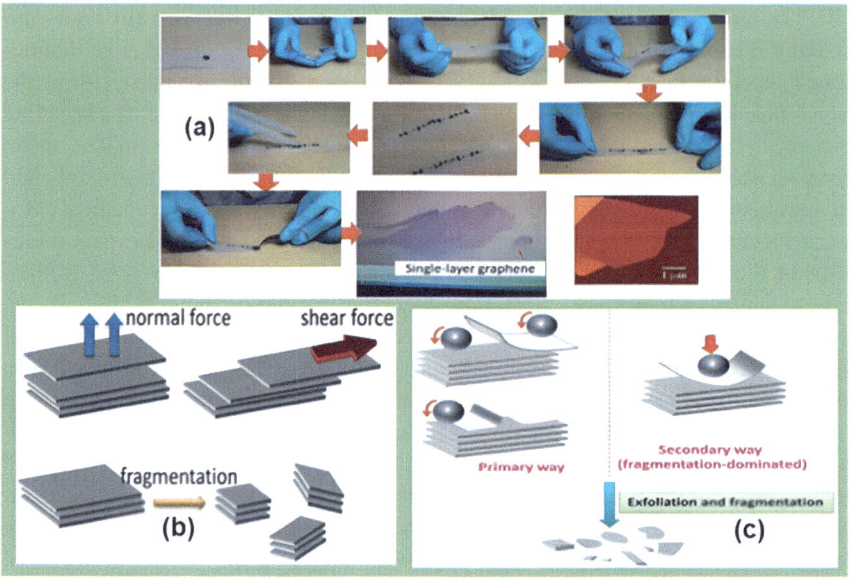

Fig. 3.6 Schematic illustration of the Scotch-tape based micromechanical cleavage of HOPG (**a**), Mechanical exfoliation of graphite into graphene flakes based on applied mechanical forces and the auxiliary route for fragmentation (**b**), Illustration of the mechanical mechanism of exfoliation via ball milling (**c**). Reprinted with permission from Yi and Shen (2015). Copyright 2015@Royal society of chemistry

technique, known for its efficacy, has been successfully adapted to produce graphene, opening up new possibilities for its large-scale production and various applications in the field of materials science and nanotechnology**.** The arc discharge technique was first developed by Kratschmer et al. to synthesize fullerene (C60) (Krätschmer et al. 1990). After that, this method was adapted for the synthesis of graphene (Wu et al. 2016). Typically, a carbon precursor at the anode and graphite at the cathode are placed inside the reaction chamber, which is filled with either a gas or liquid medium. When applying an electric current, dissociation of the precursor occurs, leading to plasma formation at extremely high temperatures ranging from 3727 to 5727 °C, sufficient for extracting graphene (Arora and Sharma 2014). After the discharge ceases, the resulting soot is collected under ambient conditions. To facilitate the graphene synthesis, an inert atmosphere was introduced during the arc discharge process using hydrogen (H_2) and helium (He) gases. In the pioneering work by Subrahmanyam et al., graphene was initially synthesized using the arc discharge method in an H_2 environment, resulting in 2–4 graphene layers (Subrahmanyam et al. 2009). Subsequently, Li et al. synthesized graphene with 2–6-layer nanosheets using a mixture of He and NH_3 as the atmospheric environment, followed by the arc discharge method. However, these methods are often expensive due to the use of specific vacuum equipment (Li et al. 2012). Wang et al. developed a low-cost graphene synthesis method to overcome large-scale graphene production challenges.

They used air as a medium instead of H_2/He gas in arc discharge to produce graphene, and the yield reached tens of grams per day with a number of layers of 2–10 graphene nanosheets. However, maintaining precise, high air pressure was necessary to ensure successful graphene growth (Wang et al. 2010a, b). Wu et al. established an arc discharge method for large-scale graphene production under different atmospheric conditions (Wu et al. 2016). Li et al. developed a cost-effective method for synthesizing graphene using petroleum asphalt in a water medium. This approach utilized inexpensive raw materials, with water serving as the medium and carbon-rich asphalt as the raw-material, making it highly inexpensive. Meanwhile, the arc-discharge technique offered notable advantages, including simplicity, time efficiency, no metal catalyst required, and the ability to produce high-quality graphene inexpensively. However, precise control is crucial for this process, and the energy-intensive nature of arc discharge synthesis might pose challenges to scaling up graphene production (Stankovich et al. 2007a). This arc discharge method provides a cost-effective solution to graphene production challenges and has the potential to be scaled up for large-scale production.

3.4.4 *Unzipping of Carbon Nanotube*

The unzipping or unrolling of carbon nanotube (CNT) entails cutting the cylindrical structure to flatten the carbon sheet into monolayers, bi-layers, or few-layers graphene structure. The unzipping or unrolling of carbon nanotubes (CNT) is a promising approach for the large-scale production of high-quality graphene nanostrips or graphene nanoribbons (GNRs). Researchers have reported numerous procedures for unzipping of the carbon nanotube (CNT) in order to produce graphene and graphene derivatives, including intercalation and exfoliation (Cano-Márquez et al. 2009), chemical attack (Kosynkin et al. 2009), metal-catalyzed cutting (Thess et al. 1996), thermal expansion (Morelos-Gómez et al. 2012), and plasma etching (Jiao et al. 2009). Unzipping a CNT often necessitates a very high strain rate (10^{8-10}/s) in order to break the C–C bond either in the axial or longitudinal direction (Tiwary et al. 2015). Moreover, this process could be applied to both single-walled carbon nanotubes (SWCNT) as well as multi-walled carbon nanotubes (MWCNT), resulting in a varied number of graphene layers with precisely defined dimensions, and the resultant products are known as graphene nanoribbons (GNRs). Kosynkin et al. developed an acid treatment method to synthesize graphene nanoribbons (GRNs) by unzipping of multiwalled carbon nanotubes (MWCNTs). They suspended MWCNT in concentrated sulfuric acid (H_2SO_4), followed by oxidative treatment with potassium permanganate ($KMNO_4$) (Fig. 3.7b). During oxidation, layers of MWCNT subsequently open up, introducing oxygen-containing functional groups like carbonyl, carboxylic, and hydroxyl along the plane and edges. Their approach, now referred to as the "experiment," enables the fabrication of uniform GNRs with widths ranging from 100 to 500 nm. Moreover, GNRs have both semiconducting and metallic properties, with presumed zigzag-shaped edges. The GNRs synthesized from multiwalled carbon nanotubes (MWCNTs), consisting of approximately 15–20 concentric and single-wall carbon nanotubes (SWCNTs) with diameters ranging from 40 to 80 nm. These graphene nanoribbons (GNRs) are highly soluble in water, alcohol, and polar protic solvents, making them an emerging material for various fields of applications. The main drawback of this approach was that it produced low-yield and destroyed a large amount of precursor material (Kosynkin et al. 2009). In order to enhance the yield, the multiwalled carbon nanotubes (MWCNTs) were intercalated with oxalic acid prior to chemical unzipping. Oxalic acid, with a size of 0.38 nm, is compatible with the interlayer spacing of MWCNTs (0.34 nm), enabling easy intercalation between the layers. Moreover, it ensures that oxalic acid can be effectively intercalated into the lattice space of MWCNTs (Elias et al. 2010). Elias et al. in their study, reported longitudinal cutting of MWCNT using Ni and Co metallic catalysts to synthesize graphene nanoribbons (GRNs). The resultant nanoribbons are of 100–500 nm in length and 10–40 nm in diameter (Elias et al. 2010) (Fig. 3.7c). Kumar et al. reported the unzipping of MWCNTs by irradiating MWCNTs using a laser pulse energy ranging from 200 to 350 mJ and revealed that the laser irradiation approach is a straight-forward method to synthesize graphene nanoribbons (GNRs) without any surface contamination. While the unzipping and unrolling of CNTs can significantly reduce GNRs width, the overall unzipping process exhibits low yield (Kumar et al. 2011). Dongjin Ko et al. developed an innovative electrochemical unzipping method for large-scale graphene nanoribbons (GNRs) production by using MWCNTS powders as a working electrode without adding any additives. After optimising the setup, sulfuric acid (H_2SO_4) was added for the intercalation between the layers, which leads to unzipped CNT (Ko et al. 2020). M. Terrones and his research group reported a method for the synthesis of nanoribbons from multi-walled carbon nanotubes (MWCNTs) unzipping by metallic intercalation. They explained that MWCNT was intercalated with lithium (Li) in liquid ammonia (liq. NH_3) and then treated with acid followed by thermal exfoliation. The resultant exfoliated MWCNT (Ex-MWCNT) consists of approximately 60% nanoribbons, along with partially opened and damaged MWCNTs, as well as graphene flakes detached from the tube walls (Kosynkin et al. 2009) (Fig. 3.7a). This resultant product is due to the need to cut longer MWCNTs, as complete unzipping along the entire length is rare. Achieving full unzipping and obtaining nanoribbons with controlled dimensions would require further improvements and unwrapping specific tube types. Hence, the presence of these by-products in the resultant mixture bounds their applications, such as in electronic devices, unless effective separation and sorting methods are developed. However, partially unzipped ex-MWCNTs could find utility in composites, where the unwrapped edges interface with a matrix, and they may also exhibit intriguing properties for nanoelectronics. Kim et al. demonstrated that under high electrical bias, MWCNTs exhibit superplasticity. By applying electrical currents and nanomanipulation, they successfully unwrapped MWCNTs to synthesize graphene nanoribbons (GNRs) of specific dimensions. This electric-based approach not only enabled precise control over GNR width and length but also facilitated simultaneous structural characterization of the produced GNRs (Kim et al. 2010a, b). Dai et al. group developed a physicochemical method, involving the embedding of MWCNTs

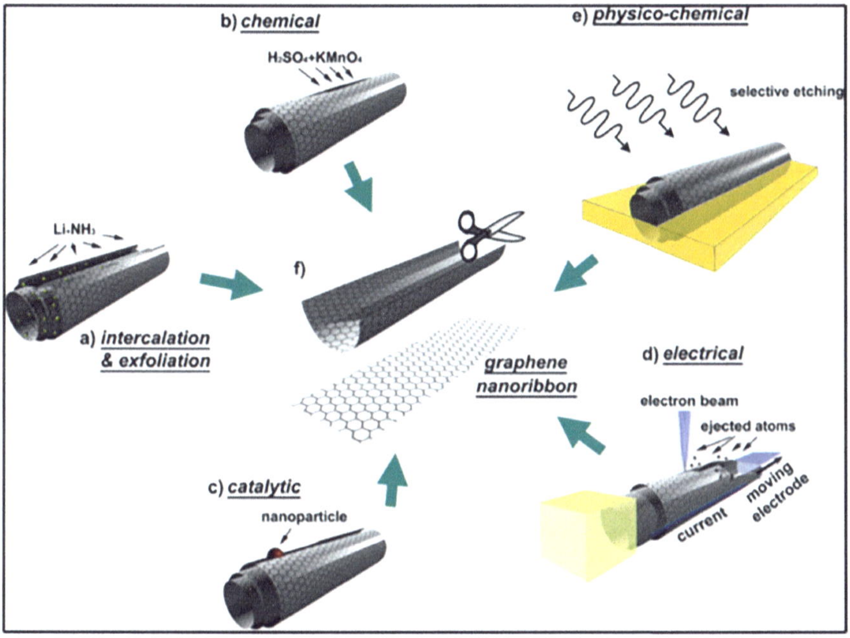

Fig. 3.7 Different methods of unzipping carbon nanotubes for the synthesis of graphene nanoribbons (GNRs) or graphene: **a** metallic intercalation of MWCNTs with Li and liquid NH_3, and followed by exfoliation using acid and heat treatments; **b** chemical method, Acid treatment using H_2SO_4 and $KMnO_4$ as oxidizing agents) **c** catalytic approach (Elias et al. 2010; Terrones 2009), **d** the electrical method, and **e** physicochemical method using Ar plasma treatment. The resultant products are either graphene nanoribbons (GNRs) or graphene sheets **f**. Reprinted with permission from Terrones et al. (2010). copyright 2010@Elesevier

in a poly (methyl methacrylate) film (Jiao et al. 2009). The film is then flipped to expose the MWCNT walls, which are selectively etched away by using Ar plasma, and the duration of the etching time determines the thickness of the resultant product. Although this method is more labour-intensive and has a relatively low yield (around 20%), the resulting GNRs have a more uniform shape compared to other unzipping methods. Notably, partial opening of MWCNTs using plasma etching was reported as early as in 2004 by S. Trasobares et al., but the resulting segments were short and the process lacked control and further development (Trasobares et al. 2004).

3.4.5 Chemical Oxidative-Exfoliation-Reduction

Chemical-based techniques are highly effective for the large-scale production of graphene and graphene's derivatives. Oxidative-exfoliation, introduces oxygen groups during oxidation, enabling them to spread throughout the carbon structure, effectively weakening the interlayer van der Waals forces, and these oxygen functional groups enhance the interlayer space between the two graphite layers. The structure and chemical composition of GO are primarily influenced by factors such as the choice of oxidizing agents, the graphite precursor, reaction conditions, and the specific synthesis procedure employed. Graphene oxide (GO) is commonly used as a precursor material for the synthesis of reduced graphene oxide (rGO) and graphene, followed by reduction by using numerous reducing agents. Moreover, various reduction methods such as chemical, thermal, electrochemical, microwave, photoreduction, and hydrothermal methods can be employed to transform GO into reduced graphene oxide (rGO). These reduction methods facilitate the removal of oxygen functionalities, thereby restoring the sp^2 carbon network and enhancing the electrical properties of the resulting rGO material.

3.4.5.1 Chemical Oxidative-Exfoliation

Oxidative exfoliation of graphite is a highly promising technique for producing graphite oxide (GO). There are various well-established methods for the synthesis of graphene oxide (GO) by oxidation of graphite followed by exfoliation, including the Brodie method, Staudenmaier methods, Hofmann methods, Hummer's methods, and the Tour method, or improved Hummer's methods. In these methods, graphite is chemically treated with a strong acid like H_2SO_4, conc. HNO_3, or H_3PO_4, followed by an oxidizer's treatment with $KMnO_4$, $KClO_3$, or $NaNO_3$. These oxidizers facilitate the oxidation at carbon edges, leading to the expansion of graphite layers. The Brodie method, documented by Benjamin Brodie in 1859, was the first approach developed for the oxidation of graphite. It entails oxidizing graphite with potassium chlorate ($KClO_3$) and concentrated HNO_3 (fuming) at 60 °C for a few days to produce multi-layer graphene oxide (GO) (Brodie 1859). The resultant product i.e., Brodie's Graphene oxide (BR-GO) is dispersible in alkaline as well as aqueous solutions and non-dispersible in acidic media due to the addition of oxygen functional groups; hence, graphene oxide is hydrophilic in nature. However, the main drawbacks of the Brodie method are its long reaction time and the release of hazardous gases such as dinitrogen tetra-oxide (N_2O_4) (Kosynkin et al. 2009) and nitrogen dioxide (NO_2) (Yu et al. 2016) during the reaction. In 1898, L. Staudenmaier introduced an improved version of Brodie's method for the oxidation of graphite to produce graphene oxide (GO). Staudenmaier's modifications aimed to address the safety concerns associated with the original process. One of the key modifications was the gradually addition of tiny amounts of the oxidizing agent, potassium chlorate ($KClO_3$), into the reaction mixture over an extended period. This gradually addition of $KClO_3$ prevented the formation of explosive by-products and minimized the release of excessive heat. Moreover, the inclusion of concentrated H_2SO_4 increased the acidity of the reaction mixture, allowing the use of less fuming HNO_3 for oxidation. This minimised the release of harmful gases such as N_2O_4 and NO_2, increasing the method's safety. Although the properties of L. Staudenmaier's graphene oxide (ST-GO) and Brodie's graphene oxide (BR-GO) were similar, however, despite these enhancements, Staudenmaier's method still had drawbacks. The addition of $KClO_3$ normally takes place over the course of a week, lengthening the procedure. Moreover, the release of toxic chlorine dioxide (ClO_2) gas by an inert gas was necessary. Overall, Staudenmaier's improvements on Brodie's method made the production of graphene oxide safer, but they also added complexity and lengthened the process. It demonstrated the potential for improving existing methods by tweaking reaction conditions to achieve better results while minimizing hazards (Somanathan et al. 2015). To overcome these problems, in 1937, Hofmann proposed a method for the oxidation of graphite. They used non-fuming conc. HNO_3 and $KClO_3$ to synthesize Hofmann graphite oxide (HO–GO) with reduced oxygen proportion ($C/O > 2.5$). By employing $KClO_3$ as a potent oxidant, reactive dioxygen is generated in situ, effectively oxidizing the graphite into GO in an acidic solution. On the other hand, HNO_3 has a strong inclination to react with aromatic carbon of graphite surfaces. However, this approach still resulted in the emission of toxic gases. Nevertheless, this method offered a less hazardous means of synthesizing graphite oxide compared to earlier proposed methods (Verma et al. 2011).

William S. Hummers and Richard E. Offeman established a new method for the oxidation of graphite, named Hummer's method (Hummers and Offeman 1958). which has since become the most widely used method for producing graphene oxide, offering a safer and faster alternative to previous methods. They demonstrated that treating graphite by using a mixture of potassium permanganate ($KMnO_4$), sulfuric acid (H_2SO_4), and a tiny amount of sodium nitrate ($NaNO_3$) to oxidize graphite resulted in the production of Hummer's graphite oxide (HU–GO) with a C/O proportion ratio of 2.25. In this process, graphite powder and $NaNO_3$ were mixed in H_2SO_4 and cooled (temperature controlled to 0 °C) using an ice bath, followed by continuous stirring. Then, $KMnO_4$ was gradually added to the reaction mixture and again stirred at 35 °C. Subsequently, water was gradually added and left to reach 98 °C for a short time. After the reaction was completed, the suspension was treated with H_2O_2 and water to eliminate any residual manganese dioxide (MnO_2) and permanganate and also to achieve neutral pH. To further purify the mixture, it was subjected to filtration, which involved washing through centrifugation as well as decantation. This process effectively removed the dissolved mellitic acid salt, also known as benzenehexacarboxylic acid or graphitic acid (Verma et al. 2011). Hummers' method offers some advantages, like quality, safety, and time, compared to other oxidation approaches. Nevertheless, its main drawbacks, such as the high reaction temperature and release of toxic gases during the reaction, hindered its wide range of applications (Jilani and Melaibari 2022). To address these concerns, extensive endeavors have focused on enhancing or modifying Hummers' method to alleviate safety risks, reduce reaction time and temperature, enhance oxidation efficiency, and lower costs. In 2010, Tour and colleagues modified Hummers' method to synthesize graphene oxide (GO) with increasing oxygen content. They replaced $NaNO_3$ by using a small amount of phosphoric acid (H_3PO_4) as a dispersive and etching agent during the oxidation process of graphite, along with large amounts of $KMnO_4$ and H_2SO_4. The resulting GO, known as TU-GO, exhibited a high yield, a more ordered structure with fewer

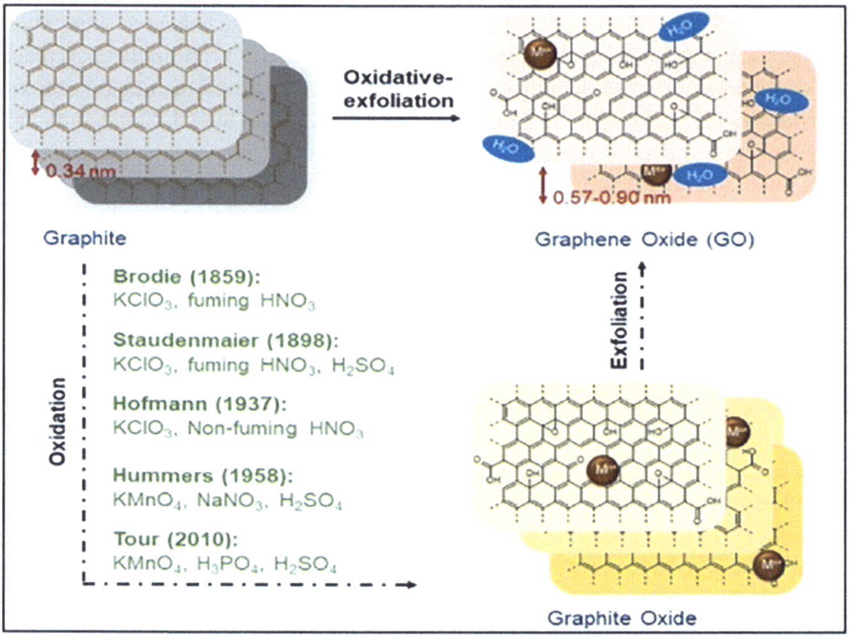

Fig. 3.8 Schematic illustration of chemical oxidative-exfoliation methodology for the oxidation of graphite to synthesize graphene oxide (GO). Reprinted with permission from Kumar et al. (2020). Copyright 2021@Elesevier

defects, and extensive oxidation, compared to other methods (BR–GO, ST–GO, HO–GO, and HU–GO). The improved method fostered the formation of well-ordered structures, which were attributed to the interaction between the two vicinal diols on the basal plane and the phosphoric acid, forming five-membered cyclic phosphate groups. Notably, Tour's modified Hummer's method enhances the concentration of hydrophilic groups in the oxidized form of graphite, i.e., GO. Tour's method for oxidation is considered safer compared to other methods as it doesn't produce toxic gases and requires low-temperature heating (50 °C). The improved Hummers' method offers additional advantages, including lower toxicity and enhanced quality of the synthesized product (Lavin-Lopez et al. 2016). Due to its numerous benefits, Tour's method is highly suitable for large-scale production of graphene oxide (GO) (Marcano et al. 2010) (Fig. 3.8).

3.4.5.2 Reduction

Graphene oxide (GO) undergoes reduction processes to restore its disrupted sp^2 bonding and retain its honeycomb lattice structure. Thermal, chemical, and electrochemical methods are commonly employed for GO reduction. During reduction, oxygenous functional groups like hydroxyl, carbonyl, and carboxylic groups in GO are typically eliminated. However, achieving a complete reduction to produce pure graphene remains a challenge (Agarwal and Zetterlund 2021; Dreyer et al. 2010). The resulting reduced graphene oxide (rGO) bears similarities to pristine graphene but with some defects and dimensional variations. The quality of rGO depends on factors such as the type of reducing agent and process conditions, including temperature, pressure, time, and voltage. Additionally, the carbon-to-oxygen (C/O) ratio provides insight on the reducing agent efficiency in GO reduction (Chua and Pumera 2014). A higher C/O ratio indicates greater de-oxygenation and results in higher-quality rGO. According to studies, rGO shows similarities with pristine graphene, especially in surface, mechanical strength, electrical conductivity, and thermal conductivity. Although both GO and rGO have surface defects, they have been widely utilized in various applications, including supercapacitors (Jaiswal et al. 2023, 2018; Saha et al. 2016), membrane separation (Pakdel et al. 2021; Voicu and Thakur 2021), biomedical fields (Devi et al. 2023; Jha et al. 2021), water treatment (Saravanan et al. 2022), and energy storage (Salahdin et al. 2022). Different reduction processes are used for synthesizing rGO, including chemical reduction, thermal reduction, electrochemical reduction, microwave irradiation, hydrothermal reduction, photoreduction, and green reduction methods.

Chemical Reduction: The chemical reduction of graphene oxide (GO) is a highly popular and appealing technique for the large-scale production of graphene and its derivatives. By employing chemical reducing agents, the method effectively removes oxygen functionalities from the intermediate GO sheets, resulting in reduced graphene oxide (rGO). This process can be carried out at room temperature or with moderate heating, requiring no sophisticated equipment. Therefore, it is considered an economical and straightforward approach for large-scale graphene production. Various reducing agents, including hydrazine (N_2H_4), hydrohalic acid (HCl), sodium borohydride ($NaBH_4$), urea (CH_4N_2O), thiourea, dimethylhydrazine, ammonia (NH_3), Hydroiodic acid with acetic acid ($C_2H_4O_2$), vitamin C, hydroxylamine hydrochloride ($NH_2OH{\cdot}HCl$), hexylamine ($C_6H_{15}N$), potassium hydroxide (KOH), sodium hydroxide (NaOH), hydrohalic acids (HCl, HBr, and HI), Na–NH_3, poly(norepinephrine), alcohol, sulfur compounds ($Na_2S_2O_4$, $NaHSO_3$, $Na_2S{\cdot}9H_2O$, $Na_2S_2O_3$, Na_2SO_3, SO_2, and $SOCl_2$), N-methyl-2-pyrrolidinone (NMP), lithium aluminium hydride ($LiAlH_4$), and iron powder, have been used for the chemical reduction of graphene oxide (GO). Among these, hydrazine has emerged as the most widely used reductant due to its effectiveness in achieving a higher reduction in graphene oxide (GO) (Guex et al. 2017; Hofmann 1934; Park and Ruoff 2009). Moon et al. introduced a novel reduction method for synthesizing highly conductive reduced graphene oxide (rGO) using hydroiodic acid (HI) as a reducing agent without the need for a nitrogen source. Their study aimed to investigate the role of nitrogen and determine the most effective reduction method by combining the traditional NH method with the HI method. The introduction of the HI method to rGONH was hypothesized to remove or reduce any nitrogen atoms present. To explore this, a step-by-step reduction approach was employed, alternating between two different reducing reagents, HI/NH and NH/HI, as depicted in Fig. 3.9. This approach allowed for a better understanding of nitrogen incorporation as well as its removal or reduction, leading to the production of highly qualified and qualitative rGOs with significantly improved electrical conductivity. The findings of this study shed light on innovative reduction techniques and their impact on the properties of graphene-based materials, offering potential applications in various fields requiring enhanced electrical conductivity (Moon et al. 2011). Ambrosi et al. conducted a comparative study on the reducing properties of $LiAlH_4$, $NaBH_4$, and N_2H_4 for graphene oxide (GO) reduction. They found that $LiAlH_4$ resulted in the highest carbon-to-oxygen (C/O) ratio of 12, while $NaBH_4$ and N_2H_4 had ratios of 9.5 and 11.5, respectively. However, the charge transfer resistance of the reduced graphene oxide (rGO) produced with $LiAlH_4$ (5.9 kΩ) was higher than that of $NaBH_4$ (4.0 kΩ) and N_2H_4 (3.3 kΩ) (Ambrosi et al. 2012). H. Korucu et al. reported the reduction of GO by using sulfur dioxide (SO_2) with or without the combination of numerous chemicals like NaBH4 NaOH, L-ascorbic acid, $Na_2B_4O_7 \cdot 10H_2O$, $Na_2S_2O_3$, OH, and NH_4OH. The combination of SO_2 with $NaBH_4$ demonstrated superior reduction performance, yielding an rGO with a carbon-to-oxygen (C/O) ratio of 11.73. This suggests that the reduction efficiency of SO_2 can be enhanced by combining it with other chemicals for GO reduction (Korucu et al. 2019). Ren et al. conducted a study on the temperature-dependent reduction of graphene oxide (GO) using hydrazine hydrate. They observed that the efficiency of GO reduction with hydrazine hydrate was primarily influenced by temperature rather than reduction time. High temperatures were found to be necessary for effective reduction, resulting in an increase in the carbon-to-oxygen (C/O) ratio of GO from 3 to 15.1 when reduced at 95 °C for 3 h. X-ray photoelectron spectroscopy (XPS) and Fourier-transform infrared (FT-IR) data analysis revealed that the reduction at temperatures above 60 °C resulted in the removal of most epoxide groups from GO, leading to the formation of a sp^2 structure and an increase in the electrical conductivity of GO from 1.5 × 10–6 to 5 S/cm. However, complete restoration of the GO structure to that of graphite was not achieved due to the presence of remaining hydroxyl groups, even at elevated temperatures (Martinez et al. 2011). Park et al. reported a study on the reduction of graphite oxide and GO using hydrazine as the reducing agent. They observed that the resulting materials, rGO and reduced graphite oxide, showed distinct chemical and structural characteristics. The reduced graphite oxide (rGO) had a C/O ratio of approximately 10.2, indicating significant removal of oxygen atoms through chemical reduction. Moreover, reduced graphite oxide had a lower C/O atomic ratio of approximately 5, suggesting a lesser degree of reduction using hydrazine (Park et al. 2012). These findings emphasize the differences in the reduction process and its impact on the resulting materials' properties. Further research is required to explore the implications of these variances for potential applications in fields like electronics, energy storage, and catalysis.

Thermal Reduction: The direct reduction of graphene oxide (GO) through thermal annealing has gained significant attention, particularly for producing reduced graphene sheets that possess similar structural and electrical properties to pristine

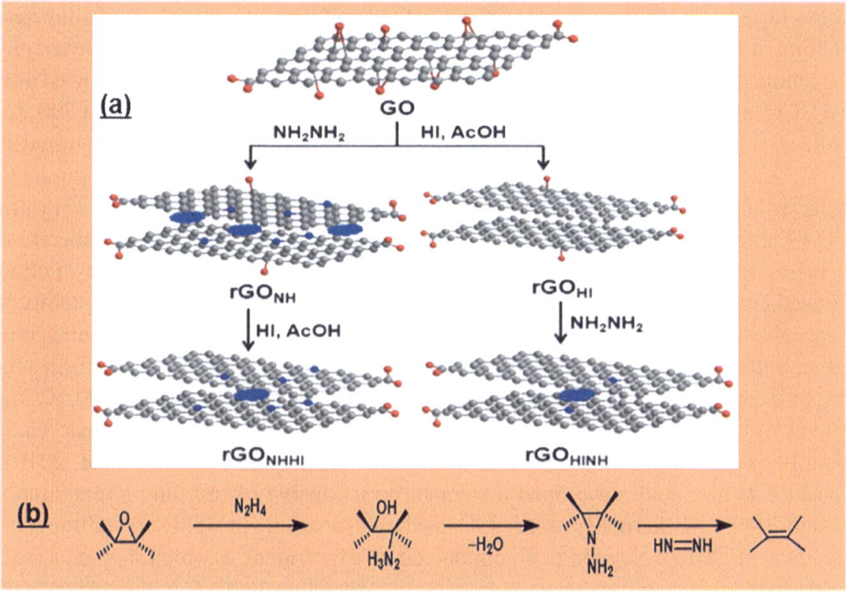

Fig. 3.9 **a** Schematic flow chart of two different step-by-step reductions of GO for the synthesis of rGOs, representing carbon (gray colour), nitrogen (blue colour), and oxygen atom (red colour), respectively. Reprinted with Copyright permission from Moon et al. (2011). Copyright 2011@Royal Society of Chemistry

graphene. This reduction approach is especially favoured for nano-electronic applications. Several studies have explored different reduction temperatures to investigate the kinetics and thermodynamics involved in the thermal reduction process and to understand the properties of the resulting reduced graphene oxide (rGO). The choice of the reduction temperature is crucial and depends on the intended application of the rGO, as it requires a delicate balance between achieving the desired level of reduction and minimizing the structural decomposition of the graphene oxide (GO) during annealing. During thermal reduction, the oxygenous functional groups (–OH, –COOH, –CO, and epoxide) attached to the basal plane of GO decompose into gases, leading to the expansion of the GO layers and inducing enough pressure to exfoliate them into reduced graphene layers. This decomposition also results in the removal of carbon atoms, releasing CO_2, and resulting in the formation of small, wrinkled, and defect-rich reduced graphene sheets. Moreover, during this exfoliation process, approx 30% GO mass is lost due to the release of CO_2, leading to structural damage as well as lattice defects in the graphene sheets. These structural defects have a significant impact on the electronic properties of the resulting product. In a study conducted by Jung et al., electrical resistivity capacities and temperature-programmed desorption (TPD) were performed to investigate the thermal stability and kinetics of graphene oxide (GO) during thermal annealing. They analyzed the electrical resistivity data and determined that the activation energy (AE) for the thermally reduced single-layered GO film was approximately 371 kcal/mol, while for the multi-layered GO film, it was around 324 kcal/mol. Moreover, TPD measurements were carried out to examine the desorption products during thermal annealing. The results showed that H_2O, CO, and CO_2 were the main desorption products up to a temperature of 300 °C, while no release of O_2 was detected. Notably, the thermal desorption of GO started at 70 °C, as confirmed by the detection of the H_2O, CO, and CO_2 desorption products (Jung et al., n.d.). Schniepp et al. reported that the carbon-to-oxygen (C/O) ratio was <7 at temperatures below 500 °C, while it increased >13 when the temperature increased from 500 °C to 750 °C (Schniepp et al. 2006). To obtain highly crystalline reduced graphene oxide (rGO) and restore its graphitic structure, annealing temperatures above 1000 °C have been explored. Such high temperatures are advantageous for complete pyrolysis of the source material (carbon), as well as for restoring its graphitic structure and reduction processes. Temperatures exceeding 1000 °C are crucial for producing large-surface-area rGO sheets with desirable intrinsic electrical properties, as mentioned in references (Negishi et al. 2016; Su et al. 2010). Recently, Ishida et al. developed a solar furnace capable of reaching temperatures around 2000 °C to thermally reduce GO in ethanol, resulting in highly crystalline rGO (Ishida et al. 2016). Negishi et al. conducted an experiment in which they observed a reduction process occurring at a temperature exceeding 1000 °C in the presence of ethanol. The experimental setup involved introducing ethanol vapour at a rate of 0.5 cm^3/min, with H_2/Ar serving as a carrier gas at 150–250 cm^3/min. This reduction process resulted in a remarkable elongation of the conjugated π-electron system in reduced graphene oxide (rGO) due to the effective restoration of its graphitic structure. The researchers also found that annealing rGO at high temperatures in ethanol led to a significant decrease in the number of oxygen functional groups. Additionally, the carrier mobilities (μ) of rGO sheets increased as the temperature raised, and rGO synthesized in the presence of ethanol exhibited higher carrier mobilities compared to those synthesized in H_2 or H_2/Ar atmospheres (Ishida et al. 2016; Su et al. 2010).

Electrochemical Reduction: The electrochemical reduction of graphene oxide (GO) offers a convenient and eco-friendly approach for the production of high-quality reduced graphene oxide (rGO) at room temperature. In this process, GO can be reduced electrochemically at room temperature in an electrochemical cell with a non-toxic buffer solution (aqueous) or supporting electrolyte. The GO film can be coated or deposited on different types of substrates, including conducting substrates like indium tin oxide, glass carbon, or gold, as well as insulating substrates like flexible plastic or glass (Ishida et al. 2016). By applying an electric potential (power), the reduction takes place, leading to the removal of oxygen groups from the GO surface. Simultaneously, films are formed on the working electrode's surface. The complete removal of oxygenous functionalities from graphene oxide (GO) to restore the pristine properties of graphene remains challenging. while the oxygen content in the resulting electrochemically reduced graphene oxide (ERGO) can be controlled by adjusting the cathodic potential and electrochemical reduction time (Doğan et al. 2013). Zhou et al. found that ERGO achieved an oxygen-to-carbon (O/C) ratio of 0.04 after 5000 s of electrochemical reduction at 0.90 V (Ag/AgCl). Similarly, Li et al. reported that an O/C ratio of 0.18 was achievable with ERGO through electrochemical reduction at 1.6 V versus SCE for 180 s (Li et al. 2013). The electrochemical reduction process's time requirement is typically analysed by recording the current versus time (i–t) curve in the constant potential reduction technique, as reported by Zhou et al. and others. These studies contribute to understanding the parameters influencing the reduction process and tailoring the oxygen content in ERGO. The performance of electrochemical reduction can be fine-tuned by adjusting the buffer solution, electrolyte, and electrolysis parameters. Different electrolytes have been explored, including organic solvents (Zhou et al. 2009), alkaline and alkali earth metal chlorides (Zhou et al. 2009), acetonitrile and propylene carbonate (Karačić et al. 2019), sodium nitrate ($NaNO_3$) (Kauppila et al. 2013), and potassium nitrate (KNO_3) (Peng et al. 2011), Potassium chloride (KCl) (Ramesha and Sampath 2009), sodium chloride (NaCl) (Wang et al. 2013a, b), sodium sulphate ($NaSO_4$) (Wang et al. 2009), and phosphate buffer solutions (PBSs, $K2HPO_4/KH_2PO_4$) (Li et al. 2013; Marrani et al. 2019). Among others, overall, electrochemical reduction represents a simple, rapid, cost-effective, and environmentally friendly technique for obtaining high-quality rGO. This method is advantageous due to its simplicity, speed, cost-effectiveness, and environmental friendliness in producing high-quality rGO. Javier A. et al. reported the electrochemical reduction of graphene oxide films in organic and aqueous electrolytes with varying proton concentrations. Their research objective was to understand the involvement of H^+ in the reduction process and determine the type and quantity of defects formed in graphene sheets during reduction (Quezada-Renteria et al. 2019).

3.4.6 Chemical Vapor Deposition (CVD)

Chemical vapor deposition is a bottom-up synthesis technique. This technique is used to produce high-quality graphene attaining controlled number of layers in a large area and its derivatives with tunable properties on a commercial scale (Brownson and Banks 2012). In CVD, graphene is generally grown by the catalytic decomposition of a carbon-containing precursor gas on a substrate inside a reaction cavity under specific conditions of pressure, temperature and gas flowing rate parameters (Brownson and Banks 2011; Yan et al. 2013). The detail of the growth process can be outlined in three main steps: gas phase decomposition, surface chemical reaction, and gas phase diffusion.

3.4.6.1 Gas Phase Decompositions

In this step, gaseous precursors enriched with carbon-sources, are thermally decomposed into hydrogen and carbon atoms or radicals. This can be proceeded by heating the precursors or applying plasma or electric fields to activate them. The decomposition reaction depends on the type and concentration of the precursors, the carrier gas, and the reaction temperature and pressure. The most common carbon source for graphene synthesis is methane (CH_4), but other hydrocarbons such as ethylene (C_2H_4), acetylene (C_2H_2), ethanol (C_2H_5OH), or benzene (C_6H_6) can also be utilized for the preparation of graphene and its derivates.

3.4.6.2 Surface Chemical Reaction

This step involves the carbon atoms or radicals where it generally adsorbs on the surface of a metal catalyst substrate, such as copper (Cu) or nickel (Ni), and form a single layer or multiple layers of graphene depending on the substrate thickness and solubility. The surface reaction is influenced by the type and morphology of the catalyst, the partial pressure of hydrogen, and the growth time. The metal catalyst can also be substituted with other substituting elements, such as boron (B), nitrogen (N), sulfur (S), or oxygen (O), to form graphene derivatives having different electronic properties.

3.4.6.3 Gas Phase Diffusion

In this step, excess carbon atoms or radicals are generally diffused from the bulk of the metal catalyst to its edges or grain boundaries, where they can be removed by hydrogen etching or desorption. This step helps to control the number of graphene layers and prevent multilayer growth. The diffusion rate depends on the specific conditions of temperature, pressure, and carbon solubility in the metal catalyst. The detailed parameters for CVD synthesis of graphene and its derivatives with various carbon precursors can be thoroughly read in the articles (Kairi et al. 2018).

A standard CVD instrument, having a quartz reaction cavity, a pump, a mass flow controller, thermocouples for measuring temperature, vacuum system, gas delivery system, energy system and computer auto-control system, are illustrated in the Fig. 3.10 (Sun et al. 2021a). Here, different combination of substrates can be utilized in the film growth process of CVD for graphene, they usually comprise Copper (Cu) (Seekaew et al. 2017), Nickel (Ni) (Yavari et al. 2011), Stainless steel (Ghaemi et al. 2016) and Iron (Fe) (An et al. 2011). In general, Methane (CH_4) and acetylene (C_2H_2) are usually used as a carbon source. There are two CVD processes, i.e. thermal CVD and plasma-enhanced CVD (PECVD), which are commonly used for the activation process of the carbon source (Cheng et al. 2015). The thermal CVD setup configuration encompasses a vacuum tube, a high-temperature heating furnace, a vacuum pump, a pressure-controlling vacuum gauge, and a mass flow controller to govern the supply of carbon and carrier gases vital for graphene synthesis. In Plasma-Enhanced CVD (PECVD), plasma induces gas source decomposition, followed by interaction with a metal substrate, leading to graphene film growth (Li et al. 2016). Various power sources such as direct current (DC) (Tzeng et al. 2013), microwave (Woehrl et al. 2014), and radio frequency (RF) (Yang et al. 2013) have served as plasma initiators. PECVD offers an advantage over thermal CVD by enabling the graphene growth

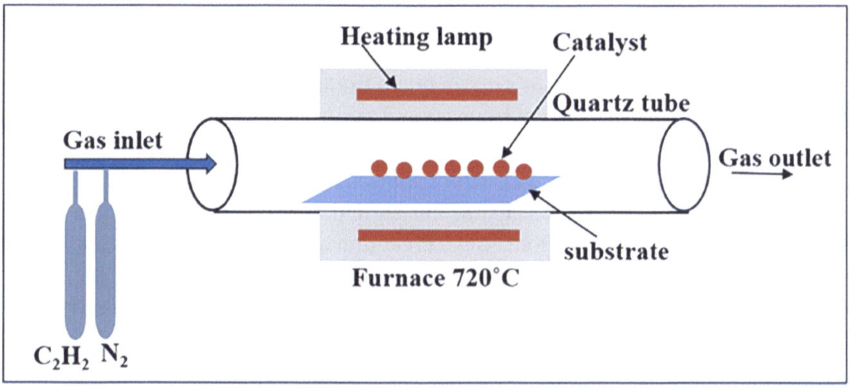

Fig. 3.10 Illustrative diagram of a standard chemical vapour deposition process

process under lower pressure and temperatures (Okigawa et al. 2015). High temperatures lead to the decomposition of the carbon source into carbon and hydrogen atoms (Magrez et al. 2007).

In this technique, hydrogen and argon gas are used in the form of carrier gases to eliminate and clean undesirable oxides on the surface of the metal catalyst. Xu. et al. (2021) reported the growth of graphene on the transition metal substrates such as Cu and Ni by the fabrication process CVD. Further, T. Mahmoudi and colleagues (Mahmoudi et al. 2018) provided a comprehensive account of the growth mechanism, described into two primary phases: (i) Generation of carbon through the pyrolysis of the gas precursor, and (ii) Development of the graphene carbon structure via the aggregation of segregated carbon on the metal catalyst's surface. By carefully controlling the synthesis parameters, such as temperature, pressure, precursor gas, and growth time, properties of graphene and its derivatives can be tailored for various applications, including electronics, energy storage, sensors, and more (Mahmoudi et al. 2018; Mbayachi et al. 2021; Sun et al. 2021b).

3.4.7 *Pyrolysis*

The term "pyrolysis" stems from the greek components "pyro" and "lysis". "Pyro" renders to fire, while "lysis" signifies separation. This terminology reflects the process of breaking down synthesized carbon atoms on a metal surface and it is employed to produce few-layer graphene structures. In general, it is refereed as the thermal decomposition of materials at elevated temperature, often in an inert environmental condition. The excessive vibration created at high temperatures causes long chain molecules to break into smaller molecules. The pyrolysis of organic based substances produces volatile products and leaves a carbon-rich solid residue of the materials

(Shams et al. 2015). Initially, this process was used for refining petroleum, now it transforms agricultural waste into valuable, eco-friendly products.

In 2011, Professor James Tour's group first introduced the concept of synthesizing graphene via the carbonization of biomass and bio-waste materials. This pioneering effort yielded high-crystallinity, low-defect graphene from sources such as food, insects, and wastewater (Ruan et al. 2011). Subsequently, a comprehensive literature review on these subject revealed advancements in the production of graphene-based nanomaterials from diverse biomasses (Prekodravac et al. 2021). This technique potentially offers a cost-effective and renewable alternative to graphene. Some typical preparation processes for graphene-like materials synthesis from biomass carbonization via pyrolysis include salt-based activation, chemical blowing, template-based confinement, coupling with hydrothermal carbonization pretreatment, post exfoliation, and some other methods (Prekodravac et al. 2021). At elevated temperatures, the thermal decomposition of silicon carbide (SiC) occurs. Under these conditions, silicon (Si) is liberated, leaving behind carbon (C) atoms that arrange themselves into a few layers of graphene, as reported by Hibino et al. (2010). This process has undergone significant advancement, especially in the continuous manufacture of graphene films on a millimeter scale at a temperature of 750 °C on a thin film of nickel coated on SiC substrate (Juang et al. 2009). The remarkable aspect of this method is its continuous graphene film production across the entire SiC-coated surface. However, it's important to note that this approach is not viable for large-scale graphene synthesis while this method offers the benefit of generating highly pure monolayer graphene.

Pyrolysis is a thermochemical process breaking down feedstock bonds without oxygen, yielding biochar, bio-oil, syngas, and more. Oxygen-free environment, 400–1200 °C or higher (Armah et al. 2022; Liu et al. 2015). The process of pyrolysis categorized as conventional (CP) and microwave-assisted (MAP). CP uses electric heating, often inefficient. MAP preferred: faster, efficient, precise (Bu et al. 2016).

Biomass, rich in carbon and renewable, presents a potential source for graphene and GLC materials. Global biomass waste of about 10 billion metric tons yearly is a growing resource. Biomass advantages: cost-effective, accessible, potentially reducing graphene production costs. Various biomass materials have been successfully transformed into graphene and carbon-derived compounds, such as wheat straw, sawdust, bamboo, etc. Biomass-derived graphene features unique properties due to aligned nanographene domains, often termed graphene-like materials (Berktas et al. 2020; Taqi-Uddeen Safian et al. n.d.; Zakaria et al. 2022).

There are also some typical preparation processes for graphene-like materials synthesis from biomass carbonization via pyrolysis. These include salt-based activation, template-based confinement, chemical blowing, coupling with hydrothermal carbonization pretreatment, post exfoliation, and some other methods. The following key steps summarize the essential stages of the pyrolysis method, providing a clear and concise overview of the pyrolysis process for the synthesis of graphene and its derivatives, as shown in Fig. 3.11. Robinson's key finding aligns with Zoraida et al. (Robinson et al. 2022). They used anthracene oil to create carbon precursors via microwave and conventional methods. Microwave tech used 60% less energy.

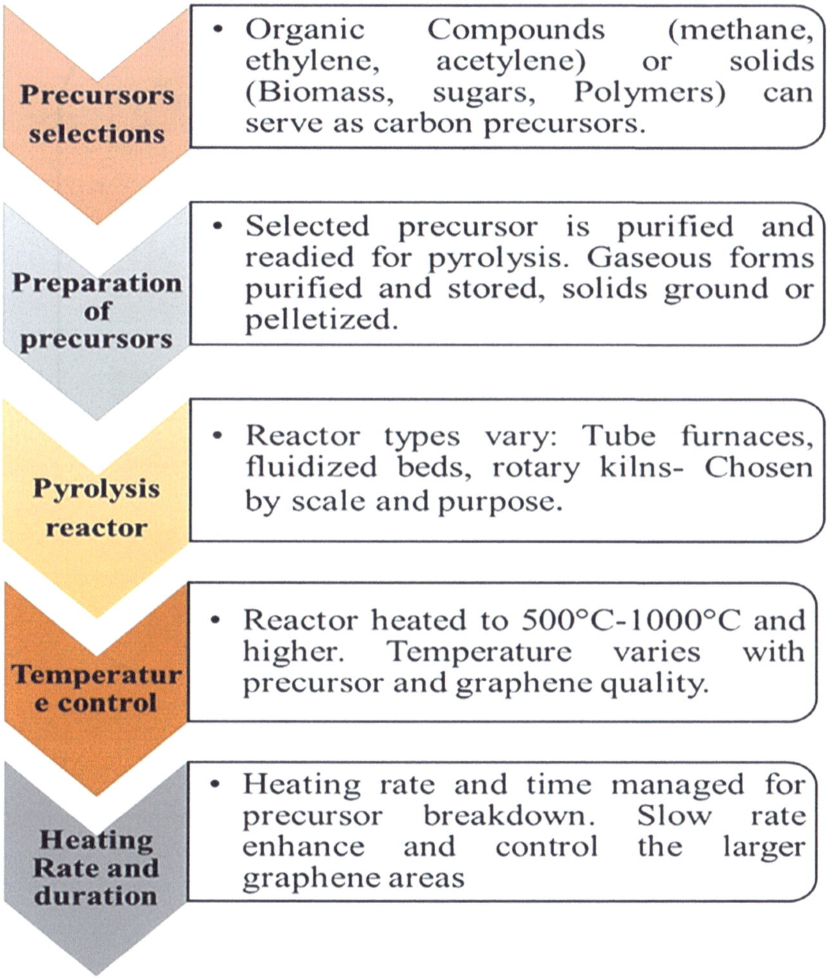

Fig. 3.11 Steps: essential phases of the pyrolysis method

Graphite from microwaved precursors had better graphitization and smaller crystallite sizes than exclusive microwaved coke. Graphene from these precursors showed improved lattice, structure, and fewer oxygenated groups. Zhang et al. (2018) made hollow carbon nanofibers on biochar surface using pine nutshell and no catalyst while Gopalkrishnan et al. (2019) reported graphene having few layers using ginger. Danafar et al. (2021) developed nanosized GO flakes using onion sheaths via pyrolysis. Later, a straightforward and commercial technique was developed to synthesize GO from agricultural waste. In recent years, a group of researchers recommended to develop green production-based techniques for the fabrication of graphene using environmentally friendly biomass resources such as chitosan, sugar, and alfalfa plant,

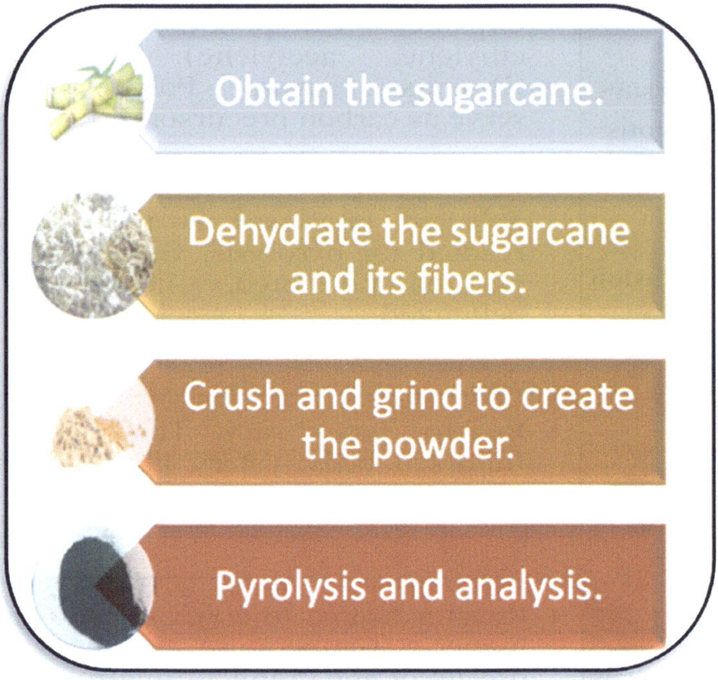

Fig. 3.12 Pyrolysis technique for carbon-based waste

as illustrated in the Fig. 3.12. They reported their important features, recent and potential applications in detail (Asif and Saha 2023; Saha and Dutta 2022).

3.4.8 Plasma Synthesis

The synthesis of graphene using the plasma process was an emerging technique that showed promise for large-scale production of high-quality graphene and its derivatives (Zafar and Jacob 2022). These high-tech synthesis practices can produce graphene without any solvents in a few seconds. The substrate-free process involves the delivery of a carbon-containing precursor into a microwave-generated Ar plasma. These cutting-edge methods primarily highlight their procedural factors for producing premium-quality graphene. This graphene is devoid of defects, consists of mono to a few layers, exhibits excellent carrier mobility, and maintains exceptional purity. The flow chart for the synthesis of graphene using plasma techniques can be displayed in Fig. 3.13.

In this process, plasma sources such as RF (13.56 MHz) and MW (2.45 GHz) are usually used for the synthesis of graphene due to their quality and manufacturing

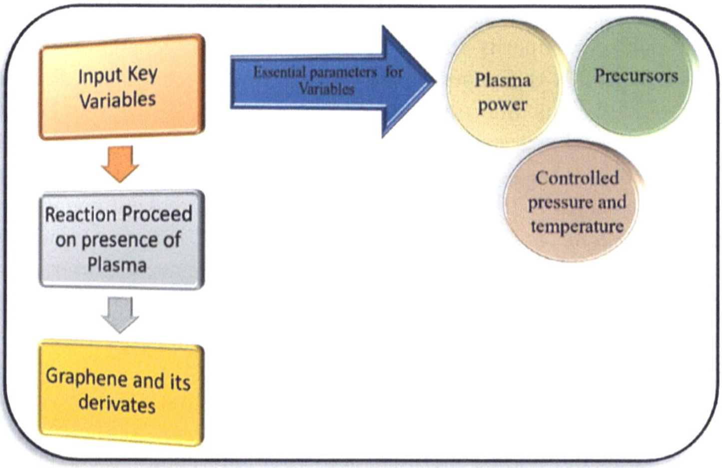

Fig. 3.13 A comprehensive overview of synthesizing Graphene by plasma process

standards. RF power of 200 W and microwave power above 1 kW are optimal for graphene manufacture. RF's effect on quality was studied by Nang and Kim et al. (Son and Ham 2017), while microwave synthesis doesn't require substrate pre-heating. Microwave plasma power from 350 to 1100 W doesn't substantially impact graphene synthesis quality (Dato et al. 2009). The flow rate and precursor composition are two essential parameters for the synthesis of graphene and its derivatives. Carbon-containing precursors like methane, toluene, and renewable sources such as sesame oil were already investigated by the researchers (Wassei et al. 2012; Zafar et al. 2022). Fragmentation of precursors is crucial, typically involving hydrocarbons mixed with hydrogen.

In past time, tea tree oil, camphor, and even edible items like honey were explored by the other researcher groups (Al-Jumaili et al. 2022; Nasir et al. 2017; Sharma et al. 2013) [247–250]. The state of precursors can be solid, liquid, or gaseous, with gaseous ones often yielding better quality graphene, reported by Deng et al. (2019). In the synthesis process, hydrogen (H_2) gas is generally employed alongside a precursor. A few efforts using argon (Ar) without H_2 compromise graphene quality (Peng et al. n.d.). Gas-to-precursor ratio significantly impacts the process of synthesis, manipulating the product type (graphite, carbon nanotube, or graphene), as well as graphene morphology and structure (Bo et al. 2013) varied CH_4 concentration and substrate temperature, observing effects on graphene sheet size and density. Optimal CH_4/H_2 gas ratios exceeding 1:20 are preferred for vertical graphene synthesis. Later, hydrogen, mixed with inert gases or used alone, acts as a catalyst, influencing grain size, layer number, and graphene formation, described by Vlassiouk et al. (2011).

In the next steps, pressure inside the chamber significantly impacts the manufacture rate of graphene. A higher vacuum pressure accelerates production in microwave plasma due to the collisions of heavy electron-particles. Sub-atmospheric pressure reduces temperature rise, slowing the rate of synthesis. Vacuuming the chamber is a crucial parameter for impurity-free, high-quality graphene generation (Muñoz and Gómez-Aleixandre 2013). Temperature also affects graphene synthesis. PECVD runs cooler due to active ions. Substrate heating varies with temperature. MW and RF-PECVD required it, while some MW systems use external heating system, reported by the researcher group Bo et al. (2013). RF and MW plasma are extensively investigated sources. Plasma serves to transfer heat and momentum, breaking precursor chains and reducing substrate pre-heating. Further, Zhang et al. (2016) noted similar stages, with a buffer layer (SiC) forming before vertical graphene nucleation from defects, benefiting growth via active sites. In recent years, Fortugno et al. (2022) studied the impact of oxygen in microwave plasma synthesis of free-standing graphene. Reactant concentration greatly influences graphene formation where low content reduces residue and enhances production. Water in the precursor promotes CO formation and graphene synthesis. Toman et al. (2021) explored the microwave power's impact on graphene structure, finding the transition from amorphous to crystalline form dependent on ethanol flow rate, microwave power, and molecule mixture.

Various plasma techniques for graphene synthesis include DC plasma, cathodic vacuum arc plasma, thermal jet plasma, and hot filament/hot wire plasma (Li et al. 2016; Singh et al. 2019; Wang et al. 2015a, b). While, these sources yield similar graphene characteristics, setup and procedures differ. For example, DC plasma employs graphite rod anode, tungsten cathode, DC power supply, and quartz chamber (Yu et al. 2011). Hot filament uses heated Tantalum filaments to dissociate gases and create graphene (Zhai et al. 2020). Radiofrequency thermal plasma jet system involves a plasma torch, generator, reactor, and precursor delivery for graphene synthesis (Fronczak et al. 2022). Micro-plasma, an emerging technique, holds promise for graphene synthesis. Operating at low temperatures, it features non-equilibrium reactivity and high energy density. Micro-plasma has been used for graphene quantum dot synthesis under atmospheric conditions, yielding dots with an average diameter of 4.9 nm and photoluminescence at 448 nm (Yang et al. 2019). Inductively coupled plasma facilitated graphene synthesis and coating on stainless steel using a robotic arm, allowing precise control over plasma exposure (Aissou et al. 2022). The plasma process offers several advantages for graphene synthesis, such as scalability, the potential for high production rates, and the ability to control the number of layers in the resulting graphene. However, challenges also exist, including the need to optimize the process parameters to achieve the desired graphene quality and scalability while keeping costs manageable.

3.4.9 Epitaxial Growth on the Silicon Carbide

Epitaxial growth on single crystalline silicon carbide (SiC) surfaces is a highly regarded method for synthesizing graphene. The term "epitaxy" originates from the greek words "epi" meaning "over" or "upon," and "taxis" meaning "order" or "arrangement." Epitaxial growth refers to depositing a single crystalline film on a similar substrate, resulting in high-quality graphene on single-crystalline SiC substrates. This process comprises two main categories: homo-epitaxial and hetero-epitaxial growth. Homo-epitaxial growth occurs when the film and substrate are the same material, while hetero-epitaxial growth involves different materials. SiC was first employed for electrical measurements on patterned epitaxial graphene. This process offers a controlled method for producing high-quality graphene layers, which are of immense interest for a wide range of electronic and optoelectronic applications due to graphene's exceptional electronic, thermal, and mechanical properties.

In the process, SiC crystal's upper layers (3CeSiC, 4HeSiC, or 6HeSiC) decompose in high-temperature vacuum annealing, leading to Si atom release. Surface C atoms rearrange, forming epitaxial graphene. Argon or slight disilane slows silicon loss, enabling higher temps for better graphene. Formation starts atop SiC, where $\approx$3 SiC bilayers decompose ($\approx$0.75 nm) to create 1 graphene layer ($\approx$0.344 nm), progressing inward (Van Bommel et al. 1975).

Initially, graphite formation on 6H–SiC (0001) surfaces was first observed in 1975 by the researcher group Van Bommel et al. (1975). Through heat treatment ranging from 1000 °C to 1500 °C in an ultrahigh vacuum, they produced graphite on both polar planes (0001) of SiC. Later in 2004, researcher group Heer et al. successfully created ultrathin graphite, comprising 1–3 monoatomic graphene layers, on the Si-terminated (0001) surface of single-crystal 6H–SiC (Heer 2010). They also investigated its electronic properties. The growth rate of graphene on SiC varies according to the specific polar crystal face. Graphene develops faster on the carbon (C)-face rather on the silicon (Si)-face (de Heer et al. 2011; Heer 2010). Larger domains (about 200 nm) of multilayered, rotationally disordered graphene are typically generated on the C-face (de Heer et al. 2011; Hass et al. 2006). While, ultra-high vacuum (UHV) annealing leads to smaller domains, roughly 30–100 nm on the C-face (Forbeaux et al. 2000; Hass et al. 2006). Si (0001)- and C (000–1)-terminated surfaces annealed at high temperatures (above 1000 °C) under ultra-high vacuum (UHV) conditions undergo graphitization due to Si evaporation (Forbeaux et al. 2000; Hass et al. 2006).

This technique holds great potential for large-scale manufacture and commercialization of graphene for electronics applications. It allows the creation of light-emitting devices, high-frequency electronics, and radiation-hard devices. Graphene on SiC has facilitated the fabrication of top-gated transistors on a wafer scale (Kedzierski et al. 2008). Especially, high-frequency transistors boasting a 100 GHz cutoff frequency have been accomplished (Lin et al. 2010), surpassing the performance of state-of-the-art Si transistors with analogous gate lengths (Schwierz 2010). Moreover, graphene on SiC has gained recognition as an exceptional resistance standard due to its manifestation of the Hall effect (quantum), despite the accompanying elevated expenses. Epitaxial graphene growth on SiC is accompanied by significant challenges that researchers are actively addressing. One major hurdle involves managing defects introduced during the growth process, which can compromise the quality of the resulting graphene layer. Researchers are exploring innovative techniques to mitigate these defects and enhance the electronic properties of graphene, enabling its optimal performance in diverse applications. Precise control over layer thickness is another critical concern, particularly for producing monolayer graphene, as the number of layers directly influences its properties. Scalability is vital for industrial applications, requiring consistent high-quality growth over larger areas while minimizing variations. Additionally, the choice of the SiC substrate polytype plays a crucial role, influencing the growth process and the properties of the resulting graphene.

The epitaxial growth of graphene on silicon carbide is a promising technique for producing high-quality graphene layers with controlled properties. This process has the potential to enable a wide range of applications in electronics, photonics, and beyond, once challenges related to defects, scalability, and control of layer thickness are overcome. Ongoing research and advancements in this field continue to improve the quality and feasibility of epitaxial graphene growth on SiC substrates.

3.5 Advantages and Disadvantages of Various Synthesis Methods of Grapheme and Its Derivates

The synthesis of graphene and its derivatives involves various methods, each with its own set of advantages and disadvantages. An overview of the advantages and disadvantages of various synthesis methods of graphene and its derivates are tabulated here:

Sr. No.	Methods	Advantages	Disadvantages
1	Electrochemical exfoliation	• Low cost and high quality • Scalable and environmentally friendly • Short reaction time	• Slight oxidation • Production of MLG
2	Mechanical exfoliation	• High quality • Monolayer production • Larger size • Defect free	• Labour intensive • Non-scalable • Low yield
3	Arc discharge	• High-quality grapheme • Suitable for small-scale Production	• Low scalability for large-scale production • Equipment and energy-intensive
4	Unzipping of carbon nanotube	• Produce grapheme nanoribbons • Unique properties	• Limited scalability
5	Chemical oxidative-exfoliation-reduction	• Low cost • High yield • Scalable	• Low quality • Functionalized • Large number of defects

(continued)

(continued)

Sr. No.	Methods	Advantages	Disadvantages
6	Chemical Vapour Deposition (CVD),	• High quality • Scalable • Large size • Excellent conductivity • Choice of carbon precursors	• Expensive • Complicated procedures • Un-efficient transfer process
7	Pyrolysis	• Produces graphene from organic precursors	• Results in varied graphene qualities • It requires post processing to improve quality
8	Plasma synthesis	• High purity • Controlled growth	• Equipment complexity • High cost • It may have limitations in certain applications
9	Epitaxial growth	• High quality • Low defects • No substrate transfers • Seamless integration	• Multi-layered graphene • Uncontrolled size • High cost

3.6 Conclusions

The chapter concludes with a comprehensive exploration of the synthesis pathways for graphene and its derivatives. They have attracted significant attention due to their exceptional properties, which have sparked intensive interests in various fields. The synthesis methods, ranging from top-down methods like electrochemical and mechanical exfoliation to bottom-up approaches like plasma synthesis and chemical vapor deposition, illuminate the dynamic landscape of graphene fabrication. As we dissect each technique, we gain a deeper understanding of their intricacies, advantages, and limitations. This insight not only addresses existing knowledge gaps but also fuels inspiration for further research and innovation in the realm of graphene materials. Overall, this chapter provides a comprehensive overview of the synthesis of graphene and its derivatives, which can serve as a valuable resource for researchers in this field.

In conclusion, the synthesis of graphene and its derivatives offers a bright outlook for innovation across industries. Overcoming challenges related to scalability, layer control, quality, environmental impact, and standardization will pave the way for their widespread adoption and transformative impact. Moreover, it's vital to note that the suitability of a synthesis technique of the graphene and its derivates depend on the specific requirements of the intended application based on considering factors such as cost, scalability, and the desired properties. Researchers generally choose methods based on a balance between these factors for their specific needs.

3.7 Promising Outlook and Challenges

This chapter provides an overview of the various synthesizing method of graphene and its derivates. Graphene, a remarkable two-dimensional carbon allotrope, has captured immense scientific and industrial interest due to its extraordinary properties. They have attracted great attention in several areas of science, including biomedical applications such as drug delivery systems, biosensors, and imaging systems. These synthesizing routes offer distinct advantages and face unique challenges. The different physicochemical properties of graphene and its derivatives guide its biocompatibility and toxicity. Despite the growing increase in the bioavailability and toxicity studies of graphene and its derivatives, there is still much to be unveiled to develop safe and effective formulations. Further research is needed to fully understand the potential of graphene and its derivatives in various applications.

References

Agarwal, V., Zetterlund, P.B.: Strategies for reduction of graphene oxide—a comprehensive review. Chem. Eng. J. **405**, 127018 (2021). https://doi.org/10.1016/j.cej.2020.127018

Aissou, T., Braidy, N., Veilleux, J.: A new one-step deposition approach of graphene nanoflakes coating using a radio frequency plasma: Synthesis, characterization and tribological behaviour. Tribol. Int. **167**, 107406(2022). https://doi.org/10.1016/j.triboint.2021.107406

Al-Jumaili, A., Zafar, M. A., Bazaka, K., Weerasinghe, J., Jacob, M.V.: Bactericidal vertically aligned graphene networks derived from renewable precursor. Carbon Trends **7**, 100157 (2022). https://doi.org/10.1016/j.cartre.2022.100157

Ambrosi, A., Chua, C.K., Bonanni, A., Pumera, M.: Lithium aluminum hydride as reducing agent for chemically reduced graphene oxides. Chem. Mater. **24**(12), 2292–2298 (2012). https://doi.org/10.1021/CM300382B

Ambrosi, A., & Pumera, M. (2014). *The CVD graphene transfer procedure introduces metallic impurities which alter the graphene electrochemical properties †*. https://doi.org/10.1039/c3nr05230c

An, H., Lee, W.J., Jung, J.: Graphene synthesis on Fe foil using thermal CVD. Curr. Appl. Phys. **11**(4 SUPPL.) (2011). https://doi.org/10.1016/j.cap.2011.03.077

Ando, T.: The electronic properties of graphene and carbon nanotubes. NPG Asia Materials|NPG Asia Mater. **1**(1), 17–21 (2009). https://doi.org/10.1038/10.1038/asiamat.2009.1

Armah, E.K., Chetty, M., Adedeji, J.A., Estrice, D.E., Mutsvene, B., Singh, N., Tshemese, Z.: Biochar: Production, Application and the Future. In Bartoli, M., Giorcelli, M., Tagliaferro, A. (eds.) Biochar (p. Ch. 2). IntechOpen (2022). https://doi.org/10.5772/intechopen.105070

Arora, N., Sharma, N.N.:. Arc discharge synthesis of carbon nanotubes: Comprehensive review (2014). https://doi.org/10.1016/j.diamond.2014.10.001

Asif, F.C., Saha, G.C.: Graphene-like carbon structure synthesis from biomass pyrolysis: a critical review on feedstock–process–properties relationship. *C* **9**(1), 31 (2023). https://doi.org/10.3390/c9010031

Avouris, P., Dimitrakopoulos, C.: Graphene: synthesis and applications (2012)

Bachmatiuk, A., Mendes, R.G., Hirsch, C., Jähne, C., Lohe, M.R., Grothe, J., Kaskel, S., Fu, L., Klingeler, R., Eckert, J., Wick, P., Rümmeli, M.H.: Few-layer graphene shells and nonmagnetic encapsulates: a versatile and nontoxic carbon nanomaterial. ACS Nano **7**(12), 10552–10562 (2013). https://doi.org/10.1021/NN4051562

Balandin, A.A., Ghosh, S., Bao, W., Calizo, I., Teweldebrhan, D., Miao, F., Lau, C.N.: Superior thermal conductivity of single-layer graphene. Nano Lett. **8**(3), 902–907 (2008). https://doi.org/10.1021/NL0731872

Balandin, A.A.: Thermal properties of graphene and nanostructured carbon materials. Nat. Mater. 10(8), 569–581. Nature Publishing Group (2011). https://doi.org/10.1038/nmat3064

Benítez-Martínez, S., Valcárcel, M.: Graphene quantum dots in analytical science (2015). https://doi.org/10.1016/j.trac.2015.03.020

Berktas, I., Hezarkhani, M., Haghighi Poudeh, L., Saner Okan, B.: Recent developments in the synthesis of graphene and graphene-like structures from waste sources by recycling and upcycling technologies: a review. Graphene Technol. **5**(3–4), 59–73 (2020). https://doi.org/10.1007/s41127-020-00033-1

Bianco, A., Cheng, H.-M., Enoki, T., Gogotsi, Y., Hurt, R. H., Koratkar, N., Kyotani, T., Monthioux, M., Park, C.R., Tascon, J.M.D., Zhang, J.: All in the graphene family—a recommended nomenclature for two-dimensional carbon materials. Carbon **65**, 1–6 (2013). https://doi.org/10.1016/j.carbon.2013.08.038

Bo, Z., Yang, Y., Chen, J., Yu, K., Yan, J., Cen, K.: Plasma-enhanced chemical vapor deposition synthesis of vertically oriented graphene nanosheets. Nanoscale 5(12), 5180–5204 (2013). https://doi.org/10.1039/c3nr33449j

Bolotin, K.I., Sikes, K.J., Jiang, Z., Klima, M., Fudenberg, G., Hone, J., Kim, P., Stormer, H.L.: Ultrahigh electron mobility in suspended graphene. Solid State Commun.commun. **146**, 351–355 (2008). https://doi.org/10.1016/j.ssc.2008.02.024

Brodeie.: brodie1859. Phil. Trans. R. Soc. **149**, 249–259 (1859). https://doi.org/10.1098/rstl.1859.0013

Brownson, D.A.C., Banks, C.E.: The electrochemistry of CVD graphene: Progress and prospects. Phys. Chem. Chem. Phys. **14**(23), 8264–8281 (2012). https://doi.org/10.1039/c2cp40225d

Brownson, D.A.C., Banks, C.E.: CVD graphene electrochemistry: the role of graphitic islands. Phys. Chem. Chem. Phys. **13**(35), 15825–15828 (2011). https://doi.org/10.1039/c1cp21978b

Bu, Q., Morgan, H.M., Liang, J., Lei, H., Ruan, R.: Chapter Two-Catalytic microwave pyrolysis of lignocellulosic biomass for fuels and chemicals. In: Li, Y., Ge, X. (eds.) Advances in Bioenergy, vol. 1, pp. 69–123. Elsevier (20160. https://doi.org/10.1016/bs.aibe.2016.09.002

Cano-Márquez, A.G., Rodríguez-Macías, F.J., Campos-Delgado, J., Espinosa-González, C.G., Tristán-López, F., Ramírez-González, D., Cullen, D.A., Smith, D.J., Terrones, M., Vega-Cantú, Y.I.: Ex-MWNTs: graphene sheets and ribbons produced by lithium intercalation and exfoliation of carbon nanotubes. Nano Lett. **9**(4), 1527–1533 (2009). https://doi.org/10.1021/NL803585S

Chae, M.-S., Kim, J., Jeong, D., Kim, Y., Hoon Roh, J., Min Lee, S., Heo, Y., Yoon Kang, J., Hoon Lee, J., Sung Yoon, D., Geun Kim, T., Tai Chang, S., Seon Hwang, K.: Enhancing surface functionality of reduced graphene oxide biosensors by oxygen plasma treatment for Alzheimer's disease diagnosis (2016). https://doi.org/10.1016/j.bios.2016.10.049

Chandra, V., Kim, K.S.: Highly selective adsorption of Hg2+ by a polypyrrole–reduced graphene oxide composite. Chem. Commun.commun. **47**(13), 3942–3944 (2011). https://doi.org/10.1039/c1cc00005e

Chen, C.H., Yang, S.W., Chuang, M.C., Woon, W.Y., Su, C.Y.: Towards the continuous production of high crystallinity graphene via electrochemical exfoliation with molecular in situ encapsulation. Nanoscale **7**(37), 15362–15373 (2015). https://doi.org/10.1039/c5nr03669k

Chen, K., Xue, D.: Preparation of colloidal graphene in quantity by electrochemical exfoliation (2014). https://doi.org/10.1016/j.jcis.2014.08.057

Cheng, L., Yun, K., Lucero, A., Huang, J., Meng, X., Lian, G., Nam, H.S., Wallace, R.M., Kim, M., Venugopal, A., Colombo, L., Kim, J.: Low temperature synthesis of graphite on Ni films

using inductively coupled plasma enhanced CVD. J. Mater. Chem. C **3**(20), 5192–5198 (2015). https://doi.org/10.1039/c5tc00635j
Choudhary, M., Shukla, S. K., Narang, J., Kumar, V., Govender, P.P., Niv, A., Hussain, C.M., Wang, R., Mangla, B., Babu, R.S.: Switchable graphene-based bioelectronics interfaces. Chemosensors **8**(8). MDPI Multidisciplinary Digital Publishing Institute (2020). https://doi.org/10.3390/CHEMOSENSORS8020045
Chua, C.K., Pumera, M.: Chemical reduction of graphene oxide: A synthetic chemistry viewpoint. Chem. Soc. Rev. **43**(1), 291–312. Royal Society of Chemistry (2014). https://doi.org/10.1039/c3cs60303b
Cooper, A.J., Wilson, N.R., Kinloch, I.A., Dryfe, R.A.W.: Single stage electrochemical exfoliation method for the production of few-layer graphene via intercalation of tetraalkylammonium cations (2013). https://doi.org/10.1016/j.carbon.2013.09.009
Danafar, H., Mohammadi, A., Mousazadeh, N., Ghaffarlou, M., Mollasalehi, A., Sharafi, A., Barsbay, M., Nosrati, H.: An innovative green approach to the production of bio-sourced and nano-sized graphene oxide (GO)-like carbon flakes. Curr. Res. Green Sustain. Chem. **4**, 100200 (2021). https://doi.org/10.1016/j.crgsc.2021.100200
Dasari, B.L., Nouri, J.M., Brabazon, D., Naher, S.: Graphene and derivatives e Synthesis techniques, properties and their energy applications (2017). https://doi.org/10.1016/j.energy.2017.08.048
Dato, A., Radmilovic, V., Lee, Z., Phillips, J., Frenklach, M.: Substrate-free gas-phase synthesis of graphene sheets. Nano Lett. **8**(7), 2012–2016 (2008). https://doi.org/10.1021/NL8011566
Dato, A., Lee, Z., Jeon, K.J., Erni, R., Radmilovic, V., Richardson, T.J., Frenklach, M.: Clean and highly ordered graphene synthesized in the gas phase. Chem. Commun.commun. **40**, 6095–6097 (2009). https://doi.org/10.1039/b911395a
Dreyer, D.R., Jia, H.P., Bielawski, C.W.: Graphene oxide: a convenient carbocatalyst for facilitating oxidation and hydration reactions. Angew Chem. Int. Ed. Engl. 49(38), 6813–6816 (2010). https://doi.org/10.1002/anie.201002160. PMID: 20602388
de Heer, W.A., Berger, C., Ruan, M., Sprinkle, M., Li, X., Hu, Y., Zhang, B., Hankinson, J., Conrad, E.: Large area and structured epitaxial graphene produced by confinement controlled sublimation of silicon carbide. Proc. Natl. Acad. Sci. **108**(41), 16900–16905 (2011). https://doi.org/10.1073/pnas.1105113108
Deng, B., Liu, Z., & Peng, H. (2019). Toward Mass Production of CVD Graphene Films. *Advanced Materials*, *31*(9). https://doi.org/10.1002/ADMA.201800996
Designed Research; K, T. F. H. F. M., Performed Research; K, M. Y. S. F. M., Pnas, J.: The evolution of electronic structure in few-layer graphene revealed by optical spectroscopy. **107**(34), 14999–15004 (2010). https://doi.org/10.1073/pnas.1004595107/-/DCSupplemental
Devi, S., Tripathi, U.K., Roy, D., Dwivedi, M.: Coherent loading-deloading mechanism in polymeric nanohybrid network structures. Langmuir **39**(6), 2281–2290 (2023). https://doi.org/10.1021/ACS.LANGMUIR.2C02918
Dideikin, A.T., Vul', A.Y. Graphene oxide and derivatives: The place in graphene family. Front. Phys. **6**. Frontiers Media S.A (2019). https://doi.org/10.3389/fphy.2018.00149
Dimiev, A.M., Eigler, S.: Graphene Oxide: Fundamentals and Applications. Wiley (2016)
Dimiev, A.M., Khannanov, A., Vakhitov, I., Kiiamov, A., Shukhina, K., Tour, J.M.: Revisiting the mechanism of oxidative unzipping of multiwall carbon nanotubes to graphene nanoribbons. ACS Nano **12**(4), 3985–3993 (2018). https://doi.org/10.1021/acsnano.8b01617
Doğan, H.Ö., Ekinci, D., Demir, Ü.: Atomic scale imaging and spectroscopic characterization of electrochemically reduced graphene oxide (2013). https://doi.org/10.1016/j.susc.2013.01.014
Dong, J., Wang, K., Sun, L., Sun, B., Yang, M., Chen, H., Wang, Y., Sun, J., Dong, L.: Application of graphene quantum dots for simultaneous fluorescence imaging and tumor-targeted drug delivery. Sens. Actuators B **256**, 616–623 (2018). https://doi.org/10.1016/j.snb.2017.09.200
Dong, B., Liu, G., Zhou, J., Wang, A., Wang, J., Jin, R., Lv, H.: Biogenic gold nanoparticles-reduced graphene oxide nanohybrid: synthesis, characterization and application in chemical and biological reduction of nitroaromatics (2015). https://doi.org/10.1039/c5ra19806b

Eda, G., Chhowalla, M.: Chemically derived graphene oxide: Towards large-area thin-film electronics and optoelectronics. Adv. Mater. **22**(22), 2392–2415 (2010). https://doi.org/10.1002/ADMA.200903689

Elias, A.L., Botello-Méndez, A.R., Meneses-Rodríguez, D., González, V.J., Ramírez-González, D., Ci, L., Muñoz-Sandoval, E., Ajayan, P.M., Terrones, H., Terrones, M.: Longitudinal cutting of pure and doped carbon nanotubes to form graphitic nanoribbons using metal clusters as nanoscalpels. Nano Lett. **10**(2), 366–372 (2010). https://doi.org/10.1021/NL901631Z

Fan, Z., Li, S., Yuan, F., Fan, L.: Fluorescent graphene quantum dots for biosensing and bioimaging. RSC Adv. 5(25), 19773–19789. R. Soc. Chem. (2015). https://doi.org/10.1039/c4ra17131d

Feng, L.L., Wu, Y.X., Zhang, D.L., Hu, X.X., Zhang, J., Wang, P., Song, Z.L., Zhang, X.B., Tan, W.: Near infrared graphene quantum dots-based two-photon nanoprobe for direct bioimaging of endogenous ascorbic acid in living cells. Anal. Chem. **89**(7), 4077–4084 (2017). https://doi.org/10.1021/acs.analchem.6b04943

Fernández-Merino, M.J., Guardia, L., Paredes, J.I., Villar-Rodil, S., Solís-Fernández, P., Martínez-Alonso, A., Tascón, J.M.D.: Vitamin C is an ideal substitute for hydrazine in the reduction of graphene oxide suspensions. J. Phys. Chem. C **114**(14), 6426–6432 (2010). https://doi.org/10.1021/JP100603H

Forbeaux, I., Themlin, J.-M., Charrier, A., Thibaudau, F., Debever, J.-M.: Solid-state graphitization mechanisms of silicon carbide 6H–SiC polar faces. Appl. Surf. Sci. 162–163, 406–412 (2000). https://doi.org/10.1016/S0169-4332(00)00224-5

Fortugno, P., Musikhin, S., Shi, X., Wang, H., Wiggers, H., Schulz, C.: Synthesis of freestanding few-layer graphene in microwave plasma: The role of oxygen. Carbon **186**, 560–573 (2022). https://doi.org/10.1016/j.carbon.2021.10.047

Fronczak, M., Keszler, A.M., Mohai, M., Jezsó, B., Farkas, A., Károly, Z.: Facile and continuous synthesis of graphene nanoflakes in RF thermal plasma. Carbon **193**, 51–67 (2022). https://doi.org/10.1016/j.carbon.2022.03.008

Gaj, J., Clapa, M., Nowak, D., Juszczak, J., Galazka, M., Pelka, M., Niedzielski, P.: Metallurgical graphene under different gas atmospheres and UV radiation for gas-sensing applications. Sens. Actuators A **312**, 112152 (2020). https://doi.org/10.1016/j.sna.2020.112152

Gao, Y., Li, Y., Zhang, L., Huang, H., Hu, J., Shah, S.M., Su, X.: Adsorption and removal of tetracycline antibiotics from aqueous solution by graphene oxide (2011). https://doi.org/10.1016/j.jcis.2011.11.015

Geim, A.K., Novoselov, K.S.: The rise of graphene. Nat. Mater. **6**(3) (2007). https://doi.org/10.1038/nmat1849

Geim, A.K.: Graphene: status and prospects. Science (2009). https://doi.org/10.1126/science.1158877

Ghaemi, F., Abdullah, L.C., Tahir, P.M., Yunus, R.: Synthesis of different layers of graphene on stainless steel using the CVD method. Nanoscale Res. Lett. **11**(1) (2016). https://doi.org/10.1186/s11671-016-1709-x

Gondosiswanto, R., Lu, X., Zhao, C.: Preparation of metal-free nitrogen-doped graphene via direct electrochemical exfoliation of graphite in ammonium nitrate. Aust. J. Chem. **68**(5), 830 (2015). https://doi.org/10.1071/CH14447

Gopalakrishnan, A., Kong, C.Y., Badhulika, S.: Scalable, large-area synthesis of heteroatom-doped few-layer graphene-like microporous carbon nanosheets from biomass for high-capacitance supercapacitors. New J. Chem. **43**(3), 1186–1194 (2019). https://doi.org/10.1039/c8nj05128c

Guex, L.G., Sacchi, B., Peuvot, K.F., Andersson, R.L., Pourrahimi, A.M., Ström, V., Farris, S., Olsson, R.T.: Experimental review: Chemical reduction of graphene oxide (GO) to reduced graphene oxide (rGO) by aqueous chemistry. Nanoscale **9**(27), 9562–9571 (2017). https://doi.org/10.1039/c7nr02943h

Güler, Ö., Güler, S.H., Selen, V., Albayrak, M.G., Evin, E.: Production of graphene layer by liquid-phase exfoliation with low sonication power and sonication time from synthesized expanded graphite. Fullerenes, Nanotubes, Carbon Nanostruct. **24**(2), 123–127 (2016). https://doi.org/10.1080/1536383X.2015.1114472

Guo, S., Dong, S.: Graphene nanosheet: Synthesis, molecular engineering, thin film, hybrids, and energy and analytical applications. Chem. Soc. Rev. **40**(5), 2644–2672 (2011). https://doi.org/10.1039/c0cs00079e

Haar, S., Bruna, M., Lian, J.X., Tomarchio, F., Olivier, Y., Mazzaro, R., Morandi, V., Moran, J., Ferrari, A.C., Beljonne, D., Ciesielski, A., Samorì, P.: Liquid-Phase Exfoliation of Graphite into Single- and few-layer graphene with α-Functionalized Alkanes. J. Phys. Chem. Lett. **7**(14), 2714–2721 (2016). https://doi.org/10.1021/ACS.JPCLETT.6B01260

Hass, J., Feng, R., Li, T., Li, X., Zong, Z., de Heer, W.A., First, P.N., Conrad, E.H., Jeffrey, C.A., Berger, C.: Highly ordered graphene for two dimensional electronics. Appl. Phys. Lett. **89**(14), 143106 (2006). https://doi.org/10.1063/1.2358299

Heer. (n.d.). 1012.1644. https://doi.org/10.48550/arXiv.1012.1644

Hernandez, Y., Nicolosi, V., Lotya, M., Blighe, F.M., Sun, Z., De, S., McGovern, I.T., Holland, B., Byrne, M., Gun'ko, Y.K., Boland, J.J., Niraj, P., Duesberg, G., Krishnamurthy, S., Goodhue, R., Hutchison, J., Scardaci, V., Ferrari, A.C., Coleman, J.N.: High-yield production of graphene by liquid-phase exfoliation of graphite. Nat. Nanotechnol. **3**(9), 563–568 (2008). https://doi.org/10.1038/nnano.2008.215

Hibino, H., Kageshima, H., Nagase, M.: Graphene growth on silicon carbide. NTT Techn. Rev. 8(8) (2010)

Hofmann.: BF01451376 (1). Kolloid-Zeitschrift **68**, 149–151 (1934). https://doi.org/10.1007/BF01451376

Huang, X., Yin, Z., Wu, S., Qi, X., He, Q., Zhang, Q., Yan, Q., Boey, F., Zhang, H.: Graphene-based materials: synthesis, characterization, properties, and applications. Small **7**(14), 1876–1902 (2011). https://doi.org/10.1002/smll.201002009

Hummers, W.S., Offeman, R.E.: Preparation of graphitic oxide (1958). https://pubs.acs.org/sharingguidelines

IEEE Staff.: In: 2016 IEEE International Conference on Manipulation, Manufacturing and Measurement on the Nanoscale (3M NANO). IEEE (2016)

Ishida, T., Miyata, Y., Shinoda, Y., Kobayashi, Y.: Anomalous restoration of graphitic layers from graphene oxide in ethanol environment at ultrahigh temperature using solar furnace. Appl. Phys. Express **9**(2), 025103 (2016). https://doi.org/10.7567/APEX.9.025103

Jacobberger, R.M., Machhi, R., Wroblewski, J., Taylor, B., Gillian-Daniel, A.L., Arnold, M.S.: Simple graphene synthesis via chemical vapor deposition. J. Chem. Educ. **92**(11), 1903–1907 (2015). https://doi.org/10.1021/ACS.JCHEMED.5B00126/SUPPL_FILE/ED5B00126_SI_001.PDF

Jafri, R.I., Rajalakshmi, N., Ramaprabhu, S.: Nitrogen doped graphene nanoplatelets as catalyst support for oxygen reduction reaction in proton exchange membrane fuel cell (n.d.). https://doi.org/10.1039/c0jm00467g

Jaiswal, R., Saha, U., Goswami, T.H., Maji, P.K.: Effect of partially reduced fullerenol-graphene hybrid nanofiller on photophysical and super capacitance properties of fluorescence conducting polymer nanocomposites. Polym. Compos.. Compos. **44**(6), 3401–3415 (2023). https://doi.org/10.1002/PC.27329

Jaiswal, R., Saha, U., Goswami, T.H., Srivastava, A., Eswara Prasad, N.: "Pillar effect" of chemically bonded fullerene in enhancing supercapacitance performances of partially reduced fullerenol graphene oxide hybrid electrode material (2018). https://doi.org/10.1016/j.electacta.2018.06.110

Jalili-Firoozinezhad, S., Hasan, M., Moghadam, M., Mohammad, C., Ghanian, H., Ashtiani, M.K., Alimadadi, H., Baharvand, H., Martin, I., Scherberich, A.: Polycaprolactone-templated reduced-graphene oxide liquid crystal nanofibers towards biomedical applications (2017). https://doi.org/10.1039/c7ra06178a

Jayasena, B., Subbiah, S.: A novel mechanical cleavage method for synthesizing few-layer graphenes (2011). https://doi.org/10.1186/1556-276X-6-95

Jha, R., Singh, A., Pramod, Sharma, K., Porwal, O., Neeraj, Fuloria, K.: Graphene-based nanomaterial system: a boon in the era of smart nanocarriers. J. Pharm. Invest. **51**, 245–280 (2021). https://doi.org/10.1007/s40005-021-00513-3

Jiao, L., Zhang, L., Wang, X., Diankov, G., Dai, H.: Narrow graphene nanoribbons from carbon nanotubes. Nature (2009). https://doi.org/10.1038/nature07919

Jilani, A., Melaibari, A.A.: MoS_2–Cu/CuO@graphene heterogeneous photocatalysis for enhanced photocatalytic degradation of mb from water. Polymers **14**(16) (2022). https://doi.org/10.3390/polym14163259

Juang, Z.-Y., Wu, C.-Y., Lo, C.-W., Chen, W.-Y., Huang, C.-F., Hwang, J.-C., Chen, F.-R., Leou, K.-C., Tsai, C.-H.: Synthesis of graphene on silicon carbide substrates at low temperature. Carbon **47**, 2026–2031 (2009). https://doi.org/10.1016/j.carbon.2009.03.051

Jung, I., Field, D.A., Clark, N.J., Zhu, Y., Yang, D., Piner, R.D., Stankovich, S., Dikin, D.A., Geisler, H., Ventrice, C.A., Ruoff, R.S.: Reduction kinetics of graphene oxide determined by electrical transport measurements and temperature programmed desorption (n.d.). https://doi.org/10.1021/jp904396j

Kairi, M.I., Khavarian, M., Bakar, S.A., Vigolo, B., Mohamed, A.R.: Recent trends in graphene materials synthesized by CVD with various carbon precursors. J. Mater. Sci. **53**(2), 851–879 (2018). https://doi.org/10.1007/s10853-017-1694-1

Karačić, D., Korać, S., Dobrota, A.S., Pašti, I.A., Skorodumova, N.V., Gutić, S.J.: When supporting electrolyte matters–tuning capacitive response of graphene oxide via electrochemical reduction in alkali and alkaline earth metal chlorides. Electrochim. Acta. Acta **297**, 112–117 (2019). https://doi.org/10.1016/j.electacta.2018.11.173

Kauppila, J., Kunnas, P., Damlin, P., Viinikanoja, A., Kvarnström, C.: Electrochimica acta electrochemical reduction of graphene oxide films in aqueous and organic solutions. Electrochim. Acta. Acta **89**, 84–89 (2013). https://doi.org/10.1016/j.electacta.2012.10.153

Kedzierski, J., Hsu, P.-L., Healey, P., Wyatt, P.W., Keast, C.L., Sprinkle, M., Berger, C., de Heer, W.A.: Epitaxial graphene transistors on SiC substrates. IEEE Trans. Electron Dev. **55**(8), 2078–2085 (2008). https://doi.org/10.1109/TED.2008.926593

Kim, J., Cote, L.J., Kim, F., Yuan, W., Shull, K.R., Huang, J.: Graphene oxide sheets at interfaces. J. Am. Chem. Soc. **132**(23), 8180–8186 (2010a). https://doi.org/10.1021/JA102777P

Kim, K., Sussman, A., Zettl, A.: Graphene nanoribbons obtained by electrically unwrapping carbon nanotubes. ACS Nano **4**(3), 1362–1366 (2010b). https://doi.org/10.1021/NN901782G

Kim, J., Song, S.H., Jin, Y., Park, H.J., Yoon, H., Jeon, S., Cho, S.W.: Multiphoton luminescent graphene quantum dots for in vivo tracking of human adipose-derived stem cells. Nanoscale **8**(16), 8512–8519 (2016). https://doi.org/10.1039/c6nr02143c

Ko, D., Choi, J., Yan, B., Hwang, T., Jin, X., Kim, J.M., Piao, Y.: A facile and scalable approach to develop electrochemical unzipping of multi-walled carbon nanotubes to graphene nanoribbons. J. Mater. Chem. A **8**(42), 22045–22053 (2020). https://doi.org/10.1039/d0ta03782f

Koo, J.H., Paek, S.M.: Microwave-assisted synthesis of Ge/GeO_2-reduced graphene oxide nanocomposite with enhanced discharge capacity for lithium-ion batteries. Nanomaterials **11**(2), 1–15 (2021). https://doi.org/10.3390/nano11020319

Korucu, H., Şimşek, B, Kocakerim, M.M., Karakaş, İ.H.: Effective reduction of graphene oxide using sulfur dioxide-containing chemical compounds. Int. J. Environ. Sci. Technol. **16**(3), 8329–8342 (2019). https://doi.org/10.1007/s13762-019-02419-y

Kosynkin, D.V., Higginbotham, A.L., Sinitskii, A., Lomeda, J.R., Dimiev, A., Price, B.K., Tour, J.M.: Longitudinal unzipping of carbon nanotubes to form graphene nanoribbons. Nature (2009). https://doi.org/10.1038/nature07872

Krätschmer, W., Lamb, L.D., Fostiropoulos, K., Huffman, D.R.: Solid C60: a new form of carbon. Nature **347**(6291), 354–358 (1990). https://doi.org/10.1038/347354a0

Kumar, P., Panchakarla, L.S., Rao, C.N.R.: Laser-induced unzipping of carbon nanotubes to yield graphene nanoribbons (2011). https://doi.org/10.1039/c1nr10137d

Kumar, V., Kumar, A., Lee, D., Park, S.: Estimation of number of graphene layers using different methods: a focused review. Materials **14**(16), 4590 (2020). https://doi.org/10.3390/ma14164590

Kyesmen, P.I., Onoja, A., Amah, A.N.: Fullerenes synthesis by combined resistive heating and arc discharge techniques. SpringerPlus **5**(1) (2016). https://doi.org/10.1186/s40064-016-2994-7

Lavin-Lopez, M.D.P., Romero, A., Garrido, J., Sanchez-Silva, L., Valverde, J.L.: Influence of different improved hummers method modifications on the characteristics of graphite oxide in order to make a more easily scalable method. Ind. Eng. Chem. Res. **55**(50), 12836–12847 (2016). https://doi.org/10.1021/ACS.IECR.6B03533

Lee, C., Wei, X., Kysar, J.W., Hone, J.: Measurement of the elastic properties and intrinsic strength of monolayer graphene. Science **321**(5887), 385–388 (2008). https://doi.org/10.1126/SCIENCE.1157996

Li, H., He, X., Kang, Z., Huang, H., Liu, Y., Liu, J., Lian, S., Tsang, C.H.A., Yang, X., Lee, S.T.: Water-soluble fluorescent carbon quantum dots and photocatalyst design. Angewandte Chemie—Int. Ed. **49**(26), 4430–4434 (2010). https://doi.org/10.1002/ANIE.200906154

Li, Y., Hu, Y., Zhao, Y., Shi, G., Deng, L., Hou, Y., Qu, L.: An electrochemical avenue to green-luminescent graphene quantum dots as potential electron-acceptors for photovoltaics. Adv. Mater. **23**(6), 776–780 (2011). https://doi.org/10.1002/ADMA.201003819

Li, Y., Chen, Q., Xu, K., Kaneko, T., Hatakeyama, R.: Synthesis of graphene nanosheets from petroleum asphalt by pulsed arc discharge in water (2012). https://doi.org/10.1016/j.cej.2012.09.123

Li, W., Liu, J., Yan, C.: Reduced graphene oxide with tunable C/O ratio and its activity towards vanadium redox pairs for an all vanadium redox flow battery (2013). https://doi.org/10.1016/j.carbon.2012.12.069

Li, M., Liu, D., Wei, D., Song, X., Wei, D., Wee, A.T.S.: Controllable synthesis of graphene by plasma-enhanced chemical vapor deposition and its related applications. Adv. Sci. **3**(11). Wiley-VCH Verlag (2016). https://doi.org/10.1002/advs.201600003

Li, K., Liu, W., Ni, Y., Li, D., Lin, D., Su, Z., Wei, G.: Technical synthesis and biomedical applications of graphene quantum dots. J. Mater. Chem. B **5**(25), 4811–4826; R. Soc. Chem. (2017). https://doi.org/10.1039/c7tb01073g

Li, L., Wu, G., Yang, G., Peng, J., Zhao, J., Zhu, J.-J.: Focusing on luminescent graphene quantum dots: current status and future perspectives (n.d.). https://doi.org/10.1039/c3nr33849e

Liao, C., Li, Y., Tjong, S.C.: Graphene nanomaterials: synthesis, biocompatibility, and cytotoxicity. Int. J. Mole. Sci. 19(11). MDPI AG (2018). https://doi.org/10.3390/ijms19113564

Lightcap, I.V., Kamat, P.V.: Graphitic design: Prospects of graphene-based nanocomposites for solar energy conversion, storage, and sensing. Acc. Chem. Res. **46**(10), 2235–2243 (2013). https://doi.org/10.1021/AR300248F

Lin, Y.-M., Dimitrakopoulos, C., Jenkins, K.A., Farmer, D.B., Chiu, H.-Y., Grill, A., Avouris, P.: 100-GHz transistors from wafer-scale epitaxial graphene. Science **327**(5966), 662 (2010). https://doi.org/10.1126/science.1184289

Lin, J., Chen, X., Huang, P.: Graphene-based nanomaterials for bioimaging (2016). https://doi.org/10.1016/j.addr.2016.05.013

Liu, W.J., Jiang, H., Yu, H.Q.: Development of biochar-based functional materials: toward a sustainable platform carbon material. Chem. Rev. **115**(22), 12251–12285 (2015). https://doi.org/10.1021/ACS.CHEMREV.5B00195

Liu, J., Kok Poh, C., Zhan, D., Lai, L., Lim, S.H., Wang, L., Liu, X., Sahoo, N. G., Li, C., Shen, Z., Lin, J.: Improved synthesis of graphene flakes from the multiple electrochemical exfoliation of graphite rod (2012). https://doi.org/10.1016/j.nanoen.2012.11.003

Liu, J., Yang, H., Zhen, S.G., Kok, C., Chaurasia, A., Luo, J., Wu, X., Kok, E., Yeow, L., Sahoo, N.G., Lin, J., Shen, Z.: A green approach to the synthesis of high-quality graphene oxide flakes via electrochemical exfoliation of pencil core3. **3**, 11745 (2013). https://doi.org/10.1039/c3ra41366g

Low, C.T.J., Walsh, F.C., Chakrabarti, M.H., Hashim, M.A., Hussain, M.A.: Electrochemical approaches to the production of graphene flakes and their potential applications (2012). https://doi.org/10.1016/j.carbon.2012.11.030

Lu, X., Zhao, C.: Controlled electrochemical intercalation, exfoliation and in situ nitrogen doping of graphite in nitrate-based protic ionic liquids. Phys. Chem. Chem. Phys. **15**(46), 20005–20009 (2013). https://doi.org/10.1039/c3cp53917b

Lu, J., Yang, J.X., Wang, J., Lim, A., Wang, S., Loh, K.P.: One-pot synthesis of fluorescent carbon nanoribbons, nanoparticles, and graphene by the exfoliation of graphite in ionic liquids. ACS Nano **3**(8), 2367–2375 (2009). https://doi.org/10.1021/NN900546B

Luk, C.M., Tang, L.B., Zhang, W.F., Yu, S.F., Teng, K.S., Lau, S.P.: An efficient and stable fluorescent graphene quantum dot-agar composite as a converting material in white light emitting diodes. J. Mater. Chem. **22**(42), 22378–22381 (2012). https://doi.org/10.1039/c2jm35305a

Luo, J., Kim, J., Huang, J.: Material processing of chemically modified graphene: Some challenges and solutions. Acc. Chem. Res. **46**(10), 2225–2234 (2013). https://doi.org/10.1021/AR300180N

MacHado, B.F., Serp, P.: Graphene-based materials for catalysis. Catal. Sci. Technol. 2(1), 54–75 (2012). https://doi.org/10.1039/c1cy00361e

Magrez, A., Seo, J.W., Kuznetsov, V.L., Forró, L.: Evidence of an equimolar C_2H_2–CO_2 reaction in the synthesis of carbon nanotubes. Angew. Chem.. Chem. **119**(3), 445–448 (2007). https://doi.org/10.1002/ange.200603764

Mahmoudi, T., Wang, Y., Hahn, Y.B.: Graphene and its derivatives for solar cells application. Nano Energy **47**, 51–65. Elsevier Ltd. (2018). https://doi.org/10.1016/j.nanoen.2018.02.047

Mao, L., Liu, C., Lu, K., Su, Y., Gu, C., Huang, Q., Petersen, E.J.: Exposure of few layer graphene to Limnodrilus hoffmeisteri modifies the graphene and changes its bioaccumulation by other organisms (2016). https://doi.org/10.1016/j.carbon.2016.08.037

Marcano, D.C., Kosynkin, D.V, Berlin, J.M., Sinitskii, A., Sun, Z., Slesarev, A., Alemany, L.B., Lu, W., Tour, J.M.: Improved synthesis of graphene oxide (2010). https://doi.org/10.1021/nn1006368

Marrani, A.G., Motta, A., Schrebler, R., Zanoni, R., Dalchiele, E.A.: Insights from experiment and theory into the electrochemical reduction mechanism of graphene oxide. Electrochim. Acta. Acta **304**, 231–238 (2019). https://doi.org/10.1016/j.electacta.2019.02.108

Martinez, U., Rojas-Carbonell, S., Halevi, B., Ren, P.-G., Yan, D.-X., Ji, X., Chen, T., Li, Z.-M.: Temperature dependence of graphene oxide reduced by hydrazine hydrate You may also like Ni-La electrocatalysts for direct hydrazine alkaline anion-exchange membrane fuel cells temperature dependence of graphene oxide reduced by hydrazine hydrate. Nanotechnology **22**, 55705–55713 (2011). https://doi.org/10.1088/0957-4484/22/5/055705

Mbayachi, V.B., Ndayiragije, E., Sammani, T., Taj, S., Mbuta, E.R., khan, A.U.: Graphene synthesis, characterization and its applications: a review. In: Results in Chemistry, vol. 3. Elsevier B.V (2021). https://doi.org/10.1016/j.rechem.2021.100163

Mendes, R.G., Bachmatiuk, A., El-Gendy, A.A., Melkhanova, S., Klingeler, R., Büchner, B., Rümmeli, M.H.: A facile route to coat iron oxide nanoparticles with few-layer graphene. J. Phys. Chem. C **116**(44), 23749–23756 (2012). https://doi.org/10.1021/JP307623H

Moon, I.K., Lee, J., Lee, H.: Highly qualified reduced graphene oxides: the best chemical reduction. Chem. Commun.commun. **47**(34), 9681 (2011). https://doi.org/10.1039/c1cc13312h

Morelos-Gómez, A., Vega-Díaz, S.M., González, V.J., Tristán-López, F., Cruz-Silva, R., Fujisawa, K., Muramatsu, H., Hayashi, T., Mi, X., Shi, Y., Sakamoto, H., Khoerunnisa, F., Kaneko, K., Sumpter, B.G., Kim, Y.A., Meunier, V., Endo, M., Muñoz-Sandoval, E., Terrones, M.: Clean nanotube unzipping by abrupt thermal expansion of molecular nitrogen: Graphene nanoribbons with atomically smooth edges. ACS Nano **6**(3), 2261–2272 (2012). https://doi.org/10.1021/NN2043252

Morozov, S.V, Novoselov, K.S., Katsnelson, M.I., Schedin, F., Elias, D.C., Jaszczak, J.A., Geim, A.K.: Giant intrinsic carrier mobilities in graphene and its bilayer (2008). https://doi.org/10.1103/PhysRevLett.100.016602

Moser, J., Barreiro, A., Bachtold, A.: View online export citation crossmark current-induced cleaning of graphene. Appl. Phys. Lett. **91**, 163513 (2017). https://doi.org/10.1063/1.2789673

Müller, S., Müllen, K.: Expanding benzene to giant graphenes: towards molecular devices. Philos. Trans. r. Soc. Math. Phys. Eng. Sci. **365**(1855), 1453–1472 (2007). https://doi.org/10.1098/rsta.2007.2026

Muñoz, R., Gómez-Aleixandre, C.: Review of CVD synthesis of graphene. Chem. Vap. Deposition 19(10–12), 297–322 (2013). https://doi.org/10.1002/cvde.201300051

Nair, R.V., Thomas, R.T., Sankar, V., Muhammad, H., Dong, M., Pillai, S.: Rapid, acid-free synthesis of high-quality graphene quantum dots for aggregation induced sensing of metal ions and bioimaging. ACS Omega **2**(11), 8051–8061 (2017). https://doi.org/10.1021/ACSOMEGA.7B01262

Nasir, S., Hussein, M.Z., Yusof, N.A., Zainal, Z.:. Oil palm waste-based precursors as a renewable and economical carbon sources for the preparation of reduced graphene oxide from graphene oxide. Nanomaterials **7**(7) (2017). https://doi.org/10.3390/nano7070182

Negishi, R., Akabori, M., Ito, T., Watanabe, Y., Kobayashi, Y.: Band-like transport in highly crystalline graphene films from defective graphene oxides OPEN. Nat. Publ. Group (2016). https://doi.org/10.1038/srep28936

Novoselov, K.S., Geim, A.K., Morozov, S.V., Jiang, D., Zhang, Y., Dubonos, S.V., Grigorieva, I.V., Firsov, A.A.: Electric field in atomically thin carbon films. Science **306**(5696), 666–669 (2004). https://doi.org/10.1126/SCIENCE.1102896

Novoselov, K.S., Jiang, D., Schedin, F., Booth, T.J., Khotkevich, V.V., Morozov, S.V., Geim, A.K.: Two-dimensional atomic crystals. Proc. Natl. Acad. Sci. u.s.a. **102**(30), 10451–10453 (2005). https://doi.org/10.1073/PNAS.0502848102

Novoselov, K.S., Geim, A.K., Morozov, S.V, Jiang, D., Zhang, Y., Dubonos, S.V, Grigorieva, I.V, Firsov, A.A.: Electric field effect in atomically thin carbon films. Phys. Rev. Lett. **404**. Kluwer (2020). www.arXiv.org/quant-ph/

Novoselov, K.S., Fal9ko, V.I., Colombo, L., Gellert, P.R., Schwab, M.G., Kim, K.: A roadmap for graphene. Nature **490** (2012). https://doi.org/10.1038/nature11458

Okigawa, Y., Kato, R., Yamada, T., Ishihara, M., Hasegawa, M.: Electrical properties and domain sizes of graphene films synthesized by microwave plasma treatment under a low carbon concentration. Carbon **82**(C), 60–66 (2015). https://doi.org/10.1016/j.carbon.2014.10.029

Pakdel, S., Majidi, S., Azamat, J., Erfan-Niya, H.: Graphene Oxide and Reduced Graphene Oxide as Nanofillers in Membrane Separation, pp. 113–144 (2021). https://doi.org/10.1007/978-3-030-72457-3_5

Park, S., Ruoff, R.S.: Chemical methods for the production of graphenes. Nat. Nanotechnol.nanotechnol. **4**(4), 217–224 (2009). https://doi.org/10.1038/nnano.2009.58

Park, H.J., Meyer, J., Roth, S., SkáKalová, V.: Growth and properties of few-layer graphene prepared by chemical vapor deposition. Carbon **48**, 1088–1094 (2009). https://doi.org/10.1016/j.carbon.2009.11.030

Park, S., Hu, Y., Hwang, J.O., Lee, E.S., Casabianca, L.B., Cai, W., Potts, J.R., Ha, H.W., Chen, S., Oh, J., Kim, S.O.: Chemical structures of hydrazine-treated graphene oxide and generation of aromatic nitrogen doping. Nat. Commun. (2012). https://doi.org/10.1038/ncomms1643

Park, M., Kim, H. S., Yoon, H., Kim, J., Lee, S., Yoo, S., Jeon, S.: Controllable singlet–triplet energy splitting of graphene quantum dots through oxidation: from phosphorescence to TADF. Adv. Mater. **32**(31) (2020). https://doi.org/10.1002/ADMA.202000936

Parveen, N., Ansari, O., Cho, M.H.: Simple route for gram synthesis of less defective few layered graphene and its electrochemical performance (2015). https://doi.org/10.1039/c5ra06404j

Parvez, K., Li, R., Puniredd, S.R., Hernandez, Y., Hinkel, F., Wang, S., Feng, X., Müllen, K.: Electrochemically exfoliated graphene as solution-processable, highly conductive electrodes for organic electronics. ACS Nano **7**(4), 3598–3606 (2013). https://doi.org/10.1021/NN400576V

Parvez, K., Wu, Z.S., Li, R., Liu, X., Graf, R., Feng, X., Müllen, K.: Exfoliation of graphite into graphene in aqueous solutions of inorganic salts. J. Am. Chem. Soc. **136**(16), 6083–6091 (2014). https://doi.org/10.1021/JA5017156

Pei, S., Cheng, H.-M.: The reduction of graphene oxide (2011). https://doi.org/10.1016/j.carbon.2011.11.010

Pei, S., Zhao, J., Du, J., Ren, W., Cheng, H.-M.: Direct reduction of graphene oxide films into highly conductive and flexible graphene films by hydrohalic acids (2010). https://doi.org/10.1016/j.carbon.2010.08.006
Peng, X.-Y., Liu, X.-X., Diamond, D., Lau, K.T.: Synthesis of electrochemically-reduced graphene oxide film with controllable size and thickness and its use in supercapacitor (2011). https://doi.org/10.1016/j.carbon.2011.04.047
Peng, K.-J., Wu, C.-L., Lin, Y.-H., Liu, Y.-J., Tsai, D.-P., Pai, Y.-H., Lin, R.: Hydrogen-free PECVD growth of few-layer graphene on an ultra-thin nickel film at the threshold dissolution temperature (n.d.). https://doi.org/10.1039/c3tc30332b
Ponomarenko, L.A., Schedin, F., Katsnelson, M.I., Yang, R., Hill, E.W., Novoselov, K.S., Geim, A.K.: Chaotic dirac billiard in graphene quantum dots. Science **320**(5874), 356–358 (2008). https://doi.org/10.1126/SCIENCE.1154663
Pop, E., Varshney, V., Roy, A.K.: Thermal properties of graphene: fundamentals and applications. MRS Bull. **37**(12), 1273–1281 (2012). https://doi.org/10.1557/mrs.2012.203
Prekodravac, J.R., Kepić, D.P., Colmenares, J.C., Giannakoudakis, D.A., Jovanović, S.P.: A comprehensive review on selected graphene synthesis methods: from electrochemical exfoliation through rapid thermal annealing towards biomass pyrolysis. J. Mater. Chem. C **9**(21), 6722–6748 (2021). https://doi.org/10.1039/d1tc01316e
Pumera, M.: Graphene-based nanomaterials and their electrochemistry. Chem. Soc. Rev. **39**(11), 4146–4157 (2010). https://doi.org/10.1039/c002690p
Pyun, J.: Graphene oxide as catalyst: Application of carbon materials beyond nanotechnology. Angewandte Chemie—Int. Ed. **50**(1), 46–48 (2011). https://doi.org/10.1002/ANIE.201003897
Qiu, J., Zhang, R., Li, J., Sang, Y., Tang, W., Gil, P.R., Liu, H.: Fluorescent graphene quantum dots as traceable, pH-sensitive drug delivery systems. Int. J. Nanomed.nanomed. **10**, 6709–6724 (2015). https://doi.org/10.2147/IJN.S91864
Quezada-Renteria, J.A., Ania, O., Chazaro-Ruiz, L. F., Rangel-Mendez, J.R.: Influence of protons on reduction degree and defect formation in electrochemically reduced graphene oxide. Orl Eans **78216** (2019). https://doi.org/10.1016/j.carbon.2019.04.109
Ramesha, G.K., Sampath, N.S.: Electrochemical reduction of oriented Graphene oxide films: an in situ Raman spectroelectrochemical study. J. Phys. Chem. C **113**(19), 7985–7989 (2009). https://doi.org/10.1021/JP811377N
Ramezani, Z., Dehghani, H.: Effect of nitrogen and sulfur co-doping on the performance of electrochemical hydrogen storage of graphene. Int. J. Hydrogen Energy **44**(26), 13613–13622 (2019). https://doi.org/10.1016/j.ijhydene.2019.03.255
Revia, R.A., Zhang, M.: Magnetite nanoparticles for cancer diagnosis, treatment, and treatment monitoring: recent advances. Mater. Today **19**(3) (2016). https://doi.org/10.1016/j.mattod.2015.08.022
Robinson, J., Binner, E., Beneroso Vallejo, D., Diaz Perez, N., Al Mughairi, K., Ryan, J., Shepherd, B., Adam, M., Budarin, V., Fan, J., Gronnow, M., Peneranda-Foix, F.: Unravelling the mechanisms of microwave pyrolysis of biomass (n.d.)
Ruan, G., Sun, Z., Peng, Z., Tour, J.M.: Growth of graphene from food, insects, and waste. ACS Nano **5**(9), 7601–7607 (2011). https://doi.org/10.1021/nn202625c
Saha, J.K., Dutta, A.: A review of graphene: material synthesis from biomass sources. Waste Biomass Valorization **13**(3), 1385–1429 (2022). https://doi.org/10.1007/s12649-021-01577-w
Saha, U., Jaiswal, R., Goswami, T.H.: A facile bulk production of processable partially reduced graphene oxide as superior supercapacitor electrode material (2016). https://doi.org/10.1016/j.electacta.2016.02.203
Salahdin, O.D., Sayadi, H., Solanki, R., Rosario, Parra, M.R., Al-Thamir, M., Abduladheem, Jalil, T., Samar, Izzat, E., Ali, Hammid, T., Luis, Arenas, A.B., Kianfar, E.: Graphene and carbon structures and nanomaterials for energy storage. Appl. Phys. A **128**, 703 (2022). https://doi.org/10.1007/s00339-022-05789-2

Sapkota, B., Benabbas, A., Lin, H.Y.G., Liang, W., Champion, P., Wanunu, M.: Peptide-decorated tunable-fluorescence graphene quantum dots. ACS Appl. Mater. Interfaces **9**(11), 9378–9387 (2017). https://doi.org/10.1021/acsami.6b16364

Saravanan, A., Senthil Kumar, P., Srinivasan, S., Jeevanantham, S., Vishnu, M., Vishal Amith, K., Sruthi, R., Saravanan, R., Vo, D.-V. N.: Insights on synthesis and applications of graphene-based materials in wastewater treatment: a review (2022). https://doi.org/10.1016/j.chemosphere.2022.134284

Schedin, F., Geim, A.K., Morozov, S.V., Hill, E.W., Blake, P., Katsnelson, M.I., Novoselov, K.S.: Detection of individual gas molecules adsorbed on graphene. Nat. Mater. **6**(9), 652–655 (2007). https://doi.org/10.1038/nmat1967

Schinwald, A., Murphy, F.A., Jones, A., MacNee, W., Donaldson, K.: Graphene-based nanoplatelets: a new risk to the respiratory system as a consequence of their unusual aerodynamic properties. ACS Nano **6**(1), 736–746 (2012). https://doi.org/10.1021/NN204229F

Schniepp, H.C., Li, J.L., McAllister, M.J., Sai, H., Herrera-Alonson, M., Adamson, D.H., Prud-homme, R.K., Car, R., Seville, D.A., Aksay, I.A.: Functionalized single graphene sheets derived from splitting graphite oxide. J. Phys. Chem. B **110**(17), 8535–8539 (2006). https://doi.org/10.1021/JP060936F

Schroeder, K.L., Goreham, R.V., Nann, T.: Graphene quantum dots for theranostics and bioimaging. Pharm. Res. 33(10):2337–2357 (2016). Springer New York LLC. https://doi.org/10.1007/s11095-016-1937-x

Schwierz, F.: Graphene transistors. Nat. Nanotechnol.nanotechnol. **5**(7), 487–496 (2010). https://doi.org/10.1038/nnano.2010.89

Seekaew, Y., Phokharatkul, D., Wisitsoraat, A., Wongchoosuk, C.: Highly sensitive and selective room-temperature NO2 gas sensor based on bilayer transferred chemical vapor deposited graphene. Appl. Surf. Sci. **404**, 357–363 (2017). https://doi.org/10.1016/j.apsusc.2017.01.286

Shams, S.S., Zhang, L.S., Hu, R., Zhang, R., Zhu, J.: Synthesis of graphene from biomass: a green chemistry approach. Mater. Lett. **161**, 476–479 (2015). https://doi.org/10.1016/j.matlet.2015.09.022

Shao, Y., Wang, J., Wu, H., Liu, J., Aksay, I. A., Lin, Y.: Graphene based electrochemical sensors and biosensors: a review. Electroanalysis **22**(10), 1027–1036 (2010). Wiley-VCH Verlag. https://doi.org/10.1002/elan.200900571

Sharma, S., Kalita, G., Hirano, R., Hayashi, Y., Tanemura, M.: Influence of gas composition on the formation of graphene domain synthesized from camphor. Mater. Lett. **93**, 258–262 (2013). https://doi.org/10.1016/j.matlet.2012.11.090

Shen.:. Biomedical applications of graphene. Theranostics **2**, 283–294 (2012)

Shin, H.J., Kim, K.K., Benayad, A., Yoon, S.M., Park, H.K., Jung, I.S., Jin, M.H., Jeong, H.K., Kim, J.M., Choi, J.Y., Lee, Y.H.: Efficient reduction of graphite oxide by sodium borohydride and its effect on electrical conductance. Adv. Func. Mater. **19**(12), 1987–1992 (2009). https://doi.org/10.1002/ADFM.200900167

Shinohara.: Graphene: An Introduction to the Fundamentals and Industrial Applications. Wiley (2015)

Singh, V., Joung, D., Zhai, L., Das, S., Khondaker, S.I., Seal, S.: Graphene based materials: Past, present and future. Prog. Mater. Sci. **56**(8), 1178–1271 (2011). Elsevier Ltd. https://doi.org/10.1016/j.pmatsci.2011.03.003

Singh, M., Sengupta, A., Zeller, K., Skoptsov, G., Vander Wal, R.L.: Effect of hydrogen concentration on graphene synthesis using microwave-driven plasma-mediated methane cracking. Carbon **143**, 802–813 (2019). https://doi.org/10.1016/j.carbon.2018.11.082

Somanathan, T., Prasad, K., Ostrikov, K., Saravanan, A., Mohana Krishna, V.: Graphene oxide synthesis from agro waste. Nanomaterials **5**, 826–834 (2015). https://doi.org/10.3390/nano5020826

Son, M., Ham, M.-H.: Low-temperature synthesis of graphene by chemical vapor deposition and its applications (2017). https://doi.org/10.1016/j.flatc.2017.07.002

Song, S.H., Jang, M.-H., Jeong, J.-M., Yoon, H., Cho, Y.-H., Jeong, W.-I., Kim, B.-H., Jeon, S.: Primary hepatocyte imaging by multiphoton luminescent graphene quantum dots. Chem. Commun.commun. **51**, 8041 (2015). https://doi.org/10.1039/c5cc01801c

Stankovich, S., Dikin, D.A., Piner, R.D., Kohlhaas, K.A., Kleinhammes, A., Jia, Y., Wu, Y., Nguyen, S.B.T., Ruoff, R.S.: Synthesis of graphene-based nanosheets via chemical reduction of exfoliated graphite oxide. Carbon **45**(7), 1558–1565 (2007a). https://doi.org/10.1016/j.carbon.2007.02.034

Stankovich, S., Dikin, D.A., Piner, R.D., Kohlhaas, K.A., Kleinhammes, A., Jia, Y., Wu, Y., Nguyen, S.T., Ruoff, R.S.: Synthesis of graphene-based nanosheets via chemical reduction of exfoliated graphite oxide (2007b). https://doi.org/10.1016/j.carbon.2007.02.034

Su, C.-Y., Xu, Y., Zhang, W., Zhao, J., Liu, A., Tang, X., Tsai, C.-H., Huang, Y., Li, L.-J.: Highly efficient restoration of graphitic structure in graphene oxide using alcohol vapors (2010). https://doi.org/10.1021/nn101691m

Subrahmanyam, K., Manna, A.K., Pati, S.K., Rao, C.: A study of graphene decorated with metal nanoparticles. Chem. Phys. Lett. **497**(1-3), 70–75 (2010). https://doi.org/10.1016/j.cplett.2010.07.091

Subrahmanyam, K.S., Panchakarla, L.S., Govindaraj, A., Rao, C.N.R.: Simple method of preparing graphene flakes by an arc-discharge method. J. Phys. Chem. C **113**(11), 4257–4259 (2009). https://doi.org/10.1021/jp900791y

Sun, X., Liu, Z., Welsher, K., Robinson, J.T., Goodwin, A., Zaric, S., Dai, H.: Nano-graphene oxide for cellular imaging and drug delivery. Nano Res. **1**(3), 203–212 (2008). https://doi.org/10.1007/s12274-008-8021-8

Sun, L., Yuan, G., Gao, L., Yang, J., Chhowalla, M., Gharahcheshmeh, M.H., Gleason, K.K., Choi, Y.S., Hong, B.H., Liu, Z.: Chemical vapour deposition. Nat. Rev. Methods Primers **1**(1), 5 (2021a). https://doi.org/10.1038/s43586-020-00005-y

Sun, L., Yuan, G., Gao, L., Yang, J., Chhowalla, M., Gharahcheshmeh, M.H., Gleason, K.K., Choi, Y.S., Hong, B.H., Liu, Z.: Chemical vapour deposition. Nat. Rev. Methods Primers **1**(1) (2021b). Springer Nature. https://doi.org/10.1038/s43586-020-00005-y

Tang, Q., Zhu, W., He, B., Yang, P.: Rapid conversion from carbohydrates to large-scale carbon quantum dots for all-weather solar cells. ACS Nano **11**(2), 1540–1547 (2017). https://doi.org/10.1021/ACSNANO.6B06867

Taqi-Uddeen Safian, M., Haron, S., Nasir, M., Ibrahim, M.: A review on bio-based graphene derived from biomass wastes (n.d.)

Terrones, M.: Nanotubes unzipped. Nature **458**(7240), 845–846 (2009). https://doi.org/10.1038/458845a

Terrones, M., Botello-Méndez, A.R., Campos-Delgado, J., López-Urías, F., Vega-Cantú, Y.I., Rodríguez-Macías, F.J., Elías, A.L., Muñoz-Sandoval, E., Cano-Márquez, A.G., Charlier, J.-C., Terrones, H.: Graphene and graphite nanoribbons: morphology, properties, synthesis, defects and applications. Nano Today **5**, 351–372 (2010). https://doi.org/10.1016/j.nantod.2010.06.010

Thess, A., Lee, R., Nikolaev, P., Dai, H., Petit, P., Robert, J., Xu, C., Lee, Y.H., Kim, S.G., Rinzler, A.G., Colbert, D.T., Scuseria, G.E., Tománek, D., Fischer, J.E., Smalley, R.E.: Crystalline ropes of metallic carbon nanotubes. Science **273**(5274), 483–487 (1996). https://doi.org/10.1126/SCIENCE.273.5274.483

Tiwary, C.S., Javvaji, B., Kumar, C., Mahapatra, D.R., Ozden, S., Ajayan, P.M., Chattopadhyay, K.: Chemical-free graphene by unzipping carbon nanotubes using cryo-milling (2015). https://doi.org/10.1016/j.carbon.2015.03.036

Toman, J., Jašek, O., Šnírer, M., Pavliňák, D., Navrátil, Z., Jurmanová, J., Chudják, S., Krčma, F., Kudrle, V., Michalička, J.: On the transition of reaction pathway during microwave plasma gas-phase synthesis of graphene nanosheets: From amorphous to highly crystalline structure. Plasma Process. Polym. **18**(8) (2021). https://doi.org/10.1002/ppap.202100008

Trasobares, S., Ewels, C.P., Birrell, J., Stephan, O., Wei, B.Q., Carlisle, J.A., Keblinski, P., Ajayan, P.M.: Carbon Nanotubes with Graphitic Wings** (n.d.). https://doi.org/10.1002/adma.200306459

Tripathi, P., Ravi, C., Patel, P., Shaz, M. A., Srivastava, O.N.: Synthesis of high-quality graphene through electrochemical exfoliation of graphite in alkaline electrolyte (n.d.)

Tsai, M.L., Wei, W.R., Tang, L., Chang, H.C., Tai, S.H., Yang, P.K., Lau, S.P., Chen, L.J., He, J.H.: Si hybrid solar cells with 13% efficiency via concurrent improvement in optical and electrical properties by employing graphene quantum dots. ACS Nano **10**(1), 815–821 (2016). https://doi.org/10.1021/ACSNANO.5B05928

Van Bommel, A.J., Crombeen, J.E., Van Tooren, A.: LEED and Auger electron observations of the SiC(0001) surface. Surf. Sci. **48**(2), 463–472 (1975). https://doi.org/10.1016/0039-6028(75)90419-7

Tzeng, Y., Chen, W.L., Wu, C., Lo, J.Y., Li, C.Y.: The synthesis of graphene nanowalls on a diamond film on a silicon substrate by direct-current plasma chemical vapor deposition. Carbon **53**, 120–129 (2013). https://doi.org/10.1016/j.carbon.2012.10.038

Vekselman, V., Feurer, M., Huang, T., Stratton, B., Raitses, Y.: Complex structure of the carbon arc discharge for synthesis of nanotubes. Plasma Sources Sci. Technol. **26**(6) (2017). https://doi.org/10.1088/1361-6595/aa7158

Verma, S., Mungse, H.P., Kumar, N., Choudhary, S., Jain, S.L., Sain, B., Khatri, O.P.: Graphene oxide: an efficient and reusable carbocatalyst for aza-Michael addition of amines to activated alkenes. Chem. Commun.commun. **47**(47), 12673–12675 (2011). https://doi.org/10.1039/c1cc15230k

Vlassiouk, I., Regmi, M., Fulvio, P., Dai, S., Datskos, P., Eres, G., Smirnov, S.: Role of hydrogen in chemical vapor deposition growth of large single-crystal graphene. ACS Nano **5**(7), 6069–6076 (2011). https://doi.org/10.1021/nn201978y

Voicu, S.I., Thakur, V.K.: Graphene-based composite membranes for nanofiltration: performances and future perspectives. Emergent Mater. **1**, 3 (2021). https://doi.org/10.1007/s42247-021-00291-6

Wan, X., Long, G., Huang, L., Chen, Y.: Graphene - A promising material for organic photovoltaic cells. Adv. Mater. **23**(45), 5342–5358 (2011). https://doi.org/10.1002/ADMA.201102735

Wang, X., Fulvio, P.F., Baker, G.A., Veith, G.M., Unocic, R.R., Mahurin, S.M., Chi, M., Dai, S.: Direct exfoliation of natural graphite into micrometre size few layers graphene sheets using ionic liquids. Chem. Commun.commun. **46**(25), 4487–4489 (2010a). https://doi.org/10.1039/c0cc00799d

Wang, D., Yan, W., Vijapur, S.H., Botte, G.G.: Electrochimica Acta Electrochemically reduced graphene oxide-nickel nanocomposites for urea electrolysis. Electrochim. Acta. Acta **89**, 732–736 (2013a). https://doi.org/10.1016/j.electacta.2012.11.046

Wang, Z., Xia, J., Zhou, C., Via, B., Xia, Y., Zhang, F., Li, Y., Xia, L., Tang, J.: Synthesis of strongly green-photoluminescent graphene quantum dots for drug carrier. Colloids Surf. B **112**, 192–196 (2013b). https://doi.org/10.1016/j.colsurfb.2013.07.025

Wang, D., Chen, J.F., Dai, L.: Recent advances in graphene quantum dots for fluorescence bioimaging from cells through tissues to animals. Part. Part. Syst. Charact. **32**(5), 515–523 (2015b). https://doi.org/10.1002/ppsc.201400219

Wang, H., Wang, Y., Cao, X., Feng, M., Lan, G.: Vibrational properties of graphene and graphene layers. J. Raman Spectrosc. **40**(12), 1791–1796 (2009). https://doi.org/10.1002/jrs.2321

Wang, H.S., Tian, S.Y., Yang, S.W., Wang, G., You, X.F., Xu, L.X., Li, Q.T., He, P., Ding, G.Q., Liu, Z., Xie, X.M.: Anode coverage for enhanced electrochemical oxidation: a green and efficient strategy towards water-dispersible graphene. Green Chem. **20**(6), 1306–1315 (2018). https://doi.org/10.1039/c7gc03345a

Wang, Z., Li, N., Shi, Z., Gu, Z.: Low-cost and large-scale synthesis of graphene nanosheets by arc discharge in air. Nanotechnology **21**(17) (2010). https://doi.org/10.1088/0957-4484/21/17/175602

Wang, B.B., Zheng, K., Cheng, Q.J., Ostrikov, K.: Plasma effects in aligned carbon nanoflake growth by plasma-enhanced hot filament chemical vapor deposition. Appl. Surf. Sci. **325**, 251–257 (2015). https://doi.org/10.1016/j.apsusc.2014.11.072

Wang, Z., Zhou, X., Zhang, J., Boey, F., & Zhang, H.: Direct electrochemical reduction of single-layer graphene oxide and subsequent functionalization with glucose oxidase (n.d.). https://doi.org/10.1021/jp906348x

Wassei, J.K., Mecklenburg, M., Torres, J.A., Fowler, J.D., Regan, B.C., Kaner, R.B., Weiller, B.H.: Chemical vapor deposition of graphene on copper from methane, ethane and propane: evidence for bilayer selectivity. Small **8**(9), 1415–1422 (2012). https://doi.org/10.1002/smll.201102276

Wee, B.H., Wu, T.F., Hong, J.D.: Facile and scalable synthesis method for high-quality few-layer graphene through solution-based exfoliation of graphite. ACS Appl. Mater. Interfaces **9**(5), 4548–4557 (2017). https://doi.org/10.1021/ACSAMI.6B11771

Wei, W., Qu, X.: Extraordinary physical properties of functionalized graphene. Small **8**(14), 2138–2151 (2012). https://doi.org/10.1002/SMLL.201200104

Woehrl, N., Ochedowski, O., Gottlieb, S., Shibasaki, K., Schulz, S.: Plasma-enhanced chemical vapor deposition of graphene on copper substrates. AIP Adv. **4**(4) (2014). https://doi.org/10.1063/1.4873157

Wu, Z.S., Ren, W., Wen, L., Gao, L., Zhao, J., Chen, Z., Zhou, G., Li, F., Cheng, H.M.: Graphene anchored with Co_3O_4 nanoparticles as anode of lithium ion batteries with enhanced reversible capacity and cyclic performance. ACS Nano **4**(6), 3187–3194 (2010). https://doi.org/10.1021/NN100740X

Wu, L., Li, W., Li, P., Liao, S., Qiu, S., Chen, M., Guo, Y., Li, Q., Zhu, C., Liu, L.: Powder, paper and foam of few-layer graphene prepared in high yield by electrochemical intercalation exfoliation of expanded graphite. Small **10**(7), 1421–1429 (2014). https://doi.org/10.1002/smll.201302730

Wu, P.C., Wang, J.Y., Wang, W.L., Chang, C.Y., Huang, C.H., Yang, K.L., Chang, J.C., Hsu, C.L.L., Chen, S.Y., Chou, T.M., Kuo, W.S.: Efficient two-photon luminescence for cellular imaging using biocompatible nitrogen-doped graphene quantum dots conjugated with polymers. Nanoscale **10**(1), 109–117 (2018). https://doi.org/10.1039/c7nr06836k

Wu, X., Liu, Y., Yang, H., Shi, Z.: Large-scale synthesis of high-quality graphene sheets by an improved alternating current arc-discharge method (2016). https://doi.org/10.1039/c6ra22273k

Wu, X., Mu, F., Wang, Y., Zhao, H.: Molecules graphene and graphene-based nanomaterials for dna detection: a review (n.d.). https://doi.org/10.3390/molecules23082050

Xu, S., Zhang, L., Wang, B., Ruoff, R.S.: Chemical vapor deposition of graphene on thin-metal films. Cell Rep. Phys. Sci. **2**(3) (2021). Cell Press. https://doi.org/10.1016/j.xcrp.2021.100372

Yan, K., Fu, L., Peng, H., Liu, Z.: Designed CVD growth of graphene via process engineering. Acc. Chem. Res. **46**(10), 2263–2274 (2013). https://doi.org/10.1021/AR400057N

Yang, C., Bi, H., Wan, D., Huang, F., Xie, X., Jiang, M.: Direct PECVD growth of vertically erected graphene walls on dielectric substrates as excellent multifunctional electrodes. J. Mater. Chem. A **1**(3), 770–775 (2013). https://doi.org/10.1039/c2ta00234e

Yang, S., Lohe, M.R., Müllen, K., Feng, X.: New-generation graphene from electrochemical approaches: production and applications. Adv. Mater. **28**(29), 6213–6221 (2016). https://doi.org/10.1002/adma.201505326

Yang, J., Luo, S., Zhou, X., Li, J., Fu, J., Yang, W., Wei, D.: Flexible, tunable, and ultrasensitive capacitive pressure sensor with microconformal graphene electrodes. ACS Appl. Mater. Interfaces **11**(16), 14997–15006 (2019). https://doi.org/10.1021/acsami.9b02049

Yavari, F., Chen, Z., Thomas, A.V., Ren, W., Cheng, H.M., Koratkar, N.: High sensitivity gas detection using a macroscopic three-dimensional graphene foam network. Sci. Rep. 1 (2011). https://doi.org/10.1038/srep00166

Yazdi, G.R., Iakimov, T., Yakimova, R.: Epitaxial graphene on SiC: a review of growth and characterization. Crystals **6**(5). MDPI AG (2016). https://doi.org/10.3390/cryst6050053

Yi, M., Shen, Z.: A review on mechanical exfoliation for the scalable production of graphene (2015). https://doi.org/10.1039/c5ta00252d

Yu, K., Bo, Z., Lu, G., Mao, S., Cui, S., Zhu, Y., Chen, X., Ruoff, R.S., Chen, J.: Growth of carbon nanowalls at atmospheric pressure for one-step gas sensor fabrication. Nanoscale Res. Lett. **6**(1) (2011). https://doi.org/10.1186/1556-276X-6-202

Yu, H., Zhang, B., Bulin, C., Li, R., Xing, R.: High-efficient Synthesis of Graphene Oxide Based on Improved Hummers Method. Sci. Rep. **6** (2016). https://doi.org/10.1038/srep36143

Zafar, M.A., Varghese, O.K., Robles Hernandez, F.C., Liu, Y., Jacob, M.V.: Single-step synthesis of nitrogen-doped graphene oxide from aniline at ambient conditions. ACS Appl. Mater. Interfaces **14**(4), 5797–5806 (2022). https://doi.org/10.1021/acsami.1c21150

Zafar, M.A., Jacob, M.V.: Plasma-based synthesis of graphene and applications: a focused review. Rev. Modern Plasma Phys. **6**(1) (2022). https://doi.org/10.1007/s41614-022-00102-3

Zakaria, N.Z.J., Rozali, S., Mubarak, N.M., Ibrahim, S.: A review of the recent trend in the synthesis of carbon nanomaterials derived from oil palm by-product materials. In: Biomass Conversion and Biorefinery. Springer Science and Business Media Deutschland GmbH (2022). https://doi.org/10.1007/s13399-022-02430-3

Zhai, Z., Shen, H., Chen, J., Li, X., Li, Y.: Metal-free synthesis of boron-doped graphene glass by hot-filament chemical vapor deposition for wave energy harvesting. ACS Appl. Mater. Interfaces **12**(2), 2805–2815 (2020). https://doi.org/10.1021/acsami.9b17546

Zhang, L.X., Sun, Z., Qi, J.L., Shi, J.M., Hao, T.D., Feng, J. C.: Understanding the growth mechanism of vertically aligned graphene and control of its wettability. Carbon **103**, 339–345 (2016). https://doi.org/10.1016/j.carbon.2016.03.029

Zhang, J., Tahmasebi, A., Omoriyekomwan, J.E., Yu, J.: Direct synthesis of hollow carbon nanofibers on bio-char during microwave pyrolysis of pine nut shell. J. Anal. Appl. Pyrolysis **130**, 142–148 (2018). https://doi.org/10.1016/j.jaap.2018.01.016

Zhou, M., Wang, Y., Zhai, Y., Zhai, J., Ren, W., Wang, F., Dong, S.: Controlled synthesis of large-area and patterned electrochemically reduced graphene oxide films. Chem. Eur. J. **15**(25), 6116–6120 (2009). https://doi.org/10.1002/CHEM.200900596

Zhou, L., Fox, L., Włodek, M., Islas, L., Slastanova, A., Robles, E., Bikondoa, O., Harniman, R., Fox, N., Cattelan, M., Briscoe, W.H.: Surface structure of few layer graphene (2018). https://doi.org/10.1016/j.carbon.2018.04.089

Zhu, Y., Murali, S., Cai, W., Li, X., Suk, J.W., Potts, J.R., Ruoff, R.S.: Graphene and graphene oxide: synthesis, properties, and applications. Adv. Mater. **22**(35), 3906–3924 (2010). https://doi.org/10.1002/adma.201001068

Zhuang, S., Nunna, B.B., Boscoboinik, J.A., Lee, E.S.: Nitrogen-doped graphene catalysts: high energy wet ball milling synthesis and characterizations of functional groups and particle size variation with time and speed. Int. J. Energy Res. **41**(15), 2535–2554 (2017). https://doi.org/10.1002/ER.3821

Chapter 4
Structure and Electrochemical Properties of Graphene, Derivatives, and Its Nanocomposites

Nkurunziza Florien, Shagun Gupta, Rachna Poria, Deepali Chaudhary, Renu Poria, Trishala Rawat, and Ankur Kaushal

Abstract Graphene is a form of carbon with a two-dimensional structure organized in a plane and looks like a honeycomb network. Graphene may assist the amount of electricity passing more than six times higher than copper. Graphene possesses good electrical, mechanical, and thermal characteristics along with other unlikeness-like magnitude relations with great specific surface area compared to other nanoparticles. Additionally, to provide an acceptable possibility of mechanical support, still, there are probabilities of being used to regulate functional features including impermeability to gases, expansion, and stability. Graphene sheets have a higher surface area-to-volume ratio with exceptional electronic transport features. These characteristics guarantee graphene to be used in various areas like solar cells, sensors, batteries, membranes, medicines, supercapacitors, and hydrogen storage. Developing composites of graphene contributes a lot to their application. The electrochemical properties of graphene rely on their surface charge and how electrons move conveniently in graphene electrodes; this was examined by using redox reaction and potassium ferricyanide. Furthermore, the unlikeness features of graphene and its derivatives (such as graphene oxide, graphene quantum dots, reduced graphene oxide, graphane, graphone, graphyne, graphdiyne, and fluorographene) bring more awareness to probable significance in several dimensions like Nanobiotechnology, biomedicine, and biosensors.

Keywords Graphene · Nanocomposite · Electrochemical · Biosensor · Other graphene derivatives

---

N. Florien · S. Gupta · R. Poria · D. Chaudhary · R. Poria · T. Rawat · A. Kaushal (✉)
Department of Bio-Sciences and Technology, Maharishi Markandeshwar (Deemed to Be) University, Mullana, Ambala 134003, India
e-mail: ankur.biotech85@gmail.com

R. Khan et al. (eds.), *Electrochemical Exfoliation of Graphene and its Derivatives*, Engineering Materials, https://doi.org/10.1007/978-981-97-2128-3_4

4.1 Introduction

Carbon, a foundational element of life, has intrigued humanity for centuries. Diamond and graphite, among its diverse allotropes, were early revelations showcasing the versatility of this extraordinary element. The distinct properties of diamond, with its hardness and brilliance, contrasted with the layers of graphite, highlighting carbon's ability to exist in varied forms. These initial discoveries set the stage for further exploration, leading to the unveiling of novel carbon allotropes like fullerenes, carbon nanotubes, and graphene. Carbon's rich history and its array of structures continue to shape our understanding and applications in fields ranging from materials science to biotechnology (Zhen and Zhu 2018). The exploration of carbon's three-dimensional structures paved the way for groundbreaking discoveries, such as zero-dimensional fullerenes and one-dimensional carbon nanotubes, transforming the landscape of materials science more than two decades ago. A notable milestone in the continuum of carbon structures is the isolation of two-dimensional graphene, achieved only in recent years (Bostwick et al. 2009). Graphene's unique properties, including exceptional electrical conductivity and mechanical strength, have opened new avenues for technological advancements and applications in various fields. This two-dimensional carbon material represents a revolutionary leap in our understanding of carbon's capabilities. In the realm of electrochemistry, carbon-based materials have gained significant prominence, particularly as electrodes, due to their myriad advantages (Tang et al. 2009a, b). Electrodes play a pivotal role in electroanalysis and electrosynthesis, and carbon's ubiquity makes it an ideal candidate. Its resource abundance, ease of manufacturing, and cost-effectiveness make carbon electrodes highly attractive for a range of applications. Moreover, carbon exhibits excellent electrical conductance, high thermal stability, and chemical robustness, providing versatility in diverse environments (Zhou and Yao 2022).The electrocatalytic properties of carbon electrodes are multifaceted, influenced by factors such as structural configuration, organization, surface chemistry, and aggregation (Tang et al. 2009a, b). Understanding and manipulating these factors is crucial for optimizing the performance of carbon electrodes in various electrochemical processes. Researchers continue to delve into the intricate interplay of these elements to enhance the efficiency and effectiveness of carbon-based electrodes. Carbon, with its rich history of elemental forms and the recent emergence of two-dimensional graphene, remains at the forefront of materials science. Carbon electrodes, leveraging the element's abundance and unique properties, hold immense promise for advancing electroanalytical techniques and electrosynthesis processes, contributing to the ongoing evolution of electrochemistry and allied fields (Tang et al. 2009a, b). Carbon exhibits a diverse array of allotropes, each with unique properties and applications. Notable examples include fullerenes, carbon nanotubes (CNTs), graphite, and graphyne. Carbon fiber (CF), amorphous carbon (AC), and charcoal further contribute to this rich allotropes landscape. Fullerenes, spherical structures of carbon atoms, and CNTs, cylindrical carbon structures, showcase exceptional strength and conductivity. Graphite, with its layered structure, contrasts with the single-layer arrangement of graphyne. Carbon allotropes like CF and AC find utility in various industries, highlighting the versatility and significance of carbon in diverse material forms (Zhen and Zhu 2018). Graphene, a revolutionary addition to the carbon family, represents a two-dimensional nanoparticle derived from graphite. The initial discovery of a single layer of carbon, which essentially constitutes graphene, can be traced back to Boehm and co-workers in 1986 (De Marchi et al. 2018). However, it gained renewed attention in 2004 when Andre Geim and Konstantin Novoselov rediscovered it using the innovative "Scotch Tape method," a simple yet effective technique for isolating graphene layers from graphite (Randviir et al. 2014). Graphene is distinctive for its two-dimensional framework, where carbon atoms are arranged in a hexagonal lattice resembling a "honeycomb crystal plane" (Sur 2012). The remarkable electronic, mechanical, and thermal properties of graphene arise from this unique atomic arrangement. A single sheet of graphene, as illustrated in Fig. 4.1, features exogenous functional groups such as carboxylic acid, aldehyde, ketone, epoxide, and hydroxyl groups, enhancing its chemical reactivity and expanding its potential applications. The characterization of graphene and its modifications on carbon electrodes involves a range of sophisticated techniques, including Transmission Electron Microscopy (TEM), X-Ray Diffraction (XRD), Scanning Electron Microscopy (SEM), X-Ray Photoelectron Spectroscopy (XPS), and Raman spectroscopy (Tang et al. 2009a, b). These methods allow researchers to visualize the structure, composition, and morphological features of graphene at the nanoscale, providing crucial insights into its properties and performance. Transmission electron microscopy enables high-resolution imaging, unveiling the atomic arrangement of graphene layers. X-ray diffraction provides information about the crystal structure, while Scanning Electron Microscopy offers topographical details. X-ray photoelectron Spectroscopy is employed to analyze the elemental composition and chemical states, and Raman spectroscopy provides valuable data on the vibrational modes and defects in graphene. Graphene, with its intriguing structure and exceptional properties, has revolutionized the field of carbon materials. From the pioneering work of Boehm and co-workers to the groundbreaking rediscovery by Novoselov and Geim, graphene's journey has been marked by innovation and ingenuity. Its diverse functionalization and thorough characterization through advanced techniques underscore its potential for applications in fields ranging from electronics to energy storage, highlighting graphene's central role in the ever-expanding realm of carbon allotropes (Tang et al. 2009a, b). Graphene has become a focal point of scientific and industrial interest owing to its extraordinary combination of properties, making it a material of immense versatility and potential. Recognized for its remarkable strength and fine structure, graphene exhibits an impressive array of exemplary characteristics. Among these are its high mechanical strength, thermal conductivity, electrical conductivity, and optical properties. This multifaceted material also boasts a substantial specific surface area, heightened sensitivity, chemical stability, and outstanding optical transmittance, signifying its ability to allow the passage of light through its layers. Notably, graphene's flexibility and transparency contribute to its appeal in various applications, enabling its integration into flexible electronic devices and transparent conductive films. The material's exceptional resistance, which surpasses that of steel by 200 times and rivals the hardness of diamond, underscores its robust nature and durability (De Marchi et al. 2018). Such mechanical resilience positions graphene as an ideal candidate for applications requiring materials that can withstand extreme conditions. Moreover, graphene exhibits elasticity, porosity, biocompatibility, a tunable band gap, and ease of chemical functionalization, further expanding its potential applications across diverse fields. The tunable band gap allows researchers to tailor graphene's electronic properties for specific applications, ranging from semiconductors to sensors. The material's biocompatibility makes it suitable for biomedical applications, and its ease of chemical functionalization enhances its adaptability for various industrial and research purposes. Graphene's unique combination of mechanical, thermal, electrical, and optical properties, coupled with its flexibility, transparency, and resistance, positions it as a revolutionary material with broad applications in fields such as electronics, materials science, energy, and healthcare. The ongoing exploration of graphene's potential continues to fuel excitement and innovation in the scientific community and beyond.

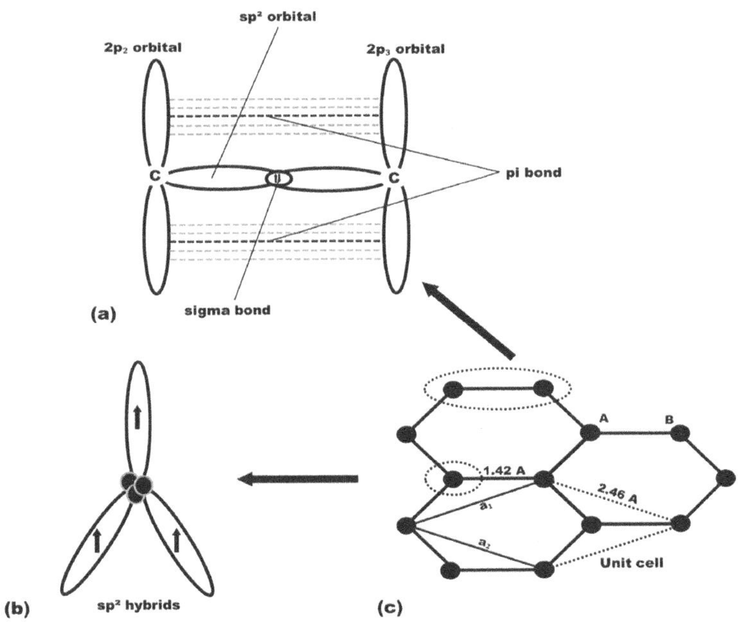

Fig. 4.1 **a** Sigma bond and pi bond formed by sp^2 hybridization, **b** The structure of sp^2 hybrids, and **c** The crystal lattice of graphene, where A and B are carbon atoms belonging to different sublattices, a1 and a2 are unit cell vectors (Tiwari et al. 2020)

4.2 Structure of Graphene

Graphene, an exceptional carbon allotrope, was initially discovered in 1986 by Boehm and co-workers, who identified a single layer of this groundbreaking material (De Marchi et al. 2018). However, its true potential came to light in 2004 when Novoselov and Geim rediscovered graphene using the innovative "Scotch Tape method." This method involved isolating graphene layers from graphite, leading to the material's resurgence and subsequent exploration of its remarkable properties. This pivotal rediscovery marked the beginning of intensive research into graphene, unlocking its extraordinary mechanical, thermal, electrical, and optical properties, with far-reaching implications across various scientific and industrial domains (De Marchi et al. 2018).

Graphene, a revolutionary carbon allotrope, derives its name from graphite and the suffix "-ene," referring to polycyclic aromatic hydrocarbons (Geim and Novoselov 2007). Its structure looks like a honeycomb or chicken wire, presenting a hexagonal lattice of carbon's single layer with sp2 hybridization (De Marchi et al. 2018; Boehm et al. 1994). This unique arrangement results in what is often described as the "hexagonal crystalline monolayer of graphite." Graphene serves as the primitive form and crystalline allotrope of carbon atoms, connected by single bonds (C–C) with a remarkable distance of 0.142 nm between two carbon atoms (Zhou et al. 2009). The hexagonal lattice showcases carbon atoms demonstrating sp2 hybridization, contributing to the material's extraordinary properties as shown in Fig. 4.1.The distinct honeycomb structure not only grants graphene its remarkable mechanical, thermal, and electrical characteristics, but also establishes its visual resemblance to a chicken wire or a honeycomb. This hexagonal arrangement is pivotal to graphene's exceptional strength, conductivity, and versatility, making it a focal point of extensive research and technological innovation across diverse scientific and industrial domains. Graphene's hexagonal crystalline monolayer is not just a structural feature; it encapsulates the material's essence, embodying the extraordinary potential that this single layer of carbon atoms holds in revolutionizing materials science and various technological applications (Zhou et al. 2009). Graphene, a single layer of carbon atoms arranged in a hexagonal lattice, exhibits extraordinary electronic, mechanical, and thermal properties due to its unique bonding and atomic structure. The bonding in graphene involves both π (pi) bonds and σ (sigma) bonds, contributing to its exceptional characteristics. The π bonds in graphene are formed by the lateral overlap of adjacent p orbitals, specifically the 2p orbitals of carbon atoms. These orbitals, referred to as 2pz orbitals, are found perpendicular to the plane of the graphene sheet. The lateral overlap results in the formation of a π bond, creating a network of delocalized electrons above and below the graphene plane. This π electron system is responsible for graphene's excellent electrical conductivity and unique electronic properties. On the other hand, the σ bonds in graphene are formed by the overlap of sp2 hybridized orbitals. In the graphene lattice, each carbon atom undergoes sp2 hybridization, involving the combination of one 2 s and two 2p orbitals (2px and 2py). The resulting three sp2 hybrid orbitals arrange themselves in a trigonal planar geometry around each carbon atom. The fourth electron from each carbon atom forms a π bond, creating the hexagonal lattice. The sp2 hybridization also leads to the formation of σ bonds between adjacent carbon atoms within the hexagonal lattice. The σ bonds in graphene are characterized by a short interatomic distance of approximately 0.142 nm, which is shorter than the interatomic distance in sp3 hybridized C–C bonds found in materials like diamond. This shorter bond length contributes to the remarkable mechanical properties of monolayer graphene. The strong σ bonds and the delocalized π bond network result in a material that is exceptionally strong, yet thin and flexible.

The strength of graphene arises from its ability to distribute and absorb stress efficiently. The sp^2 hybridized carbon atoms form a robust network of σ bonds that can withstand significant mechanical forces. Additionally, the π bond network provides flexibility, allowing graphene to bend and deform without losing its structural integrity. This combination of strength and flexibility makes graphene an ideal candidate for various applications, from reinforcing composite materials to serving as a building block for advanced electronic devices. Understanding the bonding in graphene is crucial for tailoring its properties for specific applications. For instance, introducing defects or functional groups can modify the electronic structure and reactivity of graphene, expanding its utility in areas such as catalysis, sensing, and energy storage. Researchers continue to explore novel ways to manipulate the bonding in graphene to unlock its full potential for technological advancements. The bonding in graphene, involving both π and σ bonds, plays a pivotal role in determining its unique electronic and mechanical properties. The delocalized π bond network contributes to its exceptional electrical conductivity, while the strong σ bonds provide remarkable mechanical strength. This intricate interplay of bonding in graphene has fuelled extensive research and exploration, driving innovations in materials science and technology. Graphene, with its unique hexagonal lattice structure as shown in Fig. 4.2a, comprises a single layer of carbon atoms interconnected by strong covalent bonds. Each carbon atom forms covalent bonds with three neighboring carbon atoms, resulting in exceptional stability and high tensile strength. Tensile strength refers to the ability of a material to withstand stretching forces before breaking, and graphene's robust covalent bonding imparts remarkable strength, making it one of the strongest materials known. The stability of graphene is further enhanced by its ability to coherently bind a fourth atom, enabling its use in synthesizing composite materials. This versatility stems from the material's capacity to form additional bonds without compromising its structural integrity. Any deformity in the graphene lattice may lead to the formation of double bonds ($C = C$) between two carbon atoms, and the number of these double bonds depends on the specific production procedures employed for graphene. Graphene's exceptional electrical conductivity is a result of its unique atomic arrangement and the delocalized pi electrons in the conjugated system of p orbitals. This makes graphene an excellent conductor of both electricity and heat (De Marchi et al. 2018).

The high electron mobility within the graphene lattice contributes to its prowess as a conductor, holding immense potential for applications in electronic devices as shown in Fig. 4.2a illustrating the different functional groups present in graphene,

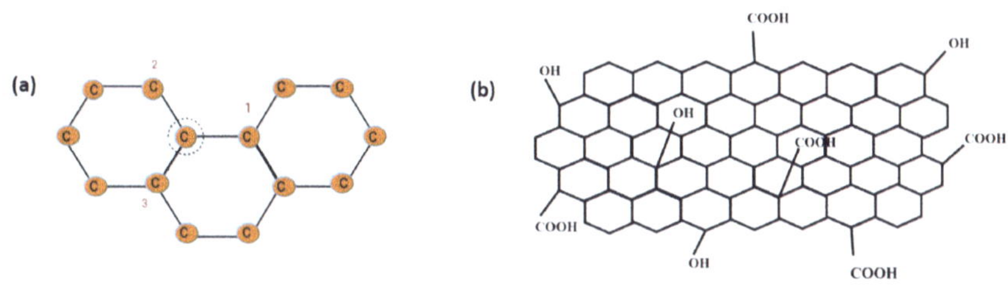

Fig. 4.2 **a** Hexagonal lattice structure of graphene (Zhen and Zhu 2018, Yang et al. 2018) **b** Structure of single sheet of graphene

and showcasing its versatility in chemical modification. These functional groups, including carboxyl, hydroxyl, and epoxy groups, make graphene amenable to various chemical reactions, expanding its utility in diverse fields. The ability to modify graphene's surface enhances its compatibility with different matrices and facilitates tailored applications. Graphene's single-layer carbon structure, covalent bonding, stability, tensile strength, and exceptional conductivity collectively position it as a transformative material. Its adaptability in forming composite materials, resistance to deformities, and versatility in functionalization make graphene a cornerstone in materials science, with wide-ranging applications from electronics to advanced materials and beyond (De Marchi et al. 2018). Graphene's exceptional conductivity is a standout feature, capable of facilitating over six times more electricity flow than copper. This remarkable property positions graphene as a potential game-changer in the realm of electrical conductivity. As a two-dimensional carbon nanomaterial, graphene, with an atomic thickness of 0.345 nm, serves as the foundational unit for various carbon allotropes, including zero-dimensional fullerenes (C60), one-dimensional carbon nanotubes (CNTs), and three-dimensional graphite (Maity et al. 2018). Derived from graphite, graphene manifests in various forms, giving rise to a family of derivatives collectively known as graphene divergent. This family includes graphene nano-platelets (GNPs), graphene oxide (GO), graphene nanoribbons (GNRs), reduced graphene oxide (rGO), graphene quantum dots (GQDs), and graphene-enabled products like graphene ink and masterbatches (Maity et al. 2018). Graphene, with its myriad derivatives, emerges as a transformative material with diverse applications. Each derivative, whether it be graphene oxide, graphene quantum dots, or reduced graphene oxide, possesses unique properties that amplify graphene's versatility. This diversity broadens graphene's application spectrum, playing a pivotal role in electronics and materials science. Enhanced electrical conductivity, mechanical strength, and tunable functionalities through various derivatives highlight graphene's transformative potential. Its impact reverberates across industries, fueling innovations in conductive materials, composites, and beyond. As researchers delve deeper into the properties of graphene derivatives, the material's ability to tailor applications and deliver multifaceted solutions positions it at the forefront of technological advancements (Maity et al. 2018). Graphene oxide (GO) is a significant derivative of graphene that has attracted considerable attention in scientific research and technological applications. Understanding its synthesis, properties, and applications provides valuable insights into the broader field of graphene-based materials. The production of GO involves the oxidation of graphite, resulting in the formation of a monolayer of graphite oxide. This process is typically achieved through chemical treatment, with a notable method known as Hummer's method. In this approach, graphite is subjected to sequential oxidation, exfoliation, and dispersion in water or suitable organic solvents, leading to the creation of individual GO sheets (Maity et al. 2018). Graphene oxide possesses distinctive properties that make it a versatile material with a wide range of applications. One notable characteristic is its transparency. Despite being a nanoparticle, GO exhibits transparency, allowing light to pass through it. This property is particularly advantageous for applications such as transparent conductive films and coatings. Additionally, GO is flexible, allowing it to conform to different surfaces and structures. Its flexibility, combined with its thin structure, makes it an excellent candidate for flexible electronics, sensors, and other flexible devices. The thinness of GO is remarkable, being 200 times thinner than steel. This property is crucial in applications where weight and thickness constraints are significant factors. Another intriguing feature of GO is its density, contributing to its impermeability to gases. Even helium, known for its small size and ability to pass through many materials, is unable to permeate GO. This impermeability makes GO useful in applications where a barrier to gases is essential, such as in packaging materials and gas separation membranes. GO's exceptional mechanical properties are noteworthy. While being thin and flexible, it exhibits outstanding mechanical strength, being 200 times stronger than steel. This strength is comparable to the resistance of diamond, highlighting GO's remarkable mechanical characteristics. This combination of strength, flexibility, and thinness makes GO an ideal candidate for reinforcement in composites, strengthening materials without compromising their weight or flexibility. Furthermore, GO is highly conductive, possessing excellent electrical conductivity. This property is exploited in various electronic and sensor applications, where GO can serve as a conductive component in devices like sensors, transistors, and batteries. In recent years, the unique properties of GO have led to its application in a diverse range of fields. In the realm of energy storage, GO-based materials have been explored for use in supercapacitors and batteries, capitalizing on their high surface area and conductivity. In the field of sensors, GO's sensitivity to changes in its environment makes it a valuable component for detecting gases, biomolecules, and other analytes. Additionally, GO has found applications in biomedicine, such as drug delivery systems and imaging agents, owing to its biocompatibility and ability to functionalize with biomolecules. Graphene oxide stands out as a remarkable derivative of graphene, showcasing a myriad of properties that contribute to its versatility and applicability across various domains. From its transparent and flexible nature to its impermeability to gases and exceptional mechanical strength, GO continues to inspire innovations in materials science, electronics, energy storage, and biomedical research. The ongoing exploration of graphene oxide and its derivatives promises continued advancements and breakthroughs in diverse scientific and technological (De Marchi et al. 2018).

4.3 Electrochemical Properties

Graphene's electrochemical properties make it an exceptional material for various applications. One of its key attributes is high electron mobility, measured at 2×10^5 cm^2/Vs, indicating efficient conductivity at room temperature (De Marchi et al. 2018). This high electron mobility is crucial for facilitating rapid and efficient electron transfer in electrochemical processes, making graphene an excellent candidate for electrodes and other electrochemical applications. Graphene's small atomic thickness, along with its high thermal conductivity, makes it an efficient conductor of heat as well. This property is advantageous in electrochemical systems where temperature management is critical. The material's high elasticity and flexibility are particularly beneficial in applications where mechanical stress and strain are involved. The great specific surface area of graphene further enhances its utility in electrochemical applications. The availability of a large surface area provides ample sites for electrochemical reactions to occur, allowing for increased efficiency in processes such as energy storage and sensing (Maity et al. 2018). Graphene's electrochemical properties extend to graphene oxide (GO), a derivative that effectively maintains electrical conductivity. This property is significant for applications where the introduction of functional groups through oxidation is necessary while retaining electrical conductivity.

In electrochemical applications involving redox reactions, graphene serves as an excellent host. It can effectively immobilize redox centers, such as those found in heme-containing metalloproteins like myoglobin, horseradish peroxidase (HRP), and cytochrome, onto the electrode surface. This ability opens avenues for various electrochemical processes, including sensing and bioelectrochemical applications. One notable example is the electrocatalytic activity of reduced graphene oxide (rGO) in the oxidation of hydrazine, as reported by. This showcases graphene's potential in catalyzing reactions, which is crucial in fuel cells, sensors, and other electrochemical devices. Graphene's electrochemical prowess, characterized by high electron mobility, thermal conductivity, elasticity, and flexibility, positions it as a formidable material in the realm of electrochemistry. Its derivatives, such as graphene oxide and reduced graphene oxide, extend its applicability to diverse electrochemical processes, offering a platform for advancements in energy storage, sensing, and catalysis (Maity et al. 2018) Graphene oxide (GO) emerges as a highly promising material for ultracapacitors due to its exceptional electrochemical capacitance and cyclic performance. These attributes make GO a preferred choice for energy storage applications. Research indicates that reduced graphene oxide (rGO) exhibits superior electrochemical capacitance and cyclic durability compared to carbon nanotubes (CNTs), further solidifying the potential of graphene derivatives in advancing ultra-capacitor technology (Maity et al. 2018). Characterization of graphene and its derivatives using techniques like Fourier-transform infrared spectroscopy (FTIR) reveals the presence of various functional groups, including carboxylic acids, alcohols, and epoxides. These functional groups enhance the biocompatibility of graphene-based materials with different biological molecules. This property opens up avenues for applications in biosensing, drug delivery, and other biomedical fields. One noteworthy feature of graphene and its derivatives is the opening of a band gap, a characteristic not present in pristine graphene. This band gap imparts high photoluminescence properties to graphene-based materials. The ability to emit light upon excitation makes them valuable in optoelectronic applications, such as light-emitting diodes (LEDs) and sensors (Ambrosi et al. 2014). The electrochemical capacitance of graphene oxide, particularly its cyclic performance, positions it as a key player in the development of ultra-capacitors. The diverse functional groups identified through FTIR characterization enhance the biocompatibility of graphene-based materials, expanding their potential applications in the biomedical field. Furthermore, the opening of a band gap in graphene derivatives adds a layer of versatility, leading to high photoluminescence properties and paving the way for advancements in optoelectronics (Ambrosi et al. 2014).

4.4 Derivatives of Graphene

Graphene, a two-dimensional structure consisting of hybridized carbon atoms arranged in a hexagonal honeycomb lattice, has captivated scientific interest since its discovery in 2004 (Satheesh and Jayavel, 2013; Banerjee et al. 2018; Ponnamma et al. 2021). This smallest material has become a focal point across diverse scientific disciplines, including biomedicine, chemistry, physics, drug delivery, electrochemical analysis, nano-manufacturing, and energy storage devices. The hexagonal lattice structure of graphene, characterized by sp2 hybridization, imparts unique properties that have sparked numerous investigations. Since 2004, researchers have delved into the potential applications of graphene due to its exceptional characteristics. The material's high electron transportation efficiency, optical transmittance, large specific surface area, and versatility make it a compelling candidate for a wide range of future applications. In biomedicine, graphene holds promise for drug delivery and therapeutic applications due to its biocompatibility and ability to interact with biological molecules (Loryuenyong et al. 2013). In the realms of chemistry and physics, graphene's electronic and optical properties propel advances in materials science and nanotechnology. The unique characteristics of graphene, such as its exceptional electrical conductivity and optical transparency, make it a key player in the development of cutting-edge materials and nanotechnological applications, (Grajek et al. 2020). In electrochemical analysis, graphene's conductivity and surface properties make it an ideal material for sensors and electrodes. Moreover, graphene's role extends to nano-manufactured devices, where its unique structure and properties contribute to the development of innovative technologies. In energy storage, graphene-based materials have been explored for applications in batteries and supercapacitors, taking advantage of their high electrical conductivity and large surface area (Tang et al. 2009a, b; Chen et al. 2012; Ibrahim et al. 2021). As the field of graphene research continues to evolve, the material's remarkable attributes open up possibilities for transformative advancements in various industries. Its role in enhancing electronic devices, contributing to medical breakthroughs, and revolutionizing energy storage highlights the broad spectrum of applications that graphene is poised to impact in the foreseeable future. The interdisciplinary nature of graphene research underscores its significance as a groundbreaking material with vast potential across scientific and technological domains (Tang et al. 2009a, b; Chen et al. 2012; Ibrahim et al. 2021). Graphene and its derivatives, owing to their unique properties, have found applications in the immobilization of biomolecules, particularly proteins. This capability makes them valuable tools in fields such as biotechnology and medicine (Seabra et al. 2014). Scientists have discovered a range of graphene derivatives, each with distinct characteristics and applications. These derivatives include graphene oxide, graphene quantum dots, reduced graphene oxide, graphone, fluorographene, graphyne, and graphdiyne (Chaudhary et al. 2021; Alam et al. 2017).

Qualitative examination of graphene and its derivatives involves various techniques, each offering insights into different aspects of their structure and composition. Scanning Electron Microscopy (SEM) is employed to estimate morphological features and identify different layers within graphene structures. Atomic Force Microscopy (AFM) examines surface roughness, morphology, and thickness at the nanoscale. Fourier Transform Infrared Spectroscopy (FTIR) is utilized to determine the presence of different functional groups in graphene derivatives. Dynamic Light Scattering (DLS) identifies particle sizes, Qualitative examination of graphene, and its derivatives involves various techniques, each offering insights into different aspects of their structure and composition. Scanning Electron Microscopy (SEM) is employed to estimate morphological features and identify different layers within graphene structures. Atomic Force Microscopy (AFM) examines surface roughness, morphology, and thickness at the nanoscale. Fourier Transform Infrared Spectroscopy (FTIR) is utilized to determine the presence of different functional groups in graphene derivatives. Dynamic Light Scattering (DLS) identifies particle sizes, and UV–VIS spectroscopy measures light absorbance in graphene and its derivatives (Chaudhary et al. 2021; Lee et al. 2019). The versatility of graphene and its derivatives is evident in their ability to occupy a significant space at the nanoscale. This characteristic has led to their application in diverse areas, particularly in drug delivery and biosensors (Kumar et al. 2023). The large specific surface area and unique physicochemical properties of graphene derivatives make them excellent candidates for interacting with biological entities, enabling precise control in drug delivery systems. The immobilization of biomolecules, especially proteins, onto graphene surfaces enhances their stability and functionality. This feature is particularly valuable in biosensor applications where the selective binding of biomolecules is crucial for sensing and detection. Additionally, the unique properties of graphene derivatives make them suitable for drug delivery systems, ensuring efficient and targeted delivery of therapeutic agents. The broad array of graphene derivatives, coupled with advanced characterization techniques, has propelled their exploration in various scientific and technological domains. From biotechnology to materials science, the unique properties of graphene derivatives continue to inspire innovation and discovery. The ability to immobilize biomolecules and occupy a significant nanoscale space opens up new possibilities for applications in medicine, sensing technologies, and beyond (Kumar et al. 2023).

4.4.1 Graphene Oxide

Graphene oxide, a derivative of graphene, is characterized by its rich composition of carbon, oxygen, and hydrogen. The synthesis of graphene oxide involves the Hummer method, where graphite powder undergoes oxidation using various oxidizing agents (Chaudhary et al. 2021; Mbayachi et al. 2021). The resulting material is highly oxygenated and features exogenous functional groups, including carboxylic acid, aldehyde, ketone, epoxide, and hydroxyl. The introduction of these oxygen-containing functional groups imparts unique properties to graphene oxide, making it highly soluble in water and various organic solvents (Gómez-Navarro et al. 2010). The dispersion of these functional groups throughout the graphene structure creates a material with remarkable versatility and compatibility with different solvents. The sp2 hybridization resulting from the formation of these functional groups enhances the stability and dispersibility of graphene oxide. This dispersion property is crucial for various applications, especially in the biomedical field, where solubility in water is often a prerequisite for interactions with biological systems. The functional groups also contribute to the reactivity of graphene oxide, making it amenable to various chemical modifications. This versatility has led to the exploration of graphene oxide in diverse areas, including drug delivery, sensors, and composite materials. Graphene oxide's synthesis through the Hummer method imparts it with a high content of oxygenated functional groups, rendering it soluble in water and organic solvents. This solubility, coupled with the dispersibility achieved through sp2 hybridization, opens up a broad spectrum of applications for graphene oxide in fields ranging from materials science to biotechnology (Gómez-Navarro et al. 2010). The synthesis of graphene oxide involves various methods, with the binding of oxygenated functional groups on graphene being a prominent approach for obtaining large quantities of graphene oxide (Boehm et al. 1994). Conversely, chemical reduction in an alkaline medium has become a widely utilized technique in contemporary graphene oxide production (Grajek et al., 2020). In Fig. 4.3, different historical approaches to graphene oxide formation are illustrated. The Brodie method, dating back to 1859, represents one of the earliest methods, followed by the Staudenmaier method in 1898. The Hofmann method, introduced in 1937, and the Hummers method, developed in 1958, are subsequent advancements in the synthesis of graphene oxide (Fig. 4.3).

Each method has its unique set of conditions and reagents, influencing the properties and characteristics of the resulting graphene oxide. The Hummers method, in particular, has gained prominence for its efficiency in producing graphene oxide on a larger scale. It involves the oxidation of graphite with strong oxidizing agents and subsequent exfoliation, resulting in graphene oxide with abundant oxygen functional groups. Chemical reduction of graphene oxide is often employed to restore some

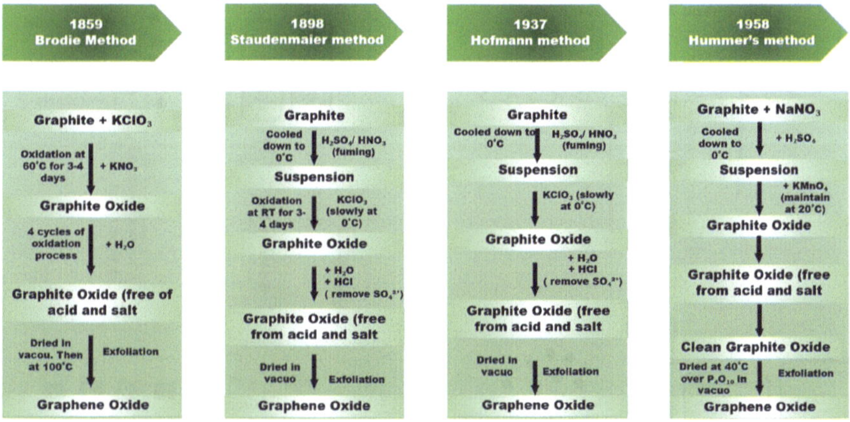

Fig. 4.3 Synthesis of graphene oxide (Lee et al. 2019)

of the graphene-like properties by removing oxygen functionalities. This reduction process, especially in an alkaline medium, plays a crucial role in tailoring the properties of graphene oxide for specific applications, ranging from electronics to biomedical fields. The historical evolution of graphene oxide synthesis methods showcases the continuous refinement and development of techniques over the years. The choice of method influences the composition and properties of graphene oxide, emphasizing the importance of selecting an appropriate approach based on the desired application.

4.4.2 *Graphene Quantum Dots*

Graphene quantum dots (GQDs), a recent addition to the nanomaterial family, are characterized by their atomical size, typically below 10 nm (Karaphun et al. 2021). Their unique features, encompassing stability, optoelectrical, electrical, and optical properties, have garnered significant attention across various scientific and technological domains (Yan et al. 2019). The synthesis of graphene quantum dots involves two primary methods: top-down and bottom-up approaches. In top-down methods, GQDs are directly produced from bulk carbon compounds through the exfoliation of liquid with techniques such as electron beam lithography. This method involves breaking down larger carbon structures into smaller quantum dots, creating nanoscale entities with specific properties (Yan et al. 2019). On the other hand, bottom-up methods involve the transformation of polymers or other precursor materials into nanoscale graphene quantum dots. These methods allow for the controlled assembly of carbon atoms, facilitating the creation of quantum dots with tailored properties. The bottom-up approach provides more control over the size, shape, and composition of GQDs, offering versatility in their applications (Yan et al. 2019). The exceptional properties of GQDs make them highly desirable in various applications. Their small size

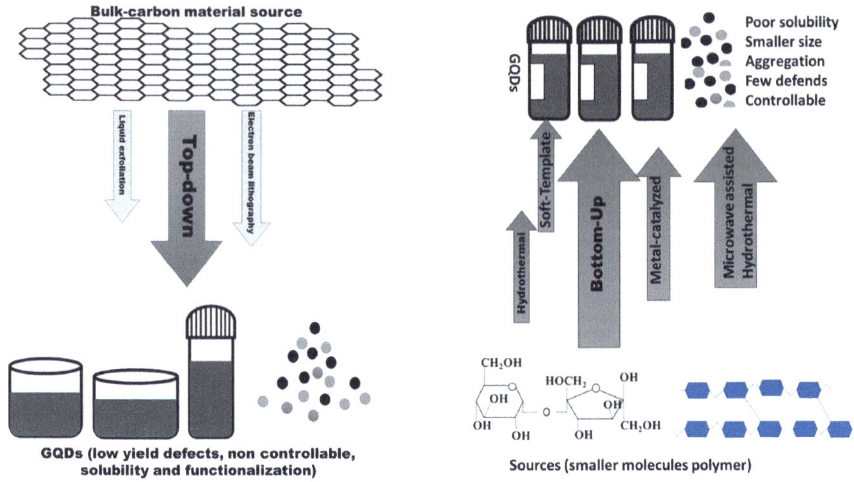

Fig. 4.4 Synthesis of GQDs by top-down and bottom-up methods (Yan et al. 2018)

imparts quantum confinement effects, leading to unique electronic and optical behaviors. The stability of GQDs is crucial for their integration into diverse technologies, and their optoelectrical and electrical properties open up possibilities for applications in sensors, electronics, and energy-related devices. GQDs have shown promise in fields such as bioimaging, sensing, drug delivery, and catalysis, owing to their size-dependent optical properties and biocompatibility. Additionally, their versatile synthesis methods contribute to their widespread adoption in research and development. The emergence of graphene quantum dots as a nanomaterial with unique properties has opened up new avenues in various scientific and technological domains. The synthesis methods, combined with the distinct characteristics of GQDs, make them valuable tools for applications ranging from electronics to biomedical research, showcasing their potential for transformative innovations in the nanomaterial landscape (Yan et al. 2019) (Fig. 4.4).

4.4.3 Reduced Graphene Oxide

Carbon's unique ability to form diverse allotropes with distinct properties is exemplified by graphene, one of the most remarkable carbon materials. Scientists have uncovered various derivatives of graphene, such as graphene oxide, graphene quantum dots, and reduced graphene oxide, expanding the potential applications of graphene in different fields (Tian et al. 2018).

Through reductive exfoliation, graphene oxide can be further transformed into reduced graphene oxide (illustrated in Fig. 4.5) (Geim and Novoselov 2007). Reduced graphene oxide has become the focus of considerable attention in diverse areas, particularly in the manufacturing of electronic devices and biosensors. Its graphenic properties, including high electric conductivity, flexibility, and a large surface area per volume, make it a highly desirable material for a range of applications (Tian et al. 2018; Tarcan et al. 2020; Robinson et al. 2008). Notably, the production of reduced graphene oxide is relatively straightforward compared to other derivatives, contributing to its widespread use and high future prospects, especially in the field of nanobiotechnology. The production of reduced graphene oxide involves various methods, each offering distinct advantages. Electrochemical reduction, thermal treatment, green reduction (utilizing environmentally friendly approaches), and chemical reduction processes are among the techniques employed to synthesize reduced graphene oxide. These processes allow for the tailoring of reduced graphene oxide properties to meet specific requirements, further enhancing its versatility in different applications. In electronic devices, reduced graphene oxide's excellent conductivity facilitates its use in components like electrodes and conductive coatings. In biosensors, its large surface area provides ample sites for the immobilization of biomolecules, enhancing sensitivity and selectivity. The ease of production and functionalization of reduced graphene oxide contribute to its growing significance in nanobiotechnology, promising innovative solutions in areas ranging from healthcare to environmental monitoring. The ability of carbon to give rise to diverse allotropes, coupled with the versatile properties of reduced graphene oxide, positions it as a key player in the advancement of various technologies. Its production ease and suitability for electronic and biosensor applications underscore its promising future impact in nanobiotechnology and beyond (Tian et al. 2018; Tarcan et al. 2020; Robinson et al. 2008).

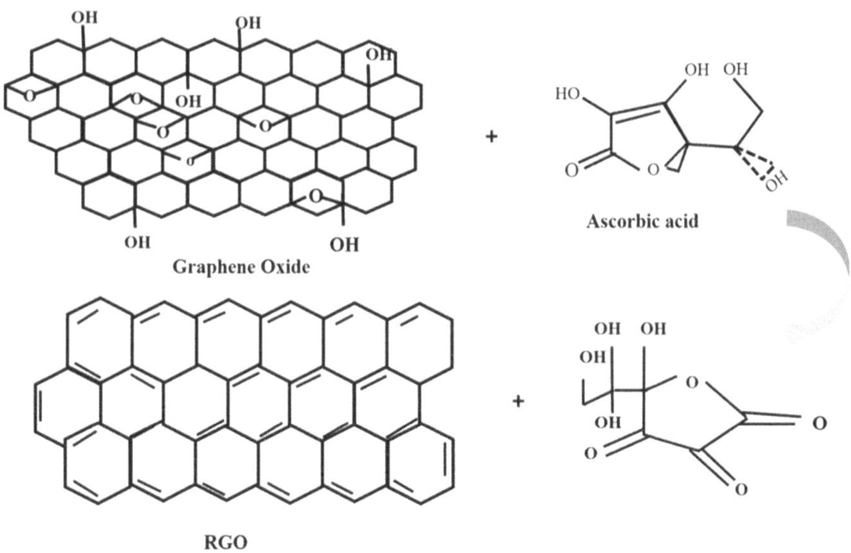

Fig. 4.5 Formation of reduced graphene oxide (Habte and Ayele 2019)

4.5 Other Graphene Derivatives

4.5.1 *Graphone*

The study conducted by Zhou et al. (2009) investigated a graphene derivative subjected to mild hydrogenation, revealing its characteristics as mildly hydrogenated compounds. Initially, pristine graphene sheets displayed no inherent magnetism. However, the research highlighted that by reducing hydrogenation in graphene, the resulting derivative exhibited proper magnetism properties. This discovery marked a pivotal understanding of the modulation of graphene's magnetic behavior through controlled hydrogenation levels. The work of Habte and Ayele (2019), Feng and Zhang (2012) further complemented this finding, emphasizing the significance of controlled hydrogenation processes in tailoring the magnetic properties of graphene derivatives. These insights into graphene's magnetic behavior have implications for applications in areas such as spintronics and magnetic sensors, offering avenues for the deliberate engineering of graphene derivatives with desired magnetic functionalities (Habte and Ayele 2019; Feng and Zhang 2012). Building upon Zhou's foundational survey in 2009, subsequent research has demonstrated that the adsorption of hydrogen atoms onto graphene serves to enhance its antiferromagnetic properties. The work conducted by Peng et al. (2014) resulted in the creation of a highly stable structure known as "triangular graphone." This distinctive arrangement features a single-handed hydrogen orientation, a design intended to amplify both electronic and magnetic properties, as elucidated by Feng and Zhang (2012). The deliberate adsorption of hydrogen atoms introduces a mechanism for precise control over the magnetic behavior of graphene derivatives. This insightful exploration and manipulation of graphene's magnetic characteristics hold significant promise for applications in advanced electronic devices and emerging technologies that leverage tailored magnetic functionalities, showcasing the transformative potential of controlled hydrogenation in the domain of graphene research (Feng and Zhang 2012).

4.5.2 *Graphyne*

Speculations about graphyne began before the 1960s, but concerted efforts by theorists to explore and discover its properties intensified in the 1980s. Graphyne is characterized by its distinctive acetylenic bond between hexagonal structures, as illustrated in Fig. 4.6 (Habte and Ayele 2019). This unique feature sets it apart from other carbon allotropes and contributes to the creation of various graphyne derivatives, such as graphdiyne and graphtriyne (Ryu et al. 2008). The acetylenic bond, formed between the benzene ring and acetylene, imparts specific electronic and structural properties to graphyne. This bond configuration not only defines the material's chemical identity but also influences its electronic and mechanical characteristics.

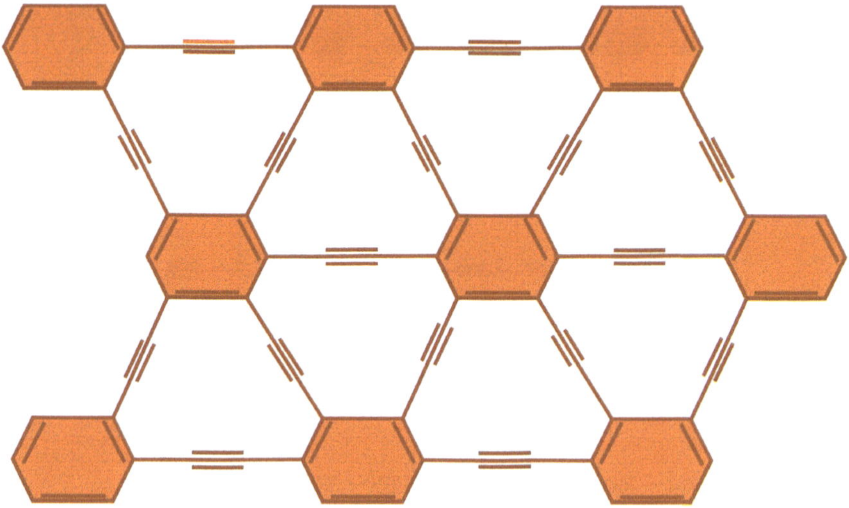

Fig. 4.6 Structure of graphyne (Huang et al. 2018)

The exploration of graphyne derivatives has unveiled diverse structures with potential applications in nanoelectronics, catalysis, and other fields. As researchers delve deeper into the properties and potential applications of graphyne and its derivatives, they continue to uncover new insights into the rich landscape of carbon allotropes (Ryu et al. 2008).

4.5.3 Graphdiyne

Continuing the exploration of carbon allotropes, researchers have further expanded on the graphyne structure by introducing a new variation. In this modified version, two acetylenic bonds connect each hexagonal structure, deviating from the original graphyne which possesses a single acetylenic bond between hexagons (illustrated in Fig. 4.7). This structural modification gives rise to a unique variation of graphyne, demonstrating the versatility and tunability of carbon allotropes. The introduction of an additional acetylenic bond between hexagons is likely to influence the electronic and mechanical properties of the material. Understanding these modifications provides valuable insights into tailoring the properties of carbon allotropes for specific applications, from electronic devices to structural materials. This ongoing research in carbon allotropes reflects the continuous exploration of novel structures with the potential to unlock new possibilities in materials science and nanotechnology (Habte and Ayele 2019). Graphyne, a carbon allotrope, represents a promising microstructure with distinctive electronic features, as noted by Habte and Ayele (2019). This unique material, explored extensively, exhibits both molecular and

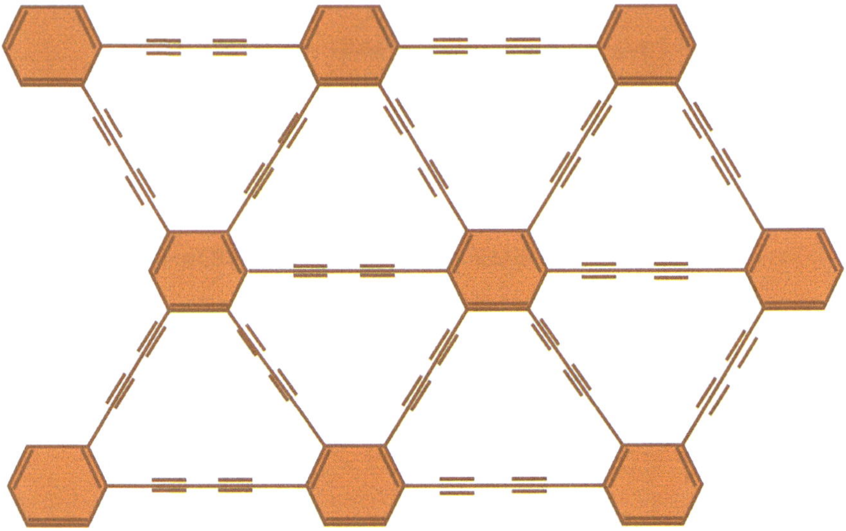

Fig. 4.7 Structure of graphdiyne (Huang et al. 2018)

mechanical features (Ryu et al. 2008). The potential of graphyne lies in its intricate structure, offering a platform for advancements in electronic, molecular, and mechanical applications, making it a subject of considerable interest in materials science. These remarkable properties make graphyne a versatile material with applications in diverse areas. Its unique molecular structure and mechanical characteristics open avenues for applications in bioremediation, where its specific properties could be harnessed for environmental cleanup. Additionally, graphyne's gas separation capabilities and catalytic properties contribute to its relevance in energy-related fields. The multifaceted utility of graphyne underscores its significance in advancing technologies and addressing challenges in various scientific and environmental domains (Jia et al. 2017).

4.5.4 Graphane

The more hydrogenated derivative of graphene, often termed graphane, represents a material that is 100% hydrogenated. In contrast to graphone, graphane is fully saturated with hydrogen atoms (Habte and Ayele 2019; Gao et al. 2019). Similar to other hydrocarbon graphene structures, it is formed by carbon and hydrogen atoms joined by covalent bonds, arranged in a two-dimensional structure. This complete hydrogenation imparts unique properties to graphane, including semi-conductivity. Despite its minor dimension, graphane finds applications in various fields (Boukhvalov 2010). The semi-conductive nature of graphane makes it promising for electronic applications, and its 2D structure enhances its utility in different technological domains, contributing to ongoing research and developments in material science and electronics (Boukhvalov 2010). Graphane, the fully hydrogenated derivative of graphene, exhibits three distinct isomers: chair, stirrup, and boat. These isomers, categorized by Sofo et al. (2007), showcase variations in stability. Among them, the chair-like and boat-like structures stand out as the most stable. In the chair-like isomer, hydrogen atoms oscillate on both sides, contributing to its heightened stability. Similarly, in the boat-like isomer, hydrogen atoms oscillate in pairs, further enhancing structural stability (Sahin et al. 2015). Understanding the stability differences among these isomers is crucial for tailoring the properties of graphane and optimizing its applications in diverse scientific and technological realms (Sahin et al. 2015).

4.6 Nanocomposites of Graphene

Nanocomposites, also referred to as hybrid materials, are formed by combining multiple nanomaterials to enhance their collective potential (Flores et al. 2009). This involves the integration of various materials, including other nanoparticles, metals, nanotubes, and polymers, each contributing distinct features. The synergy achieved through this fusion results in a novel material with unique and beneficial properties. Nanocomposites exhibit a wide range of applications across different fields, leveraging the individual strengths of their constituent materials. This interdisciplinary approach enables the creation of advanced materials with tailored characteristics, fostering innovations in areas such as materials science, electronics, and biomedical research (Boehm et al. 1994). Graphene and its derivatives, such as graphene oxide (GO) and reduced graphene oxide (rGO), play a pivotal role in the production of graphene-based nanocomposites. Their reactive oxygen species make them versatile materials for combining with various inorganic nanoparticles, including quantum dots, metal oxides, polymers, metals, and semiconductor nanoparticles. This amalgamation results in the creation of advanced compounds with enhanced selectivity, sensitivity, and low detection limits. These graphene-based nanocomposites find diverse applications, particularly in the field of biosensors (Dhand et al. 2013).

The unique properties of graphene and its derivatives, coupled with the tailored characteristics of inorganic nanoparticles, contribute to the development of highly efficient sensing platforms, paving the way for innovations in medical diagnostics, environmental monitoring, and other areas requiring precise and reliable detection technologies. Table 4.1 illustrates different derivatives and nanocomposites, their sources, methods of preparations, and their applications. Polymer nanocomposites incorporating graphene have garnered significant attention from scientists globally, owing to the exceptional graphenic features they impart. The inclusion of graphene in polymers enhances electrical conductivity, thermal conductivity, thermal stability, and surface area per volume. This has propelled the extensive utilization of graphene-related polymer nanocomposites in diverse sectors, including the manufacturing of energy-providing materials, defense mechanisms, marine devices, aerospace, robotics, and biomedical sciences (Ibrahim et al. 2021; Krishnan et al. 2019).

The integration of graphene into polymers, often involving just 1% of graphene, results in nanocomposites exhibiting superior mechanical and thermal characteristics compared to carbon nanotubes and nanoparticles, respectively (Bilisik and Akter 2022). This enhancement in properties is pivotal for various applications, ensuring improved performance in specific areas such as structural integrity, thermal management, and overall functionality. Several methods are employed to design and fabricate these graphene-polymer nanocomposites. Techniques such as the electrospinning method, coating, melt mixing, and extrusion methods allow for precise control over the dispersion of graphene within the polymer matrix. This controlled dispersion ensures uniform distribution and interaction, contributing to the advanced properties of the resulting nanocomposites (Liu et al. 2022). The significance of these graphene-based polymer nanocomposites lies in their multifaceted applications across different industries. In the energy sector, they contribute to the development of efficient materials for energy storage and conversion. In defense mechanisms, their enhanced mechanical properties make them valuable for structural components. Marine devices benefit from their improved corrosion resistance and thermal stability, while aerospace and robotics benefit from their lightweight yet strong characteristics. In biomedical sciences, the tailored properties of graphene-polymer nanocomposites open avenues for applications in drug delivery, tissue engineering, and biosensors. The versatility of these materials, coupled with their advanced properties, underscores their potential for addressing complex challenges and driving innovations across a spectrum of scientific and technological domains (Liu et al. 2022).

4.7 Conclusion

In summary, since the groundbreaking discovery of graphene in 2004 by Andre Geim and Konstantin Novoselov, extensive research efforts have been devoted to understanding its structure, derivatives, and profound significance across diverse scientific domains. Researchers worldwide have conducted numerous studies to unravel the unique properties of graphene, paving the way for its applications in various fields. To ensure the efficacy and safety of graphene-based materials in applications ranging from electronics to biomedicine, rigorous characterization techniques have been employed. Techniques such as Transmission Electron Microscopy (TEM), X-Ray Diffraction (XRD), Scanning Electron Microscopy (SEM), X-Ray Photoelectron Spectroscopy (XPS), and Raman spectroscopy have played crucial roles in elucidating the structural, chemical, and electronic properties of graphene and its derivatives. These characterization methods are essential for guaranteeing the quality and reliability of graphene-based materials, particularly in contexts where they directly impact human well-being. The ongoing multidisciplinary research surrounding graphene continues to drive innovations and shape the future of advanced materials.

Table 4.1 Illustrates different derivatives and nanocomposites, their sources, methods of preparations and their applications

Derivatives and nanocomposites	Raw materials	Preparation methods	Applications	References
Graphene oxide	Graphite powder	Hummer's method	Membrane and coatings Stimulus responsive materials Corrosion resistant	Chaudhary et al. (2021), Mbayachi et al. (2021)
Graphene quantum dots	Starch, Carbon fibers	Hydrothermal cutting, Solvothermal method	Optoelectronics	Smith et al. (2019)
Reduced graphene oxide	Ascorbic acid	Hummer's method	Membrane and coatings Stimulus responsive materials Corrosion resistant	Pan et al. (2010), Wang et al. (2012)
Poly ethylenedioxythiophene-graphene	EDOT	Electrochemically deposition	Energy storing instruments	Chu et al. (2012)
(PEDOT:PSS)/Graphene	PEDOT:PSS	In situ polymerization	Energy harvesting systems	Yoo et al. (2014)
Graphene/MnO_2/CNTs	MnO_2, Carbon nanotubes	Chemical co-precipitation method	Energy storing instruments as a robust electrode	Cheng et al. (2012)
Graphene/PANI	PANI	Electro polymerization	Paper electrode	Cong et al. (2013)
TiO_2/Graphene/PPy	TiO_2, PPy	Electrodeposition	Supercapacitor	Jiang et al. (2015)
Graphene/MnO_2	MnO_2	Electrodeposition	Asymmetric supercapacitor	Zang et al. (2014)

Up to now, it has addressed much attention in technological and biomedical aspects like diagnostics, drug delivery, biosensing, and physical and electrochemical sciences. A lot of publications have written about this atomic structure particle, in every study made about graphene revealed that this smallest particle has unique features such as high electrical conductivity, and a great surface-to-volume ratio which provides a huge fantastic future perspective. Referring to the above chapter, we recommend an open view to the scientists for more discoveries about this nanoparticle.

References

Alam, S.N., Sharma, N., Kumar, L.: Synthesis of graphene oxide (GO) by modified hummers method and its thermal reduction to obtain reduced graphene oxide (rGO). Graphene **6**(1), 1–18 (2017)

Ambrosi, A., Chua, C.K, Bonanni, A., Pumera, M.: Electrochemistry of graphene and related materials. Chem. Rev. **114**(14), 7150–7188 (2014). https://doi.org/10.1021/cr500023c

Banerjee, A.N.: Graphene and its derivatives as biomedical materials: future prospects and challenges. Interface Focus **8**(3), 20170056 (2018)

Bilisik, K., Akter, M.: Graphene nanocomposites: A review on processes, properties, and applications. J. Ind. Text. **51**(3_suppl), 3718S-3766S (2022)

Boehm, H.P., Setton, R., Stumpp, E.: Nomenclature and terminology of graphite intercalation compounds (IUPAC Recommendations 1994). Pure Appl. Chem. **66**(9), 1893–1901 (1994). https://doi.org/10.1351/pac199466091893

Bostwick, A., McChesney, J., Ohta, T., Rotenberg, E., Seyller, T., Horn, K.: Experimental studies of the electronic structure of graphene. Progress Surf. Sci. **84**(11–12), 380–413 (2009). https://doi.org/10.1016/j.progsurf.2009.08.002

Boukhvalov, D.W.: Stable antiferromagnetic graphone. Physica E E **43**(1), 199–201 (2010)

Chaudhary, K., Kumar, K., Venkatesu, P., Masram, D.T.: Protein immobilization on graphene oxide or reduced graphene oxide surface and their applications: Influence over activity, structural and thermal stability of protein. Adv. Coll. Interface. Sci. **289**, 102367 (2021)

Chen, Y., Zhang, B., Liu, G., Zhuang, X., Kang, E.T.: Graphene and its derivatives: switching ON and OFF. Chem. Soc. Rev. **41**(13), 4688–4707 (2012)

Cheng, Y., Lu, S., Zhang, H., Varanasi, C.V., Liu, J.: Synergistic effects from graphene and carbon nanotubes enable flflexible and robust electrodes for high-performance supercapacitors. Nano Lett. **12**, 4206–4211 (2012)

Chu, C.-Y., Tsai, J.-T., Sun, C.-L.: Synthesis of PEDOT-modifified graphene composite materials as flflexible electrodes for energy storage and conversion applications. Int. J. Hydrogen Energy **37**, 13880–13886 (2012)

Cong, H.-P., Ren, X.-C., Wang, P., Yu, S.-H.: Flexible graphene-polyaniline composite paper for high-performance supercapacitor. Energy Environ. Sci. **6**, 1185–1191 (2013)

De Marchi, L., Pretti, C., Gabriel, B., Marques, P.A., Freitas, R., Neto, V.: An overview of graphene materials: properties, applications and toxicity on aquatic environments. Sci. Total Environ. **1**(631), 1440–1456 (2018). BibTeX EndNote RefMan RefWorks, https://doi.org/10.1155/2014/890246

Dhand, V., Rhee, K.Y., Kim, H.J., Jung, D.H.: A comprehensive review of graphene nanocomposites: research status and trends. J. Nanomater.nanomater. **2013**, 158–158 (2013)

Feng, L., Zhang, W.X.: The structure and magnetism of graphone. AIP Adv. **2**(4), 042138 (2012)

Flores, M.Z., Autreto, P.A., Legoas, S.B., Galvao, D.S.: Graphene to graphane: a theoretical study. Nanotechnology **20**(46), 465704 (2009)

Gao, X., Liu, H., Wang, D., Zhang, J.: Graphdiyne: synthesis, properties, and applications. Chem. Soc. Rev. **48**(3), 908–936 (2019)

Geim A.K., Novoselov K.S.: The rise of graphene. Nat. Mater. 6(3), 183–191 (2007). https://doi.org/10.1038/nmat1849

Gómez-Navarro, C., Meyer, J.C., Sundaram, R. S., Chuvilin, A., Kurasch, S., Burghard, M., et al.: Atomic structure of reduced graphene oxide. Nano Lett. **10**(4), 1144–1148 (2010)

Grajek, H., Jonik, J., Witkiewicz, Z., Wawer, T., Purchała, M.: Applications of graphene and its derivatives in chemical analysis. Crit. Rev. Anal. Chem. **50**(5), 445–471 (2020)

Habte, A.T., Ayele, D.W.: Synthesis and characterization of reduced graphene oxide (rGO) started from graphene oxide (GO) using the tour method with different parameters. Adv. Mater. Sci. Eng. (2019)

Huang, C., Li, Y., Wang, N., Xue, Y., Zuo, Z., Liu, H., Li, Y.: Progress in research into 2D graphdiyne-based materials. Chem. Rev. **118**(16), 7744–7803 (2018)

Ibrahim, A., Klopocinska, A., Horvat, K., Abdel, H.Z.: Graphene-based nanocomposites: synthesis, mechanical properties, and characterizations. Polymers (basel) **13**(17), 2869 (2021). https://doi.org/10.3390/polym13172869.PMID:34502909;PMCID:PMC8434110

Jia, Z., Li, Y., Zuo, Z., Liu, H., Huang, C., Li, Y.: Synthesis and properties of 2D carbon graphdiyne. Acc. Chem. Res. **50**(10), 2470–2478 (2017)

Jiang, L.-L., Lu, X., Xie, C.-M., Wan, G.-J., Zhang, H.-P., Youhong, T.: Flexible, free-standing TiO2 -graphene-polypyrrole composite films as electrodes for supercapacitors. J. Phys. Chem. C **119**, 3903–3910 (2015)

Karaphun, A., Phrompet, C., Tuichai, W., Chanlek, N., Sriwong, C., Ruttanapun, C.: The influence of annealing on a large specific surface area and enhancing electrochemical properties of reduced graphene oxide to improve the performance of the active electrode of supercapacitor devices. Mater. Sci. Eng. B **264**, 114941 (2021)

Krishnan, S.K., Singh, E., Singh, P., Meyyappan, M., Nalwa, H.S.: A review on graphene-based nanocomposites for electrochemical and fluorescent biosensors. RSC Adv. **9**(16), 8778–8881 (2019)

Kumar, S., Prakash, J., Verma, A., Jasrotia, R., Kandwal, A., Verma, R., et al.: A review on properties and environmental applications of graphene and its derivative-based composites. Catalysts **13**(1), 111 (2023)

Lee, X.J., Hiew, B.Y.Z., Lai, K.C., Lee, L.Y., Gan, S., Thangalazhy-Gopakumar, S., Rigby, S.: Review on graphene and its derivatives: Synthesis methods and potential industrial implementation. J. Taiwan Inst. Chem. Eng. **98**, 163–180 (2019)

Liu, C., Tan, D., Chen, X., Liao, J., Wu, L.: Research on graphene and its derivativesin oral disease treatment. Int. J. Mol. Sci. **23**(9), 4737 (2022)

Loryuenyong, V., Totepvimarn, K., Eimburanapravat, P., Boonchompoo, W., Buasri, A.: Preparation and characterization of reduced graphene oxide sheets via water-based exfoliation and reduction methods. Adv. Mater. Sci. Eng. (2013)

Mbayachi, V.B., Ndayiragije, E., Sammani, T., Taj, S., Mbuta, E.R.: Graphene synthesis, characterization and its applications: a review. Results Chem. **3**, 100163 (2021)

Moon, I.K., Lee, J., Ruoff, R.S., Lee, H.: Reduced graphene oxide by chemical graphitization. Nat. Commun.commun. **1**(1), 1–6 (2010)

Pan, D., Zhang, J., Li, Z., Wu, M.: Adv. Mater. **22**, 734 (2010); Abdolhosseinzadeh, S., Asgharzadeh, H., Kim, H.S.: Fast and fully–scalable synthesis of reduced graphene oxide. Sci. Rep. **5**, 10160 (2015)

Peng, Q., Dearden, A.K., Crean, J., Han, L., Liu, S., Wen, X., De, S.: New materials graphyne, graphdiyne, graphone, and graphane: review of properties, synthesis, and application in nanotechnology. Nanotechnol. Sci. Appl. 1–29 (2014)

Ponnamma, D., Yin, Y., Salim, N., Parameswaranpillai, J., Thomas, S., Hameed, N.: Recent progress and multifunctional applications of 3D printed graphene nanocomposites. Compos. B Eng. **204**, 108493 (2021)

Randviir, E.P., Brownson, D.A., Banks, C.E.: A decade of graphene research: production, applications and outlook. Mater. Today **17**(9), 426–432 (2014). https://doi.org/10.1016/j.mattod.2014.06.001

Robinson, J.T., Perkins, F.K., Snow, E.S., Wei, Z., Sheehan, P.E.: Reduced graphene oxide molecular sensors. Nano Lett. **8**(10), 3137–3140 (2008)

Ryu, S., Han, M.Y., Maultzsch, J., Heinz, T.F., Kim, P., Steigerwald, M.L., Brus, L.E.: Reversible basal plane hydrogenation of graphene. Nano Lett. **8**(12), 4597–4602 (2008)

Sahin, H., Leenaerts, O., Singh, S.K., Peeters, F.M.: Graphane. Wiley Interdisciplinary Rev.: Comput. Mol. Sci. **5**(3), 255–272 (2015)

Satheesh, K., Jayavel, R.: Synthesis and electrochemical properties of reduced graphene oxide via chemical reduction using thiourea as a reducing agent. Mater. Lett. **113**, 5–8 (2013)

Seabra, A.B., Paula, A.J., de Lima, R., Alves, O.L., Durán, N.: Nanotoxicity of graphene and graphene oxide. Chem. Res. Toxicol.toxicol. **27**(2), 159–168 (2014)

Smith, A.T., LaChance, A.M., Zeng, S., Liu, B., Sun, L.: Synthesis, properties, and applications of graphene oxide/reduced graphene oxide and their nanocomposites. Nano Mater. Sci. **1**(1), 31–47 (2019). https://doi.org/10.1016/j.nanoms.2019.02.004

Sofo, J.O., Chaudhari, A.S., Barber, G.D.: Graphane: a two-dimensional hydrocarbon. Phys. Rev. B **75**(15), 153401 (2007)

Su, C., Loh, K.P.: Carbocatalysts: graphene oxide and its derivatives. Acc. Chem. Res. **46**(10), 2275–2285 (2013)

Sur, U.K.: Graphene: a rising star on the horizon of materials science. Int. J. Electrochem. (2012). https://doi.org/10.1155/2012/237689, https://shodhganga.inflibnet.ac.in:8443/jspui/handle/10603/359974?mode=full, http://hdl.handle.net/10603/359974

Tang, L., Wang, Y., Li, Y., Feng, H., Lu, J., Li, J.: Preparation, structure, and electrochemical properties of reduced graphene sheet films. Adv. Func. Mater. **19**(17), 2782–2789 (2009b)

Tang, L., Wang, Y., Li, Y., Feng, H., Lu, J., Li, J.: Preparation, structure, and electrochemical properties of reduced graphene sheet films. Adv. Funct. Mater. **19**(17), 2782–2789 (2009). https://doi.org/10.1002/adfm.200900377

Tarcan, R., Todor-Boer, O., Petrovai, I., Leordean, C., Astilean, S., Botiz, I.: Reduced graphene oxide today. J. Mater. Chem. C **8**(4), 1198–1224 (2020)

Tian, P., Tang, L., Teng, K.S., Lau, S.P.: Graphene quantum dots from chemistry to applications. Mater. Today Chem. **10**, 221–258 (2018)

Tiwari, S.K., et al.: Graphene research and their outputs: Status and prospect. J. Sci.: Adv. Mater. Dev. (2020). https://doi.org/10.1016/j.jsamd.2020.01.006

Wang, M., Yan, C., Ma, L.: Graphene nanocomposites. Compos. Their Prop. (2012). https://doi.org/10.5772/50840

Yan, Y., Gong, J., Chen, J., Zeng, Z., Huang, W., Pu, K., et al.: Recent advances on graphene quantum dots: from chemistry and physics to applications. Adv. Mater. **31**(21), 1808283 (2019)

Yang, G., Li, L., Lee, W.B., Ng, M.C.: Structure of graphene and its disorders: a review. Sci. Technol. Adv. Mater. **19**(1), 613–648 (2018). https://doi.org/10.1080/14686996.2018.1494493

Yoo, D., Kim, J., Kim, J.H.: Direct synthesis of highly conductive poly (3,4-ethylenedioxythiophene): Poly (4-styrenesulfonate) (PEDOT: PSS)/graphene composites and their applications in energy harvesting systems. Nano Res. **7**, 717–730 (2014)

Zhang, Z., Xiao, F., Qian, L., Xiao, J., Wang, S., Liu, Y.: Facile Synthesis of 3D MnO_2-Graphene and Carbon Nanotube-Graphene (2014)

Zhen, Z., Zhu, H.: Structure and properties of graphene. In: Graphene, pp. 1–12. Academic (2018). https://doi.org/10.1016/B978-0-12-812651-6.00001-X

Zhou, Q., Yao, H.: Recent development of carbon electrode materials for electrochemical supercapacitors. Energy Rep. **8**, 656–661 (2022). https://doi.org/10.1016/j.egyr.2022.09.167

Zhou, X., Huang, X., Qi, X., Wu, S., Xue, C., Boey, F.Y., et al.: In situ synthesis of metal nanoparticles on single-layer graphene oxide and reduced graphene oxide surfaces. J. Phys. Chem. C **113**(25), 10842–10846 (2009)

Chapter 5
Electrochemical Exfoliation a Green Approach: Waste to Wealth

Shikha Saxena, Gurmehar Singh, and Deepshikha Pande Katare

Abstract Graphene discovery in 2004 by Geim and co-workers was a major breakthrough in the field of material sciences. The creditability of graphene, which is a closely packed honeycomb two-dimensional lattice, owes to its exceptional electrical, mechanical and thermal properties. The applicability of this unique compound includes different devices for storing energy, causing conversion, generating proficient-performance composites, electronics at molecular level, field-emission, etc. That is why it has been attracting the attention of researchers since the beginning. Earlier, mechanical and chemical exfoliation was used for the preparation of graphene, but the process had a few loopholes which cannot be ignored. The problems range from high costs and environmental hazards to the production of toxic gases. To overcome these, recently new techniques came into picture like dry ball milling, electrochemical exfoliation, ultrasonication. However, out of many, electrochemical exfoliation is considered as a green, safe and substrate-free approach for graphene synthesis. It involves the usage of raw materials like pencil core, graphene pellets, graphene rods, palm oil leaf extract, etc. Therefore, the approach leads to the usage of waste or less important raw materials for more usable ones. In this chapter, different aspects of the electrochemical exfoliation process are highlighted with special reference to its green approach.

Keywords Graphene · Electrochemical exfoliation · Waste · Conversion devices

---

S. Saxena · G. Singh
Amity Institute of Pharmacy, Amity University, Noida, India

D. P. Katare (✉)
Proteomic and Translational Research Lab, Centre for Biomedical Biotechnology, Amity Institute of Biotechnology, Amity University, Noida, India
e-mail: dpkatare@amity.edy

R. Khan et al. (eds.), *Electrochemical Exfoliation of Graphene and its Derivatives*, Engineering Materials, https://doi.org/10.1007/978-981-97-2128-3_5

5.1 Introduction

5.1.1 Graphene

Carbon atoms are arranged in a single, hexagonally-shaped sheet to make graphene (Geim and Novoselov 2007; Hass et al. 2008). The C–C bond present at in-plane position, is the solitary powerful bonds in stuff, as well as the bonds in the out-of-plane position, that donates to a delocalized criss-cross of electrons, are what enable graphene to conduct electrons and provide weak interactions between graphene layers or between graphene and substrate. These carbon bonds are sp^2 hybridised. Graphene has garnered a lot of research interest in the engineering and scientific fields because of its distinctive structural characteristics and extraordinary physical properties (Chowdhury et al. 2013).

A particular property of graphene includes the fact that how the carriers having charge behave as relativistic particles without any mass and may move across the surroundings with little dispersion. Graphene's distinctive activity has led to a wide range of unusual occurrences (Lei et al. 2022). In the beginning, it should be noted that graphene is a 2D zero-bandgap semiconductor with a little valence-conduction band overlap. Additionally, it has a strong ambipolar electric field event, enabling measuring of charge carrier consolidation up to 1013 cm^{-2} as well as room-temperature transportability of 10,000 cm^{-2} s^{-1} (Lei et al. 2022). Third, it has been shown that graphene displays a rare half-integer quantum Hall effect (QHE) considering the pair of electron and hole carriers. This phenomenon is persuaded by changing the potential chemically with the electric field outcome (Zhang et al. 2005). Additionally, graphene absorbs visible light at a 2.3% rate, making it extremely transparent. A single-layer sheet at room temperature has a thermal conductivity, or k, measurement of 5000 W mK^{-1} (You and Rotkin 2015). Additionally, the mechanical strength of graphene is excellent. Using atomic force microscopy and nano-indentation, the inherent mechanical characteristics of free-standing mono-layer graphene membranes were examined. One of the widely used materials, has a breakage strength of 42 N m^{-1} and value for Young's modulus as 1.0 TPa.

Graphene and its by-product have been produced, grown, or synthesised using a variety of processes up until this point (Torres et al. 2014). Graphene sheets may be grown from the bottom up as an alternative to mechanically evaporating bulk graphite. By using chemical vapour deposition (CVD), graphene sheets layers have been produced on metal surfaces like Ni and Cu (Arco et al. 2010; Bekarevich et al. 2013). On carbon-containing surfaces like SiC, carbon segregation may also transform into graphene layers by high-temperature annealing. As these techniques does not allow large scale solution stationed systems. As a consequence, sheets of reduced graphene oxide (rGO) (Qi et al. 2010), chemically functionalized graphene, have been prepared by first oxidising and then exfoliating graphite oxide (CFG). Sophisticated processes use a variety of reduction agents, akin as hydrazine, powerful alkaline media, Fan et al. (2008) vitamin C, Dua et al. (2010) with bovine serum albumin (BSA). The Hummers process, which involves reacting graphite with a solution of strong sulfuric acid and potassium permanganate ($KMnO_4$), produces graphene oxide (GO) sheets as the precursor to rGO (H_2SO_4) (Eda et al. 2010). Unfortunately, due to partial reduction and the existence of many defects that interfere with the sp^2 network, rGO exhibits electric conductivity and it has many orders less than the pure graphene value (Liu et al. 2008). The arc discharge approach (Wu et al. 2010) and synthesis of graphene in solution involving chemical and discharge approach from non-graphitic materials, like the reaction of ethanol with sodium tracked by the method of pyrolysis (Choucair et al. 2009) as well as the production made from polyaromatic hydrocarbons similar to graphene by organic means. Some more graphene-related substances include zero-dimensional graphene quantum dots, Ponomarenko et al. (2008), Ritter and Lyding (2009), one-dimensional graphene nanoribbons (GNRs), and graphene nano-meshes, which have also been manufactured or synthesised.

There are several ways to create graphene, including top-down and bottom-up strategies. The chemical reduction strategy is effective, straightforward and frequently utilised for mass manufacturing of graphene, but it uses toxic and harsh chemicals that are harmful to both mankind and the environment. Yet, among other electrochemical exfoliations, it is thought of as a safe, secure and substrate-free method for producing graphene. It uses raw materials including pencil cores, graphene rods, pellets and extract from palm oil leaves, among others. As a result, the strategy results in the substitution of less valuable or wasted raw materials with more useful ones. Popular because it saves time, poses no risk of detonation, emits no harmful vapours and prevents environmental pollution. Latest publications validate the fact that if we follow green methods with high speed of synthesis of GO, then it has more benefits than the conventional Hummers technique; however, the Hummers method is still widely used.

5.2 Graphene-Based Materials

Because it is hydrophobic and doesn't dissolve much in most solvents, pure graphene is a substance. But the solubilisation of graphene is what the processing of graphene composites is mostly concerned with (Fan et al. 2014; Park et al. 2008; Li et al. 2008). Different functional groups have been chemically modified, functionalized, or functionalized noncovalently to the carbon backbone of graphene to increase its solubility (Bai et al. 2009; Stankovich et al. 2006a, b; Chen et al. 2011; Qi et al. 2010).

Dispersing chemicals are required to scatter hydrophobic graphene sheets or graphite particles in water. Reduction of graphene oxide (GO) under controlled conditions using hydrazine, keeping the pH of the solution at about 10 using ammonia, chemically condensed graphene oxide (rGO), also mentioned that chemically altered graphene, lead to homogeneous aqueous suspension. This is likely because rGO has oxygen-containing groups. Besides, with a value of 0.5 mg/mL, rGO's consequent solubility in water is very poor (Kim and Jeong 2010) A homogenous aqueous suspension may be produced by the sequential chemical reaction of GO sheets in water with

KOH and subsequently with hydrazine. Additionally, GO was given p-phenyl-SO_3H groups, as well as after reduction, the weakly sulfonated rGO could be efficiently diffused by prevailling water at modest concentrations (2 mg/mL) in the pH range of 3–10 (Liang et al. 2009). Functional molecules interact with oxygenated groups like carboxyl groups at the edges, and also groups like epoxy, –OH and carbon–carbon unsaturated groups in the basal plane of GO or rGO; on the surface of graphene as part of the covalent alteration of the material.

Non-Covalent functionalization has a more detrimental effect on the edifice of graphene and its byproducts than covalent functionalization, which is created on the van der Waals force or the collaboration between rGO as well as stabilisers. It also allows for the alteration of the electrical properties and solubility of these materials. Hydrogen peroxide was used to reduce poly(sodium 4-styrenesulfonate) and graphene oxide (GO) dispersion in water, yielding the first noncovalently functionalized stable graphene dispersion (PSS). The rGO sheets were held in place in this experiment by the hydrophobic PSS backbone of the material, which was kept immersed in water by the hydrophilic sulfonate side groups.

Additionally, different biomolecules like single-strand DNA (ssDNA) have been used to maintain water-based suspensions of single-layer graphene sheets during the functionating process of graphene having van der Waals force as a base.

5.3 Graphene–Polymer Composites

Because of its remarkable machine-driven strength, outstanding temperature stability and high conductivity, graphene and its derivatives are indispensable filler constituents for polymer composites. Graphene-polymer composites' features and performances are affected not only by the quality of the filler graphene and polymer matrix, but also by the filler's dispersion, bonding to the matrix as well as filler-to-matrix ratio. These features are mostly determined by production methods. Solution mixing is one of the methods employed in the production of graphene-polymer composites (Stankovich et al. 2006a, b), melt blending (Son et al. 2010) and in-situ polymerization, which are analogous to conventional polymer processing. One of the most popular techniques for creating polymer composites is solution mixing since it is simple, doesn't call for any specific equipment and enables mass manufacturing. One of the major issues with solution forming mixture is the solubility or scattering of graphene sheets in the polymer solution. For instance, it is simple to combine water-soluble polymers like poly(vinyl alcohol) (PVA) with aqueous GO dispersions at varied concentrations (Sahu et al. 2021). Making graphene-polymer composites in organic solvents is difficult, however, since GO and rGO are poorly soluble in these environments.

5.4 Graphene–Semiconductor Nanomaterial Composites

Applications in the field of electronics, optics, and power storage and translation, like solar cells, Lithium-containing batteries, and supercapacitors, have increased the demand for graphene-semiconductor nanomaterial composites. The production and support of TiO_2 and other semiconductor nanomaterials use graphene-based templates such as ZnO, SnO_2, (Sahu et al. 2021), MnO_2, Co_3O_4, Fe_3O_4, Fe_3O_5, NiO, Cu_2O, CdS, RuO_2, CdS, and CdSe. These graphene-semiconductor nanomaterial composites can be synthesised by crystallisation during process, mixing of solution, microwave-based growth, electrochemical method of deposition, deposition using vapour, and other similar techniques rapid and efficient approach for creating composites of graphene and semiconductor NPs is solution mixing.

5.5 Graphene Synthesis

Figure 5.1 demonstrates a graphic illustration of the two main graphene manufacturing pathways, which are (i) top-down and (ii) bottom-up procedures. Graphene is created via top-down processes by breaking down bulk materials like graphite into their smallest components. Liquid-phase exfoliation, plasma etching, electrochemical exfoliation, ball milling, and chemical reduction are a few of the important top-down procedures. In contrast, the bottom-up synthesis technique consequences in the manufacture of a hexagonal configuration of graphene layers after the breakdown of pioneers having carbon (both gaseous and liquid). Some of the focused bottom-up methods often employed for the creation of graphene are chemical vapour deposition (CVD), thermal pyrolysis, and epitaxial growth. In the subsections below, commonly used production techniques utilising methods top-down and bottom-up will be briefly discussed.

5.5.1 Chemical Exfoliation

When compared to other popular methods like epitaxial growth, micromechanical cleavage, and Hummer's methods, chemical exfoliation is thought to be an incredibly efficient and economical top-down synthesis procedure. Exfoliation is used in the chemical exfoliation process to generate graphene from bulk graphite (Sahu et al. 2021). The number of layers and lateral dimension of the resulting graphene may be greatly adjusted by choosing an appropriate starting material of graphite (Wu et al. 2010).

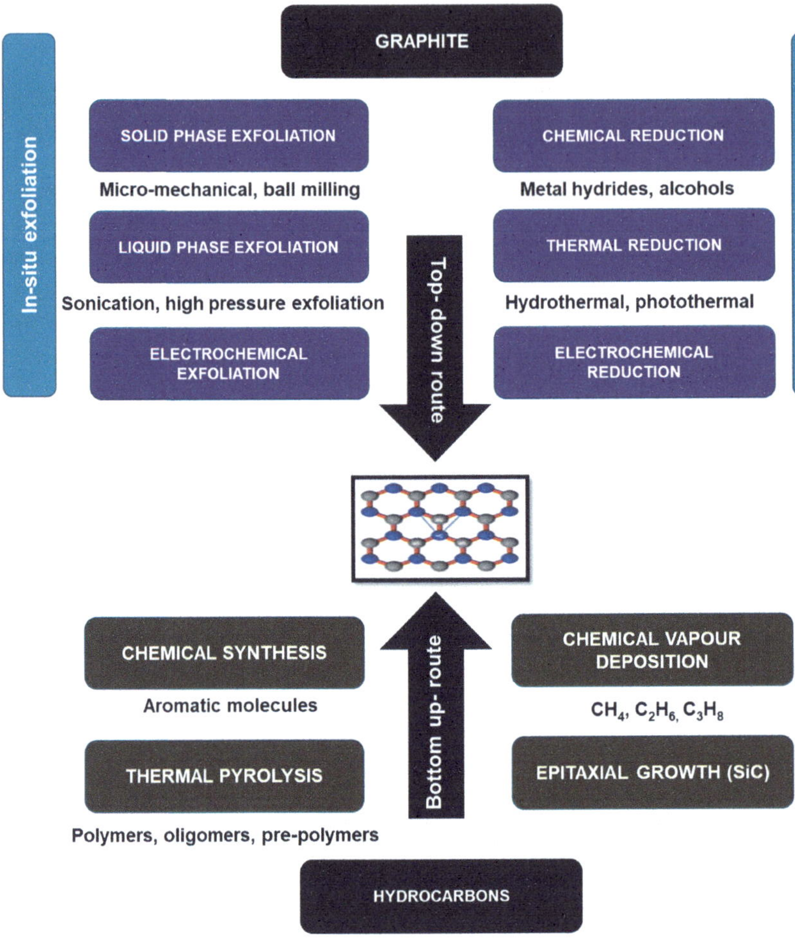

Fig. 5.1 Graphic depiction of most commonly used synthesis methods of graphene

Using reducing solvents or oxidation, graphite layers are separated during chemical exfoliation. Chemical exfoliation offers a huge promise for the synthesis of GO/rGO for more widespread commercial applications, but it is less profitable for producing graphene on a large scale than other commonly used methods due to the usage of hazardous chemicals and high production temperatures. As a result, additional study and research are needed to make this approach more practical for large-scale manufacturing while also improving its cost-effectiveness and environmental friendliness.

5.5.2 Mechanical Exfoliation

Graphene may be produced by mechanical exfoliation by means of a quantity of different top-down methods, including breakage using micromechanical method, ball milling, sonication technique, intercalation and liquid-mediated exfoliations.

Graphene's distance between layers and bonding energy is 3.34 A and 2 eV/nm^2, respectively and monolayers of graphene that are meticulously followed by weak van der Waals forces to form graphite can break easily and produce excellent-quality graphene flakes. However, this process is time-consuming and labour-intensive, restricting its usage to laboratory-based research. New methods, like ball milling and sonication, have been established to overcome these problems and are proving to be effective methods for the large-scale manufacturing of graphene. However, mechanical exfoliation has drawbacks like the process takes a lot of time and does not produce graphene sheets with uniform thickness. In addition, it is not an ascendable technique for producing high-quality graphene sheets, and a significant amount of investigation and expansion effort is still needed.

5.5.3 Electrochemical Exfoliation

Most recently, electrochemical exfoliation has shown to be a successful top-down method for producing graphene. Electrochemical exfoliation has a number of benefits over other synthesis methods, including a quick production process, minimal instrument cost and the ability to produce and modify high-quality graphene (Prekodravac et al. 2021). Fixed voltages are given to graphite electrodes throughout this process, which causes the van der Waals forces (5.9 kJ mol^{-1}) between graphite layers to diminish and eventually cause layer separation (Pingale et al. 2020). Figure 5.2 (Prakoso et al. 2020) shows a plan for the electrochemical exfoliation of graphene. Exfoliation is classified as either anodic (applying a positive bias) type or cathodic (applying a negative bias) type based on how graphene layers form on graphite electrodes in response to the applied voltage.

Aqueous solutions are typically utilised for performing graphite exfoliation, and while doing so, there are three crucial steps in the exfoliation process:

- Oxygen and hydroxyl radicals are synthesised by the electrolysis of water.
- Radicals of oxygen and hydroxyl migrate, and graphite edges open.
- Species of electrolyte intercalation, followed by gas generation for graphite expansion.

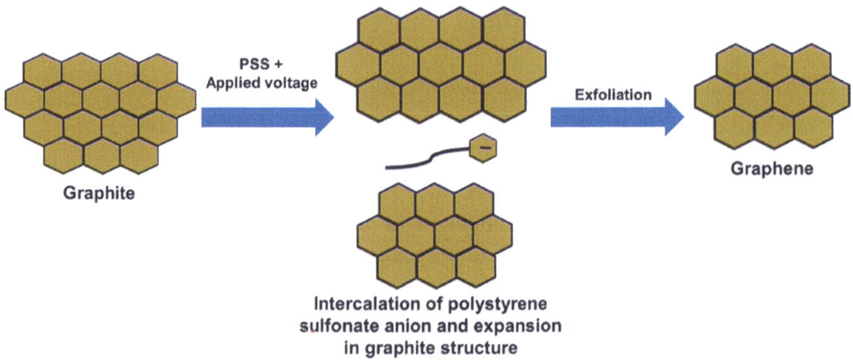

Fig. 5.2 Design of graphene synthesis via electrochemical exfoliation

5.5.4 Liquid Phase Exfoliation

High-quality graphene can be synthesised via top-down processes like CVD and epitaxial growth, but these approaches are inefficient for mass production.

5.5.5 Epitaxial Growth

Epitaxial growth on silicon carbide (SiC) is an efficient way for producing high-quality graphene with a uniformly large surface area and thickness. The very high strengthening temperature (1400 °C) and high Air pressure are responsible for the manufacture of highly organised and spotless epitaxial graphene films.

A novel synthesis technique for making high-quality, large-sized graphene in a low-vacuum situation. This technique slows down the pace at which graphene grows on the SiC support, lowering manufacturing costs (Table 5.1).

5.5.6 Problems Associated with These Methods

The methods of Brodie, Staudenmaier and Hummers had been used to manufacture graphite oxide for a considerable amount of time prior to the enhancement of oxidation and exfoliation efficacy by researchers. Chemical oxidation, however, results in the emission of numerous hazardous gases (e.g., NO_x and ClO_2) with reaction periods of hours or even days. In addition, the presence of highly reactive Mn_2O_7 intermediates in the Hummers method poses a detonation risk. Massive quantities of water are required to eradicate excess H_2SO_4 and $KMnO_4$ after oxidation, resulting in severe environmental pollution. Furthermore, heavy metal ions and Mn^{2+} impurities can be detected on GO sheets.

Table 5.1 Advantages and disadvantages of various synthesis techniques for graphene

Technique	Advantages	Disadvantages
Chemical exfoliation	Less price More yield Scalable	Defects rate high Functionalised Poor quality
Mechanical exfoliation	Monolayer creation Excellent quality Devoid of defects Big size	Low yield Non-scalable Labour intensive
Electrochemical exfoliation	Scalable High quality Low cost Environmentally friendly Less reaction time	Production of MLG Slight oxidation
Liquid phase exfoliation	Low cost Ascendable High quality Minor experimental conditions	Less monolayer yield More time required Nonuniform flakes Less size
Epitaxial growth	No transferring of substrate Unified incorporation Less defects Good quality	High cost Uncontrolled size Multi-layered graphene

Recent extensive use of the electrochemical exfoliation technique for the production of graphene nanosheets for various applications can be attributed to the technique's numerous advantages over others. Recent industrial manufacturing of few-layer graphene nanosheets establishes that the rate of graphene exfoliation surges with the concentration of intercalates, resulting in a change in colloidal conductivity. Graphene oxide is more environmentally benign than graphene because it contains functional oxygen-containing groups. In comparison to the conventional Hummers' method for the synthesis of GO, the electrochemical exfoliation of graphite is considered to be more efficient, environmentally friendly and crystallinity-enhancing. The electrochemical exfoliation method can be used to synthesise graphene from graphite extracted from spent batteries. The FTIR and XRD spectra confirmed the formation of a graphene structure.

5.6 Synthesis of Graphene Using Electrochemical Exfoliation

5.6.1 Using Pencil Electrode

The exfoliated pencil graphite electrode was used in the electrochemical synthesis of graphene oxide and graphene using an AC adapter to apply a high voltage between two pencil electrodes immersed in Na_2SO_4 aqueous solution. By use of chronoamperometry (CA), GO was successfully electrochemically reduced to rGO. This further led to the generation of the Tafel curve with the use of a linear sweep voltammetry (LSV).

5.6.2 Using Direct Electrodeposition

In this, two electrodes can be used to perform the electrochemical exfoliation. Since graphite is used as the anode and primary source of carbon, consistency is maintained with a platinum cathode. Different volume ratios of deionized water and DMF/H_2O solutions are taken and spiked separately and the material is electrochemically exfoliated. Processing of these generated dispersions is done with ultrasonic treatment (at a power of 250 W). After that, a polytetrafluoroethylene (PTFE) membrane was used to filter out the dispersions, and the residuals on the membrane were removed using vacuum filtering and deionized water. Overnight, the powder was dried in a vacuum (Losey et al. 2017). Un-exfoliated graphite flakes are removed by ultrasonication. Graphene paper was made by filtering graphene dispersions via a vacuum membrane, then peeling off the membrane and drying in a vacuum oven for a whole day.

5.7 Applications of Graphene in Various Fields

The applications of Graphene give a bigger picture as a whole and usage is based on its extraordinary mechanical, electrical and thermal properties. It is employed in electronic products, energy conservation, sensory products, coverings, composites, biomedical devices and many others. The surface zone is more biocompatible make it a striking substantial for drug distribution and tissue engineering applications. Graphene's exclusive possessions make it a promising material for many future technologies. Figure 5.3 shows applications of graphene in various fields.

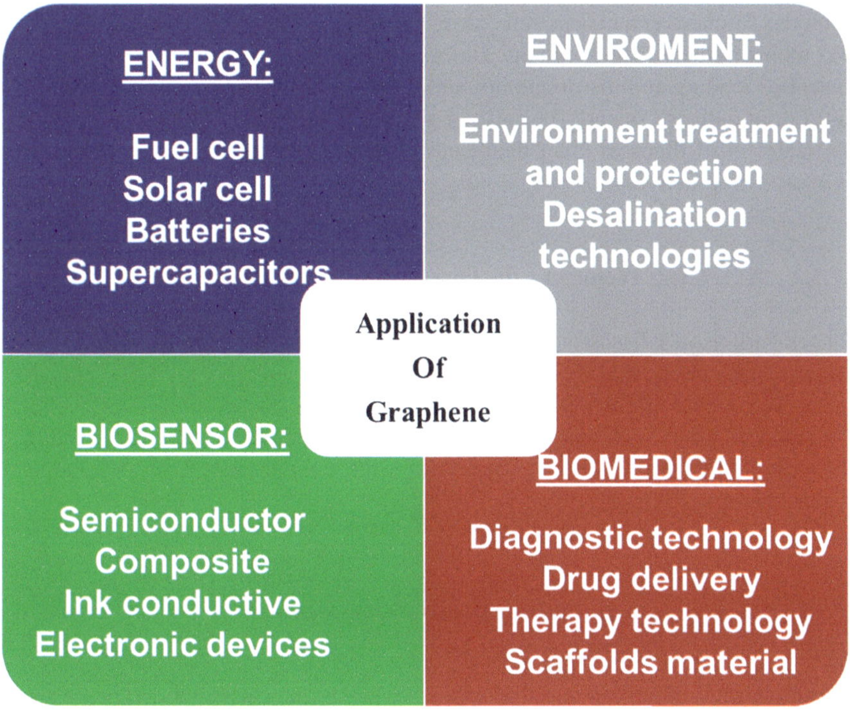

Fig. 5.3 Application of graphene

5.7.1 Drug/Gene Delivery and Cancer Therapy

Gene delivery is a capable technique to treat some hereditary ailments by transporting external DNA into cells. Graphene can also sense cancer cells in the initial phases of the illness. Furthermore, it can break them from increasing any additional load of cancer by overriding the correct construction of the tumour or causing autophagy which causes death of cancer cells.

5.7.1.1 As a Nanocarrier in Drug Delivery

GO is a perfect nanocarrier for effective medication and gene delivery and is created by vigorously oxidising graphite using Hummers' technique. The typical thickness of the GO utilised for drug administration is 1–2 nm, and its size ranges from a few nanometers to several hundred (Loh et al. 2010; Kovtyukhova et al. 1999; Sun et al. 2008). Large and planar sp^2 hybridised carbon domains, a highly precise surface area (2630 m^2/g) and oxygen-containing groups give GO its distinctive structural characteristics, which also make it exceptionally biocompatible, physiologically stable and capable of loading medicines or genes via chemical conjugation or physisorption methods. To overcome cancer cells' drug resistance, it is common clinical practise to combine numerous medications. Unfortunately, there isn't much research on the controlled loading and distribution of various pharmaceuticals using nanomaterial-based drug delivery systems because of technological difficulties. Other drugs for the treatment of illnesses other than cancer are also included in the exploration of GO-based drug delivery.

5.7.1.2 As a Gene Vector in Life-Threatening Diseases

Cancer, Parkinson's disease and cystic fibrosis are just a few of the genetically-based diseases that may be addressed by gene therapy. For gene therapy to be effective, a gene vector that inhibits DNA deprivation by nucleases and encourages cellular acceptance of DNA with more transfection efficiency is required (Yang et al. 2007). The lack of effective and secure gene vectors is the main obstacle to the development of gene therapy.

5.7.2 Biosensing and Biomolecule Detection

Biosensing and detection of biomolecules including thrombin, ATP, oligonucleotides, amino acids and dopamine may be accomplished using a variety of graphene derivatives, including pure graphene, GO, chemically concentrated GO (rGO) and doped graphene. Several distinct types of GO-based biosensors have been developed so far. The very effective fluorescence quenching capability of graphene has led to the development of a variety of novel biosensors based on FRET. Using graphene's unique electronic properties, FET-style biosensors have been created. The planned self-assembly of graphene biomolecules allows for the development of very sensitive biosensors for the uncovering of DNA and other chemicals. The use of graphene oxide (GO) as a matrix for matrix-assisted laser desorption/ionisation time of flight mass spectrometry (MALDI-TOF MS) has been described; GO's large surface area, high electrical conduction and exceptional size to load diverse biomolecules via chemical or physical interactions have run to the expansion of new biosensors based on an electrochemical concept (He et al. 2010).

5.7.3 In the Biological Imaging Field

Researchers are now looking at the viability of GO in optical and magnetic modalities for biological imaging. Dai et al. used the NIR fluorescence of GO to perform the first investigation on cellular absorption of PEG-modified GO loaded with medicinal medicines.

Lately, scientists have started creating smaller GO (with a size of 10 nm or fewer), also known as graphene quantum dots (GQDs), using a "bottom-up" strategy and chemical oxidation of graphite. These GQDs have inherent fluorescence and can be used for bioimaging. By hydrothermal cutting of GO, Pan et al. (2010) created GQDs with a blue emission colour. These as-obtained GQDs are substantially less fluorescent than II–VI type QDs and ordinary organic dyes (quantum yield less than 10%), which limits their usefulness in bioimaging applications. To address this issue, Eda et al. and his co-worker (Su et al. 2011) exposed the chemically created GQDs to hydrazine vapour for a brief period of time before observing much-improved fluorescence.

5.8 Impact of Different Parameters on Graphene Oxide Yield

Research on graphene has made great strides in recent years, and methods for synthesising graphene oxide electrochemically have been significantly improved (GO). A graphene derivative, such as GO, has caught the attention of scientists and researchers from all over the world because of its hydrophilic and readily functionalized characteristics. The electrochemical method is well-liked because it minimises environmental contamination, reduces the risk of explosion and saves time. Recent studies have shown that the green, quick and mass electrochemical synthesis of GO has more benefits than the conventional Hummers approach, but it is still important to investigate the impacts of reaction parameters. In this article, we cover recent research on the effects of different reaction conditions on the synthesis of GO sheets.

A description of the development, present difficulties and solutions for GO electrochemical synthesis techniques is also provided. We believe that this evaluation will provide those who desire to scale up the production of GO with some solid suggestions.

5.8.1 *Types of Electrolytes*

One of the most important essential variables affecting the yield of GO is the kind of electrolyte. Because each electrolyte has different oxidation capacities, each one creates a different quantity of GO. According to reports, molten salts, ionic solutions and inorganic solutions are common options for electrolytes used in the electrochemical process to create GO. Sulfuric acid (H_2SO_4), on the other hand, is a potent and straightforward proton donor ($pKa = 10$), which demonstrates its benefit as an electrolyte by refining the functionality of GO. After reaching a specific threshold oxidation degree, the composition and structure of GO do not significantly alter with the accumulation of an additional oxidising moiety. Thus, due to its quick exfoliation rate, SO_4^{2-} comparable size to the space between graphite layers, and the generation of SO_2 and O_2 gases, which further accelerate the exfoliation rate, H_2SO_4 is a very common electrolyte used to exfoliate graphite and synthesise GO. Quick exfoliation is caused by the powerful released gases, which increased the graphite layer when a low concentration of H_2SO_4 was applied, Wu et al. were able to produce few-layer graphene with a low oxidation. GO was effectively produced with simply 1 M $HClO_4$ and 1 M HNO_3, despite reports of the effects of adding 1 M H_2SO_4, 1 M $HClO_4$ and 1 M HNO_3 (Su et al. 2011).

When 1 M H_2SO_4 was employed, no GO peak was visible because of the rapid electrochemical exfoliation. Using inorganic salts ($(NH_4)_2SO_4$, Na_2SO_4, K_2SO_4, etc.), Parvez et al. showed the electrochemical exfoliation of graphite into little oxidation grade graphene sheets with a large quantity (>85%, 3 layers), and lateral size up to 44 m. However, the acidic electrolytes produced a blend of graphene and GO with mediocre quality and a larger lateral size. Consequently, utilising an inorganic salt or an acidic solution as the electrolyte will reduce or increase the degree of graphite oxidation, respectively.

5.8.2 Electrolyte Concentration

One of the main determinants of the yield of GO is also the concentration of the electrolyte. In other words, the concentration of the electrolyte affects how many intercalating ionic species are present. The graphene sheets will intercalate, expand, exfoliate, and oxidise at a quicker pace when there are more ionic species present. Also, as a reflection of modifications in the surface functioning, an increase in the electrolyte concentration may lead to an improvement in GO dispersion, a coarser surface shape and a decrease in GO sizes. Several researches showed that a higher electrolyte content boosted the degree of oxidation, which increased the output of GO. Yet, a minor quantity of water present in a highly concentrated electrolyte makes it inappropriate for GO production since it is the primary catalyst for the creation of oxygen-containing functional groups at the basal and margins of graphene. Hence, less GO will be produced at greater electrolyte concentrations. To put it another way, a high electrolyte concentration encourages the intercalation rate because it contains a lot of anions, which serve as intercalants, whereas a low concentration encourages oxidation and exfoliation because it contains a lot of water molecules. According to Dalal et al. (50), exfoliated graphite was mostly collected when the concentration of H_2SO_4 was raised and no GO was synthesised. The process was unsustainable after the 1.0 M intercalation/exfoliation, which occurred solely at the starting stage and was followed by the fastest intercalation at 2.0 M. Just a tiny portion of the sample had mild intercalation when the concentration was less than 0.50 M, whereas the majority of the sample remained unaffected. It was shown how to adjust the H_2SO_4 concentration in the electrolyte to control the degree of graphite oxide oxidation when extremely oxidised GO with a C/O ratio of less than 2 was attained in 40–60 weight per cent of H_2SO_4, whereas the highest oxidation level with a C/O ratio of 1.5–1.8 was attained in 50 weight percent of H_2SO_4.

5.8.3 Operating Temperature

As compared to other crucial reaction factors, the operating temperature had the least impact. The high-temperature resonance, according to the researchers, is brought on by graphene's negative thermal expansion coefficient. Yet, it has been shown that room temperature is the preferred setting for the high synthesis of GO, but this temperature might have some unfavourable impacts on the process. When additional energy was given to the ions by heat to intercalate into graphite, one of the evident results of the side was that the interlayer gap of graphite was increased. Moreover, once the interlayer gap was slightly raised and ion intercalation was also generated, the van der Waals interaction decreased. Thus, more research is urgently required to address the heating problems that decreased the amount of GO synthesis.

5.9 Green Approaches of Graphene—Few Examples

5.9.1 Electrochemical Exfoliation of Pencil Core

Pencil cores were electrochemically exfoliated in aqueous electrolytes to produce graphene oxide flakes in a straightforward, environmentally friendly, and economically advantageous manner (H_2SO_4 or H_3PO_4). In order to get GO flakes during the electrochemical exfoliation process, pencil cores were used as both the anode and cathode in a two-electrode electrochemical cell that was set up for the task. The two electrodes first had a static potential following a ramp-up. By changing the bias, it was possible to exfoliate both pencil electrodes. Without additional post-processing, the exfoliated graphene oxide flakes were recovered by vacuum filtering after being rinsed with ethanol and DI water. To shed further light on the shape and crystalline structure of the as-prepared exfoliated GO flakes, TEM, FESEM, AFM, Raman and XRD were used. In particular, it may be used to identify defects, ordered and disordered structures and layers of graphene. Raman spectroscopy is an essential and valuable technique for assessing the structure and quality of graphene. Overall, the electrochemical exfoliation of pencil cores has been used in this study to manufacture high-quality graphene oxide flakes in a straightforward, environmentally friendly and economically advantageous manner. Excellent electrocatalytic activity and toxicity tolerance for the oxygen reduction process in an alkaline solution are displayed by the exfoliated GO flakes.

5.9.2 Synthesis of Graphene Nanosheets

A single atomic plane taken from graphite, an unrolled single-walled carbon nanotube, or an extended flat fullerene molecule can all be used to describe graphene. The method helps to generate easy-going and good-quality graphene nanosheets with contaminating reduced material. The first and most important stage in this process is to create high-quality 2D graphene sheets since any remaining flaws would have a negative effect on graphene's electrical properties. Figure 5.4 shows the synthesis of graphene nanosheets from the waste material. By using electrochemical reduction of the exfoliated GO at a graphite electrode, we describe for the first time a simple and quick method for producing high-quality graphene nanosheets on a large scale. The more negative potential can overcome the energy barriers for the decrease of oxygen functionalities (–OH, C–O–C on the plane and –COOH on the edge), which allows the exfoliated GO to be efficiently reduced. While making graphene using Hummer's modified approach, we begin with exfoliated graphite oxide, which is created by oxidising graphite flakes in an acidic solution. This proves that surface-oxygenated species at GO are rapidly and permanently reduced, and that exfoliated GO may be reduced electrochemically at negative potentials (Novoselov et al. 2004; Charrier et al. 2002; Berger et al. 2004). It is reported to synthesise graphene nanosheets quickly and environmentally friendly utilising exfoliated graphite oxide (GO) as a precursor. Because of the rapid reduction rate, the electrochemically reduced graphene does, however, have certain flaws. By doing the electrochemical reduction at a high temperature or annealing the products, which are currently being done in our lab, these flaws might be avoided. The current technique may be utilised to create graphene nanosheets on a large scale and modify electrodes for use in electrocatalysis and biosensors that are being developed in the lab (Gomez De Arco et al. 2010). Electrochemically reduced graphene nanosheets were studied using transmission electron microscopy (TEM) and high-resolution transmission electron microscopy (HRTEM).

5.9.3 Palm Oil Leaves Extract

Preparation of graphite oxide: With a few modest adjustments, the process described was used to oxidise graphite. In essence, 2 g of graphite was introduced to H_2SO_4 for 1 h, and then 2 g of $NaNO_3$ was added gradually in an ice bath-equipped controlled temperature environment. Next, using a spatula, 9 g of $KMnO_4$ was added very gradually, giving a minute time gap between additions. The solution was then diluted with 184 ml of deionized (DI) water while being held at a regulated temperature of no higher than 80 °C. The mixture was then heated to 98 °C, let it to cool to ambient temperature, and then more H_2O_2 solution was added. To remove any $KMnO_4$ residue, the solution was then centrifuged and rinsed with HCl and DI water. Figure 5.5 shows the preparation of plant aqueous extract.

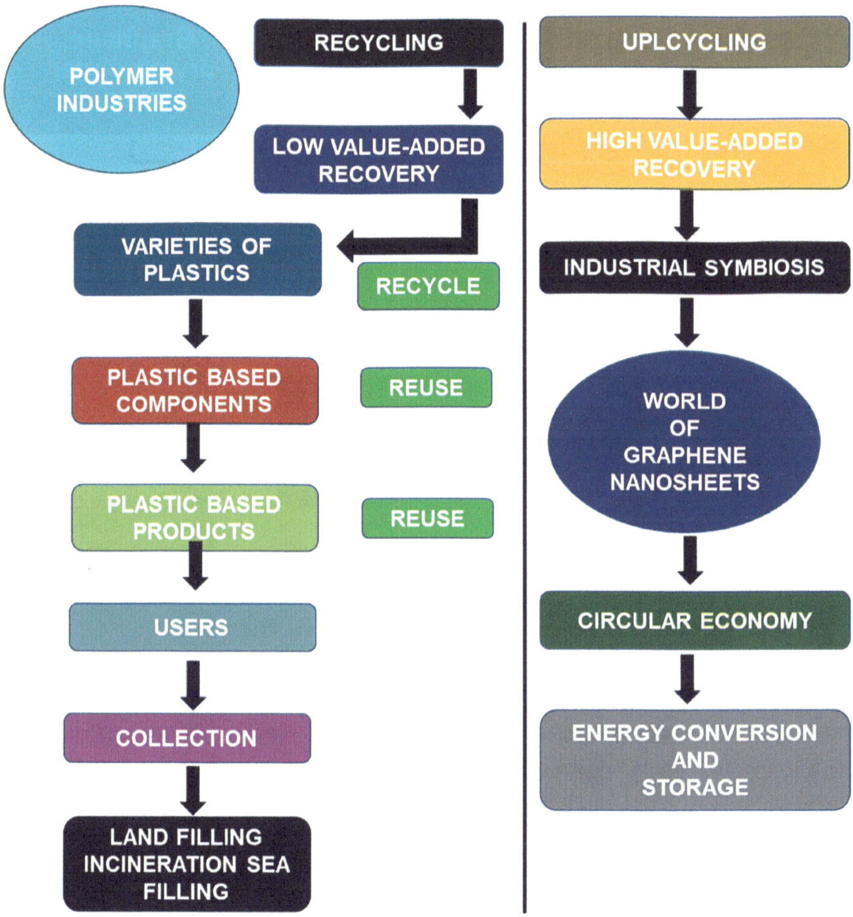

Fig. 5.4 The process representation for the waste plastics to get industrial symbiosis and circular economy for the conservation of EEE, i.e., ecology, economy and energy

Synthesis of rGO: In order to exfoliate the graphite oxide and create a GO solution, the graphite oxide powder was sonicated in distilled water for 1 h using a probe sonicator. The obtained palm oil leaf extract solution was then combined in 1:1 with the graphene oxide solution (Perumal et al. 2022). Finally, reduced GO (rGO) black powder was gathered and prepared for characterisation. Energy-dispersive X-ray (EDX) was used before and after the reduction procedure, and EDX characterisation was used to assess the oxygen and carbon content for both GO and rGO. Figure 5.6 shows a schematic diagram of the extraction of graphene from palm oil extract.

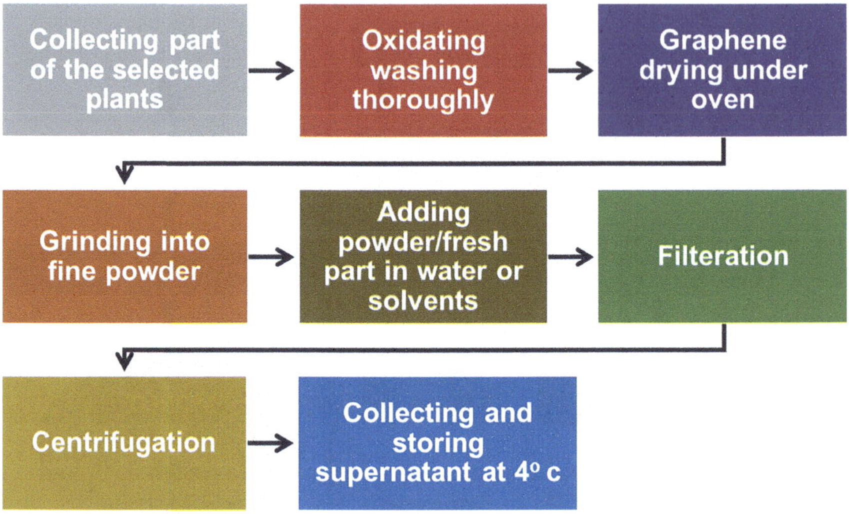

Fig. 5.5 Preparation of plant aqueous extracts for the green synthesis of rGO using plant extracts

5.10 Graphene Glass Hybrid—A Reactive Barrier

Graphene Glass Hybrid (GGH) was created by pyrolysis using a green process in which low-quality dates were utilised as a sustainable carbon source and glass trash was employed as the supporting material (GW). By batch and continuous studies, the GGH was shown to be a functional reactive material that is both affordable and acceptable for application in permeable barrier technology to treat groundwater that has been polluted with (Pb) lead and (TC) tetracycline. The GGH created revealed microporous structures that resembled 2D graphene sheets and were placed on an inorganic glass substrate. These structures were capable of removing Pb and TC from the aqueous media. Through the use of Fourier Transform Infrared (FTIR), Transmission Electron Microscopy (TEM), X-ray Diffraction (XRD), Raman spectroscopy, electron diffraction measurements, X-ray photoelectron spectroscopy, Energy-Dispersive X-ray (EDX) and surface area measurements, the GGH composite that was thus created was characterised. The results showed that the Langmuir model provided a better match for the sorption data than the Freundlich model; from the standpoint of kinetics, the pseudo-second-order model provided a better fit for the experimental data than the pseudo-first-order model. A number of column tests were conducted to evaluate the GGH reactive medium's performance in continuous mode (Graimed et al. 2022).

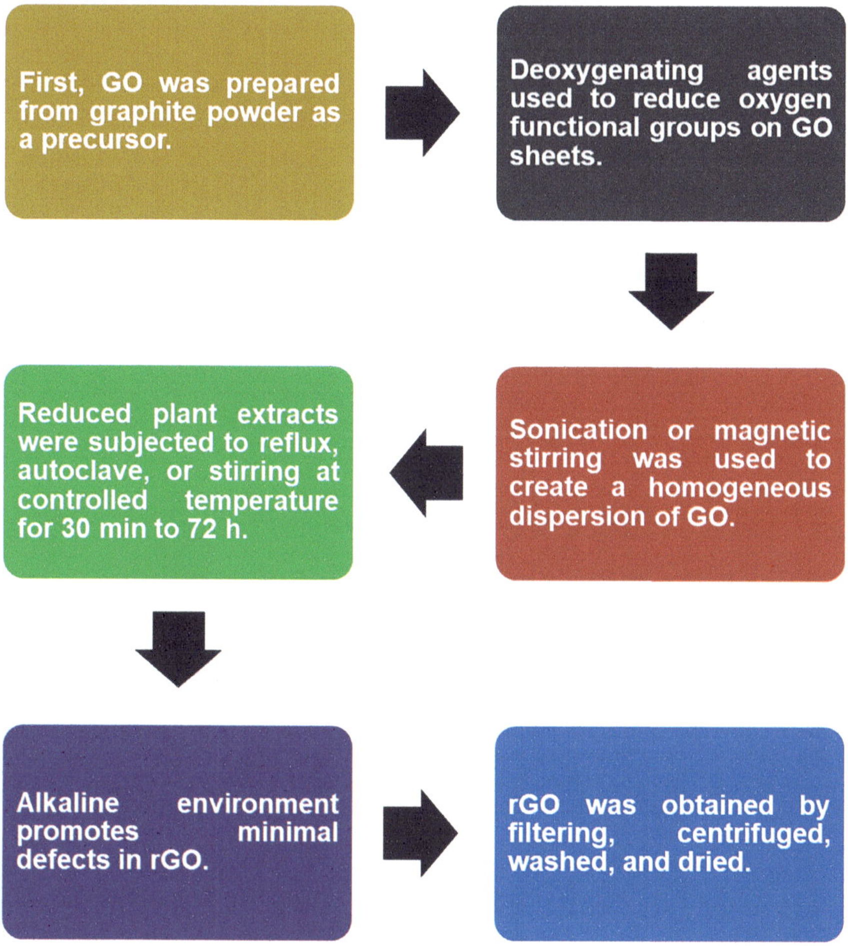

Fig. 5.6 Synthesis of graphene from palm oil leaf extract

Synthesis of Graphene Oxide:

By using glass waste to graphitize date syrup, the Graphene Glass Hybrid (GGH) was created. The glass trash was collected, rinsed, crumpled, sifted (with 1–2 mm range mesh) and then added to the date syrup in a 2:5 ratio. After that, date juice and glass scraps were combined. A furnace would be used to heat the prepared mixture. From the glass scraps, graphene oxide was extracted (Fig. 5.7).

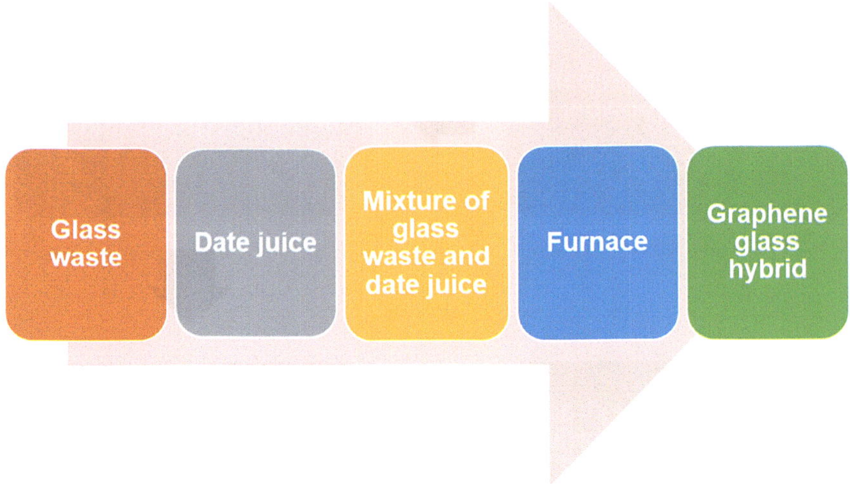

Fig. 5.7 Synthesis of graphene oxide from glass waste

5.11 Conclusion

The chemical composition of graphene is exceptional, and it has exceptional electrical, optical, thermal and mechanical capabilities. In order to address the growing demands for thin film processing, composite incorporation and device integration, its derivatives, such as epitaxially grown graphene, graphene oxide (GO), reduced GO (rGO) and exfoliated graphite, have been generated utilising a variety of synthetic processes. Unfortunately, managing the size and quantity of layers of graphene flakes when utilising exfoliation-based synthesis techniques is quite difficult. As evidenced by trend analysis data, exfoliation-based synthesis approaches have received more attention from scientists in recent years. Since graphene's discovery, the field of material scientists has been faced with the enormous challenge of producing superior-quality graphene at a reasonable cost with a controlled number of flaws and layers using environmentally friendly methods. This is because the performance of devices made of graphene is highly dependent on these production parameters. In summary, our work has demonstrated a simple, sustainable and cost-effective method for producing high-quality graphene oxide flakes by electrochemically exfoliating pencil cores. The exfoliated GO flakes exhibit excellent electrocatalytic activity and toxicity tolerance for the oxygen reduction process in an alkaline solution. With the help of the present technology, electrodes for electrocatalysis and biosensors that are being developed in the lab might be modified and graphene nanosheets could be produced on a big scale. Graphene's manufacture via several exfoliation techniques (including solid, liquid and electrochemical) and the accompanying difficulties are covered, as are the material's many potential uses. Synthesis of graphene from pencil core and palm oil leaf extract has advanced in recent years, and this chapter details such developments at 4 °C.

References

Bai, H., Xu, Y., Zhao, L., Li, C., Sh, G.: Non-covalent functionalization of graphene sheets by sulfonated polyaniline. Chem. Commun. (camb.)Commun. (camb.) **13**, 1667–1669 (2009)

Bekarevich, R.V., Miura, S., Ogino, A., Rahachou, A.U., Nagatsu, M.: The effect of substrate on the low-temperature carbon nanomaterials growth by microwave excited surface-wave plasma chemical vapor deposition. J. Phys.: Conf. Ser. **417**(012042), 1–3 (2013)

Berger, C., Song, Z., Li, T., Li, X., Ogbazghi, A.Y., Feng, R., Dai, Z., Marchenkov, A.N., Conrad, E.H., First, P.N., De Heer, W.A.: Ultrathin epitaxial graphite: 2D electron gas properties and a route toward graphene-based nanoelectronics (2004). arXiv:cond-mat/0410240

Charrier, A., Coati, A., Argunova, T., Thibaudau, F., Garreau, Y., Pinchaux, R.: Solid-state decomposition of silicon carbide for growing ultra-thin heteroepitaxial graphite films. J. Appl. Phys. **92**(5), 2479–2484 (2002)

Chen, D., Li, L., Guo, L.: An environment-friendly preparation of reduced graphene oxide nanosheets via amino acid. Nanotechnology **22**(32), 325601 (2011)

Choucair, M., Thordarson, P., Stride, J.A.: Gram-scale production of graphene based on solvothermal synthesis and sonication. Nat. Nanotechnol.nanotechnol. **4**(1), 30–33 (2009)

Chowdhury, S.M., Lalwani, G., Zhang, K., Yang, J.Y., Neville, K., Sitharaman, B.: Cytotoxicity of graphene nanoribbons. In: ECS Meeting Abstracts, p. 1218–1218. MA (2013)

Dua, V., Surwade, S.P., Ammu, S., Agnihotra, S.R., Jain, S., Roberts, K.E, Park, S., Ruoff, R.S., Manohar, S.K.: All-organic vapor sensor using inkjet-printed reduced graphene oxide. Angewandte Chemie (International ed. in English) **49**(12), 2154–2157 (2010)

Eda, G., Lin, Y.Y., Mattevi, C., Yamaguchi, H., Chen, H.A., Chen, I.S., Chen, C.W., Chhowalla, M.: Blue photoluminescence from chemically derived graphene oxide. Adv. Mater. (Deerfield Beach, Fla.) **22**(4), 505–509 (2010)

Fan, Z., Li, Y., Li, X., Fan, L., Zhou, S., De-Cai, F., Yang, S.: Surrounding media sensitive photoluminescence of boron-doped graphene quantum dots for highly fluorescent dyed crystals, chemical sensing and bioimaging. Carbon **70**, 149–156 (2014). https://doi.org/10.1016/j.carbon.2013

Geim, A.K., Novoselov, K.S.: The rise of graphene. Nat. Mater. **6**(3), 183–191 (2007)

Gomez De Arco, L., Zhang, Y., Schlenker, C.W., Ryu, K., Thompson, M.E., Zhou, C.: Continuous, highly flexible, and transparent graphene films by chemical vapor deposition for organic photovoltaics. ACS Nano **4**(5), 2865–2873 (2010). https://doi.org/10.1021/nn901587x

Graimed, B.H., Tark Abd Ali, Z.: Green approach for the synthesis of graphene glass hybrid as a reactive barrier for remediation of groundwater contaminated with lead and tetracycline. Environ. Nanotechnol. Monit. Manag. **18**, 100685 (2022)

Hass, J., de Heer, W.A., Conrad, E.H.: The growth and morphology of epitaxial multilayer graphene. J. Phys.

He, Q., Sudibya, H.G., Yin, Z., Wu, S., Li, H., Boey, F., Huang, W., Chen, P., Zhang, H.: Centimeter-long and large-scale micropatterns of reduced graphene oxide films: fabrication and sensing applications. ACS Nano **4**(6), 3201–3208 (2010)

Khoa, N.D., Anh, V.M., Duc, N.D.: Nonlinear dynamic response and vibration of functionally graded nanocomposite cylindrical panel reinforced by carbon nanotubes in thermal environment. J. Sandwich Struct. Mater. **35**(4), 110–122 (2019)

Kim, H.J., Kim, I., Lee, K.W., Yoon, D.H., Han, S.D.: Preparation and characterization of rGO electrode for electrochemical gas sensor. In: ECS Meeting, vol. 28. MA (2020)

Kim, H., Jeong, Y.G.: Polylactide/exfoliated graphite nanocomposites with enhanced thermal stability, mechanical modulus, and electrical conductivity. J. Polym. Sci., Part b: Polym. Phys. **48**(8), 850–858 (2010)

Kovtyukhova, N.I., Ollivier, P.J., Martin, B.R., Mallouk, T.E., Chizhik, S.A.: Chem. Mater. **11**(3), 771–778 (1999)

Lei, Y., Ossonon, B., Perreault, J., Tavares, A.: Electrochemical characterization of graphene-like materials obtained by electrochemical exfoliation of graphite. In: ECS Meeting Abstracts, p. 813. MA (2020)

Lei, Y., Ossonon, B.D., Perreault, J., Tavares, A.C.: Electrochemical characterization and application of graphene oxide materials obtained by electrochemical exfoliation of graphite. ECS Meeting Abstracts, MA2022-01(12), 842 (2022). https://doi.org/10.1149/ma2022-0112842mtgabs
Li, M., Jeong, Y.G.: Polyethylene terephthalate)/exfoliated graphite nanocomposites with improved thermal stability, mechanical and electrical properties. Compos. A Appl. Sci. Manuf. **42**(5), 560–566 (2011)
Li, D., Müller, M.B., Gilje, S., Kaner, R.B., Wallace, G.G.: Processable aqueous dispersions of graphene nanosheets. Nat. Nanotechnol.nanotechnol. **3**(2), 101–105 (2008)
Liang, J., Yi, H., Zhang, L., Wang, Y., Ma, Y., Guo, T., Chen, Y.: Molecular-level dispersion of graphene into Poly(vinyl alcohol) and effective reinforcement of their nanocomposites. Adv. Func. Mater. **19**, 2297–2302 (2009)
Liu, N., Luo, F., Wu, H.: One-step ionic-liquid-assisted electrochemical synthesis of ionic-liquid-functionalized graphene sheets directly from graphite. Adv. Funct. Mater. **18**(10), 1518–1525 (2008)
Loh, K.P., Bao, Q., Eda, G., Chhowalla, M.: Graphene oxide as a chemically tunable platform for optical applications. Nat. Chem. **2**(12), 1015–1024 (2010)
Losey, M.W., Kelly, J., Badgayan, N., Sahu, S., Sreekanth, P.R.: Electrodeposition. Elsevier eBooks (2017). https://doi.org/10.1016/b978-0-12-803581-8.10137-7
Mattevi, C., Eda, G., Agnoli, S., Miller, S., Mkhoyan, K.A., Celik, O.: Evolution of electrical, chemical, and structural properties of transparent and conducting chemically derived graphene thin films. Adv. Funct. Mater. **19**(16) (2009)
Nasirpouri, F., Alipour, K., Daneshvar, F., Sanaeian, M.: Electrodeposition of anticorrosion nanocoatings. In: Corrosion Protection at the Nanoscale, pp. 473–497 (2020)
Novoselov, K.S., Geim, A.K., Morozov, S.V., Jiang, D., Zhang, Y., Dubonos, S.V., Grigorieva, I.V., Firsov, A.A.: Electric field effect in atomically thin carbon films. Science (New York, N.Y.) **306**(5696), 666–669 (2004)
Pan, D., Zhang, J., Li, Z., Wu, M.: Hydrothermal route for cutting graphene sheets into blue-luminescent graphene quantum dots. Adv. Mater. (Deerfield Beach, Fla.) **22**(6), 734–738 (2010)
Park, S., An, J., Piner, R.D., Jung, I., Yang, D., Velamakanni, A.: Aqueous suspension and characterization of chemically modified graphene sheets. Chem. Mater. **20**(21), 6592–6594 (2008)
Perumal, D., Albert, E.L., Abdullah, C.A.: Green reduction of graphene oxide involving extracts of plants from different taxonomy groups. J. Compos. Sci. **6**(2), 58 (2022)
Pingale, A., Owhal, A., Katarkar, A., Belgamwar, S., Rathore, J.: Facile synthesis of graphene by ultrasonic-assisted electrochemical exfoliation of graphite. Mater. Today: Proc. (2020). https://doi.org/10.1016/j.matpr.2020.10.045
Ponomarenko, L.A., Schedin, F., Katsnelson, M.I., Yang, R., Hill, E.W., Novoselov, K.S., Geim, A.K.: Chaotic Dirac billiard in graphene quantum dots. Science (New York, N.Y.) **320**(5874), 356–358 (2008)
Prakoso, B., Ma, Y., Stephanie, R., Hanif Hawari, N., Suendo, V., Judawisastra, H., Zong, Y., Liu, Z., Sumboja, A.: Facile synthesis of battery waste-derived graphene for transparent and conductive film application by an electrochemical exfoliation method. RSC Adv. **10**, 10322–10328 (2020)
Prekodravac, J.R., Kepić, D.P., Colmenares, J.C., Giannakoudakis, D.A., Jovanović, S.P.: A comprehensive review on selected graphene synthesis methods: from electrochemical exfoliation through rapid thermal annealing towards biomass pyrolysis. J. Mater. Chem. C Mater. **9**(21) (2021)
Qi, X., Pu, K.Y., Li, H., Zhou, X., Wu, S., Fan, Q.L., Liu, B,. Boey, F., Huang, W., Zhang, H.: Amphiphilic graphene composites. Angewandte Chemie (International ed. in English) **49**(49), 9426–9429 (2010)
Ritter, K.A., Lyding, J.W.: The influence of edge structure on the electronic properties of graphene quantum dots and nanoribbons. Nat. Mater. **8**(3), 235–242 (2009)
Sahu, K.T, Ranjan, P., Kumar, P.: Chemical exfoliation synthesis of boron nitride and molybdenum disulfide 2D sheets via modified Hummers' method. Emerg. Mater. **4**(3), 645–654 (2021)

Son, J.Y., Shin, Y.H., Kim, H., Jang, H.M.: NiO resistive random access memory nanocapacitor array on graphene. ACS Nano **4**(5), 2655–2658 (2010)

Stankovich, S., Dikin, D.A., Dommett, G.H., Kohlhaas, K.M., Zimney, E.J., Stach, E.A., Piner, R.D., Nguyen, S.T., Ruoff, R.S: Graphene-based composite materials. Nature **442**(7100), 282–286 (2006a)

Stankovich, S., Piner, R.D., Chen, X., Wu, N., Nguyen, S.T., Ruoff, R.S.: Stable aqueous dispersions of graphitic nanoplatelets via the reduction of exfoliated graphite oxide in the presence of poly(sodium 4-styrenesulfonate). J. Mater. Chem. **16**(2) (2006b)

Štengl, V., Henych, J., Bludská, J., Ecorchard, P., Kormunda, M.: A green method of graphene preparation in an alkaline environment. Ultrason. Sonochem.. Sonochem. **24**, 65–71 (2015)

Su, C.Y., Lu, A.Y., Xu, Y., Chen, F.R., Khlobystov, A.N., Li, L.J.: High-quality thin graphene films from fast electrochemical exfoliation. ACS Nano **5**(3), 2332–2339 (2011)

Sun, X., Liu, Z., Welsher, K., Robinson, J.T., Goodwin, A., Zaric, S., Dai, H.: Nano-graphene oxide for cellular imaging and drug delivery. Nano Res. **1**(3), 203–212 (2008)

Torres, L., Gomez Armas, L., Carlos Seabra, A.: Optimization of micromechanical cleavage technique of natural graphite by chemical treatment. Graphene **03**(01) (2014)

Wu, Y., Wang, B., Ma, Y., et al.: Efficient and large-scale synthesis of few-layered graphene using an arc-discharge method and conductivity studies of the resulting films. Nano Res. **3**, 661–669 (2010). https://doi.org/10.1007/s12274-010-0027-3

Yan, X., Cui, X., Li, B., Li, L.S.: Large, solution-processable graphene quantum dots as light absorbers for photovoltaics. Nano Lett. **10**(5), 1869–1873 (2010)

Yan, S., Zhang, Q., Wang, W.-J., Li, B.-G.: Preparation of CO_2-switchable graphene dispersions and their polystyrene nanocomposite latexes by direct exfoliation of graphite using hyperbranched polyethylene surfactants. **7**, 4881 (2016)

Yang, Z.R., Wang, H.F., Zhao, J., Peng, Y.Y., Wang, J., Guinn, B.A., Huang, L.Q.: Recent developments in the use of adenoviruses and immunotoxins in cancer gene therapy. Cancer Gene Ther. **14**(7), 599–615 (2007)

You, D., Rotkin, S.V.: Non-Equilibrium Thermodynamics approach for QED heat conductance between graphene and SIC substrate. ECS Meeting Abstracts, MA2015-01(9), 851 (2015). https://doi.org/10.1149/ma2015-01/9/851

Yu, A., Ramesh, P., Itkis, M.E., Bekyarova, E., Haddon, R.C.: Graphite nanoplatelet-epoxy composite thermal interface materials. J. Phys. Chem. C **111**, 7565–7569 (2007)

Zhang, Y., Tan, Y.W., Stormer, H.L., Kim, P.: Experimental observation of the quantum Hall effect and Berry's phase in graphene. Nature **438**(7065), 201–204 (2005)

Chapter 6
Mechanism of Synthesis for Graphene and Its Derivatives by Electrochemical Exfoliation

Atul Chaskar, Pravin Walke, Devyani Wakchoure, and Dhriti Shenoy

Abstract The first two-dimensional atomic crystal that is now known to us is graphene. Many of its material characteristics, including mechanical stiffness, strength, and elasticity, as well as exceedingly high electrical and thermal conductivity, are incomparable. There have been numerous discoveries in the study of graphene in recent years. One of the most recent areas of material research is the chemistry of graphene and its derivatives. The production of graphene and its derivatives in large quantities utilizing low-cost techniques is highly desirable and important for practical application. There are many methods for synthesizing materials, including chemical exfoliation, mechanical exfoliation, and chemical synthesis. Electrochemical exfoliation is considered a promising method among the several that have been used to make graphene materials since it offers high yield, great performance, low cost, and excellent upscalability. This chapter provides a concise and thorough assessment of current advancements in electrochemical exfoliation approaches, a strategic design for exfoliations, mechanisms, and graphene electrochemistry.

Keywords Graphene · Electrochemical exfoliation · Synthesis · Graphene oxide · Graphene derivative

6.1 Introduction

Nanomaterials have gained a lot of interest from various branches of the research community for their easy synthesis, and exceptional mechanical, magnetic, electronic, and chemical properties. Each nanomaterial has its unique properties depending on its chemical composition, dimension, shape, and morphology. One such class of nanomaterials is carbon-based nanomaterials. Carbon and its various

A. Chaskar (✉) · P. Walke · D. Wakchoure · D. Shenoy
National Centre for Nanosciences and Nanotechnology, University of Mumbai, Mumbai 400098, India
e-mail: achaskar25@gmail.com

Department of Chemistry, Institute of Chemical Technology Mumbai, Mumbai 400019, India

R. Khan et al. (eds.), *Electrochemical Exfoliation of Graphene and its Derivatives*, Engineering Materials, https://doi.org/10.1007/978-981-97-2128-3_6

allotropes have been widely explored in the past few decades, which has been emblazoned with ground-breaking discoveries from fullerene, graphene, and carbon nanomaterials to polymers. Among all these discoveries, the Nobel Prize-winning experiment regarding 2D material graphene from graphite started a new era of "Graphene". Graphene is a monolayer of graphite, having a 2D morphology where carbon atoms have a honeycomb-like arrangement. The carbon atoms here are sp^2 hybridized with long-range π-conjugation. Due to this, it shows excellent thermal and mechanical properties and has zero bandgap like semiconductors making it an electroactive material. Its transparency makes it a suitable candidate for touchscreen and solar cell applications. Due to the frequent electron–electron collisions in graphene, the electron–phonon scattering is actually weak. The electron–electron interaction also seems to affect the local equilibrium which is seen to be above the temperature of liquid nitrogen. In contrast to the metallic behaviour predicted for doped graphene, it was found that when electron transport through graphene constrictions was analyzed below 150 K, the resistance decreased as temperature increased, a property of insulators. Apart from this, in biological environments graphene has shown higher operational stability, improved catalytic activity, a larger surface-to-volume ratio for enzyme/drug loading, and a rapid electron transfer rate for enzyme/drug immobilization (Singh et al. 2022a, b).

Since 2004, graphene and related 2D materials have been highly explored and studied for their applications in various fields from drug delivery to environmental remediation, from energy storage to optoelectronics, and from agriculture to sensing. Graphene and its derivatives are also being explored for its potential application in designing rockets, space suits, etc. Some of the graphene derivatives studied widely are graphyne, graphadiynes, Heteroatom-doped graphene (graphene oxide, fluorographene, etc.), graphane, etc. Graphyne is a stable non-natural carbon allotrope having a 2D structure. It is formed by the insertion of acetylenic linkage $-C \equiv C-$, which combines sp and sp^2 hybridized carbon atoms, while graphadiynes are formed by the insertion of diacetylene bond $-C \equiv C-C \equiv C-$ between benzene rings of graphene molecule. The combination of the p hybridized carbon atom with the sp^2 hybridized carbon atom gives it unique properties such as larger π-conjugation, well-distributed porous structures, etc. (Ivanovskii 2013; Gao et al. 2019). The addition of heteroatoms in place of hydrogen to the one-atom-thick layer of graphene has been shown to change the properties of the graphene. The addition of heteroatoms such as oxygen, fluorine, nitrogen, etc., leads to lattice defects in the structure of pristine graphene and are used to tune the magnetic, optical, thermal, and mechanical properties of graphene (Kumar 2021).

Due to the increasing applications of graphene and related 2D materials, the need for large-scale synthesis of these materials with low cost and high quality is being explored. The top-down approach of nanomaterial synthesis like the scotch tape method has been widely studied as they are cheaper, less complex, and time-consuming than bottom-up approaches of epitaxial growth and chemical vapour deposition. Electrochemical exfoliation is one such emerging top-down approach for the synthesis of nanomaterials. The milder, more convenient, and easily controllable conditions, as compared to other synthesis methods, make electrochemical methods more favourable. The electrochemical reaction occurs on the surface of the layered bulk material at the electrode which is driven by the applied potential, exfoliating the single or multi-layered 2D nanosheets of graphene and its derivatives in a quick and controlled manner (Yang 2019; Yu et al. 2015; Su et al., 2011; Prekodravac et al. 2021). This chapter will provide a comprehensive overview of electrochemical exfoliation (EE) as a promising top-down approach for the large-scale synthesis of graphene and its derivatives. It will discuss the mechanism of EE, various types of EE, and the factors affecting the quality of the synthesized materials. The chapter will also focus on the different graphene derivatives that can be synthesized through EE, such as graphene oxide, fluorographene, and heteroatom-doped graphene.

6.2 Electrochemical Exfoliation of Graphene: Techniques and Mechanism

Electrochemical exfoliation is one of the emerging techniques used to make graphene. When compared to techniques like mechanical exfoliation, molecular assembly, and chemical vapour deposition (CVD), which are all types of wet chemical procedures, the electrochemical approach, which is a wet chemical procedure, could turn out less expensive for mass production. Additionally, electrochemical activation can be used in production to avoid using harsh chemicals, which could make the process of product purification easier. Electrochemically synthesized graphene can have a tunable level of oxidation and is of relatively good quality with few hole defects. Graphene produced electrochemically is anticipated to find use in industry for coatings, nanocomposites, energy storage, electronics, and a wide range of other uses (Yang et al. 2016; Ambrosi et al. 2014).

6.2.1 Experimental Setup and Mechanism of Exfoliation

An electrochemical setup used for graphite exfoliation typically comprises a graphite working electrode, counter electrode, reference electrode, electrolyte, and power supply as represented in Fig. 6.1. Graphite powders, rods, flakes, foil, or highly oriented pyrolytic graphite (HOPG) are utilized as the working electrode, whereas most frequently used counter electrodes include pt wire, mesh, plate, rods, or graphite. Graphite flakes can be affixed to conductive carbon tapes to create a conductive surface, which serves as the working electrode. They can also be compressed directly to make graphite plates or bonded to a tungsten wire with a silver pad. The typical experimental setup has the working electrode and counter electrode submerged in electrolytes with a specified distance between them. The application of either positive or negative voltage to the graphite working electrode depends on the intended exfoliation mechanism. Developing a controllable and scalable setup for commercial use is crucial as the effective and continuous transfer of electricity to each graphene layer in graphite remains challenging (Yang 2019; Yu et al. 2015).

Mechanism: The electrochemical exfoliation process is categorized into cathodic and anodic exfoliation mechanisms based on the type of voltage applied. An aqueous solution is always used to conduct anodic exfoliation of bulk layered materials. During anodic exfoliation, anions and co-intercalating species from the reaction mixture are intercalated into graphite. The positive bias allows for the intercalation of electrolyte ions or ions created during electrolysis into the layer-structured anode. This increases the distance between graphene sheets and helps with the eventual exfoliation of the sheets (Yang 2019; Yu et al. 2015). Munuera et al. described a low-cost and facile method for electrochemically exfoliating graphite in an aqueous medium by an anodic exfoliation mechanism. The process involves using solutions with ions such as sulfates, hydroxyls, chlorates, nitrates, and chlorides. Aqueous sulfate ions are particularly effective in aiding the anodic exfoliation of graphite to graphene, driven by mechanisms such as ion intercalation, oxidation, mechanical expansion, and gas evolution reaction. The relatively high applied voltage can initiate the process of anodic oxidation of water, which results in the production of OH^- ions and oxygen radicals. Water and certain molecules can enhance the intercalation of sulfate ions, which expand the edges of graphene layers in graphite during anodic exfoliation. The pressure exerted by anions on the graphene layers as a result of their kinetic energy eventually causes the layers to separate and form multilayer graphene. It is clear that the sonication procedure caused additional layer separation, and the newly formed layers had a bigger surface area. Sulfate ions, which are larger and

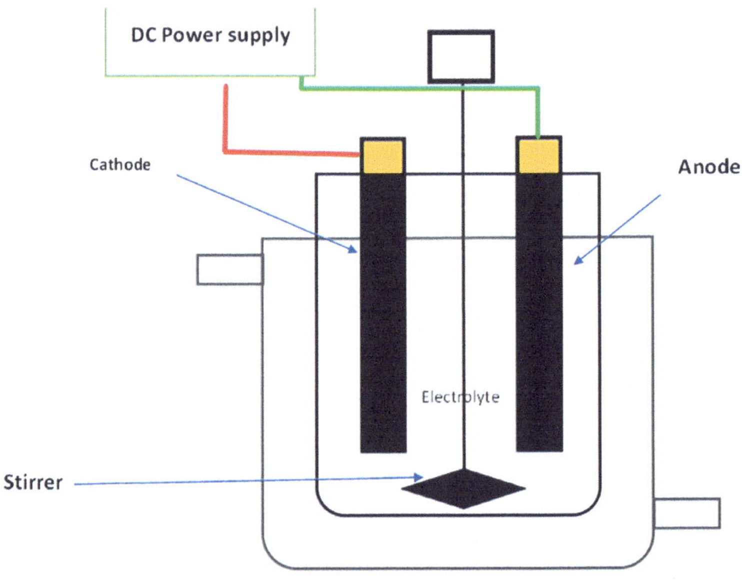

Fig. 6.1 Experimental setup for synthesis of graphene

heavier, intercalate and separate the layers held together by van der Waals forces, leading to the exfoliation of graphite (Munuera et al. 2017). Bandara et al. described a straightforward approach to enhance the yield of graphene nanosheets during the anodic exfoliation of graphite foil. Even if the structural quality of the graphene produced through anodic exfoliation tends to be higher, many times the yield is frequently low. A method to increase the yield of exfoliated graphene nanosheets involves pre-treating graphite foil with concentrated H_2SO_4 before anodic treatment in an aqueous electrolyte. This increase in yield was considered to be caused by the material undergoing more extensive intercalation during the electrolytic process. The low yield was likely caused by incomplete graphite anode intercalation during exfoliation. The structure of graphite foil has a lot of voids and interstitial spaces, therefore, methods that promote ion intercalation should be anticipated to result in higher yields. Graphite is exposed to a concentrated acid, it can fill voids and interstices with H_2So_4 molecules, making it hygroscopic. Anodic treatment of such pre-filled graphite in an aqueous electrolyte can lead to initial intercalation of SO_4^{2-} anions, promoting the further entry of anions and extensive delamination of the anode (Bandara et al. 2022).

Under a negative bias, some cations also intercalate layer-structured materials, these intercalations would cause the cathode to expand and exfoliate as well as increase the gap between the adjacent interlayers. Unlike anodic ones, to produce pristine 2D materials, bulk layered materials are subjected to cathodic exfoliation in an organic solution of alkali metal ions or quaternary ammonium cations under highly negative charge. A negative bias applied to the graphite working electrode attracts positively charged ions in the solution (such as Li^+) and any co-intercalating molecules. These intercalating species force the graphene sheets open, resulting in expansion and exfoliation. The quality of the exfoliated 2D materials as well as the intercalating process could be considerably impacted by solvent properties, size of anion, bias voltage, or current density (Yang 2019; Yu et al. 2015; Wu et al. 2017). Wu et al. investigated the cathodic exfoliation of graphite flakes and concluded that confining sandwich-structured graphite flakes to a restricted area facilitated uniform exfoliation and sufficient intercalation of electrolyte ions. The exfoliation efficacy of graphene is contingent not only upon the cationic radius, but also upon the nature of the solvent of the electrolyte, as suggested by the experimental findings. For successful electrochemical exfoliation, it is essential to achieve adequate intercalation of both charged ions and solvent. Numerous TAA^+ Tetraalkylammonium cations with varied alkyl chains, such as Tetramethylammonium perchlorate $TMAClO_4$, tetraethylammonium perchlorate $TEAClO_4$, tetrapropylammonium perchlorate $TPAClO_4$, tetrabutylammonium percholate $TBAClO_4$, tetrapentylammonium perchlorate $TPeAClO_4$, and tetrahexylammonium perchlorate $THAClO_4$ were investigated as the electrolyte salts in the search for an appropriate intercalating agent for cathodic exfoliation. These cations have high ionic diameters, allowing for a greater expansion of the interlayer space after intercalation. Additionally, many organic aprotic solvents with stronger polarity (including Acetone, dimethyl formamide (DMF), *N*-methyl-2-pyrrolidone (NMP), dimethyl sulfoxide (DMSO), and propylene carbonate PC) were chosen to investigate the ideal solvent to allow cations to intercalate into graphite electrodes effectively due to their improved capacity to stabilize charges. The selection of $TBAClO_4$/PC (tetrabutylammonium perchlorate/propylene carbonate) as the electrolyte allowed for effective intercalation and the quick expansion of graphite flakes within a minute due to the interaction between PC and TBA cations. To prevent the premature detachment of graphite flakes during the intercalation process, a porous metal mesh was employed to confine sandwich-structured graphite in a restricted area, which served as an exfoliated electrode. Initially, TBA ions in PC solvent migrate towards the graphite electrode surface and intercalate between graphite lattices under an electric field. As more TBA^+ ions intercalate over time, the GIC's layer gap gradually expands until it reaches saturation. The TBA layer diminishes the interlayer contact between adjacent graphite layers. TBA would disintegrate into gases at a greater potential, which causes GIC to expand its lattice structure very quickly. Vermicular thin sheets with an open porosity structure are produced from tightly packed graphite flakes. The study investigated the synergistic interactions of TAA cations and solvent molecules for exfoliation efficacy. Electrochemical intercalation of TBA in propylene carbonate resulted in efficient cation intercalation, uniform graphite exfoliation, and a high yield of few-layered graphene (Zhang et al. 2021).

Most of the non-aqueous electrolytes utilized in cathodic exfoliation protocols need a lengthy intercalation period, and post treatment is employed to exfoliate the intercalated graphene layers further. A novel technique would be deemed desirable if it employed electrolytes that are environmentally benign and aqueous-based, and also featured a decreased exfoliation duration. The utilization of aqueous electrolytes containing alkali metals is employed to achieve the desired outcome. It has been experimentally demonstrated that alkali metal cations in chloride-based aqueous electrolytes possess the ability to intercalate and undergo cathodic exfoliation of graphite, leading to the formation of graphene. The alkali metal cations that are utilized for this purpose include Li^+, Na^+, K^+, Rb^+, and Cs^+. However, the rate of exfoliation and yield are influenced by various factors such as the concentration of dissolved cations in the electrolyte, the size of the cations, and the applied voltage. Since van der Waals forces and electrostatic interactions are weak interlayer bonding in graphite, it takes a lot of energy to overcome them. Under the influence of the applied cathodic potential, the ionic species are driven to intercalate into the interlayer space between the graphene sheets. The exfoliation of graphite into individual graphene sheets then occurs as a result of gas evolution, which also increases the interlayer spaces inside the material and overcomes van der Waals forces. The successful exfoliation of graphite in an aqueous electrolyte depends on the critical interplay between a suitably high cathodic voltage and a high concentration of cations (Dalal et al. 2021). Dali et al. demonstrated the generation of high-quality graphene nanosheets through cathodic exfoliation of graphite in water-based electrolytes. In contrast to anodic exfoliation, cathodic exfoliation requires pre-expanded edges and interlayer spaces in graphite materials, such as graphite foil, to facilitate the effective intercalation of aqueous cations without the need for oxidizing conditions. The selection of the intercalating ion is crucial for successful intercalation in water. In terms of the electrolyte, quaternary ammonium-based cations with an intermediate size, such as tetrapropylammonium or hexyl trimethylammonium, exhibited the highest ability to expand the graphite cathode and yield graphene nanosheets in significant amounts. In the absence of oxidative processes at the graphite electrode during the electrolytic treatment, the interlayer gaps located along the periphery of the graphite are predominantly nonpolar and un-expanded. It will also prevent hydrated electrolyte cations from entering and leading to electrode expansion. Hydrated cations from the electrolyte can intercalate more easily due to the selective entry sites provided by the interlayer voids at the graphite edges. The latter serves as a molecular wedge and causes the interlayer space to enlarge more and more as gets farther away from the borders of the graphite particle. The reduction of water molecules which hydrate the cations produces hydrogen and increased gas pressure in interlayer spaces may also contribute to expansion. This process promotes further intercalation of hydrated cations, leading to the cleavage of layers and the production of enlarged products that can yield graphene nanosheets (García-Dalí et al. 2019).

6.3 Factors Affecting the Synthesis of Graphene and Its Derivative

6.3.1 Type of Graphite Material

The graphite structure is a key factor in the electrochemical exfoliation of graphite. To achieve successful exfoliation, crystalline graphite such as highly oriented pyrolytic graphite (HOPG), natural graphite flakes, graphite paper, or graphite rods are typically utilized (Liu and Aziz 2022). Efficient ion intercalation is required to increase the interlayer distance during the process of electrochemical exfoliation of graphite. Microstructures of graphite electrodes, including graphite particle sizes, defects, layer structures and thickness, composition, and proper pre-treatment, are expected to have an impact on ion intercalations. In order to prevent oxidative reaction-related damage during electrochemical exfoliation, graphite electrodes should be exfoliated rapidly during the graphene synthesis process (Liu et al. 2019).

6.3.2 Temperature

Hossain et al. reported the effect of temperature and the addition of H_2O_2 on produced graphene during electrochemical exfoliation of graphite. It was observed that as temperature increases, the defect or oxygen functional group present on the sheet decreases. Water is reduced at the negative electrode, releasing hydroxyl ions (OH–), and when H_2O_2 is added, peroxide ions (O_2^{2-}) are formed. During the electrochemical exfoliation of graphite, the expansion of carbon layers is promoted by the interaction of OH– and O_2^{2-} with graphite edges and grain boundaries. Moreover, the process leads to the production of SO_2 and O_2 gases due to residual SO_4^{2-} ions and water oxidation. The gases produced during the electrochemical exfoliation process, such as SO_2 and O_2, intercalate between the layers of graphite and exert a pressure that promotes the separation of adjacent layers. As the exfoliation temperature increases, there is a significant reduction in the amount of oxygen functional groups, which are a type of defect, resulting in a lower number of defects. The number of defects and the amount of oxygen functional groups can be altered by varying the temperature and the amount of H_2O_2 used (Hossain and Wang 2016). However, it has been noted that room temperature is preferred for the high synthesis of GO (Graphene Oxide), though the rise in temperature can have some impact on the process. The increase in interlayer spacing in graphite can be attributed to the enhanced thermal energy of intercalating ions, which promotes their intercalation into the graphite structure. The exfoliated synthesis yield increased by 4.5 times when compared to a room-temperature synthesis (Liu and Aziz 2022). In summary, the temperature and addition of H_2O_2 during electrochemical exfoliation of graphite have significant effects on the resulting properties of graphene materials. Higher temperatures promote the expansion of carbon layers and reduce the amount of oxygen functional groups present in the material. The addition of H_2O_2 can also alter the number of defects and oxygen functional groups. However, it is important to note that room temperature is preferred for the synthesis of GO, although some impact on the process may occur with increasing temperature. Understanding the impact of temperature and H_2O_2 can lead to improved production of high-quality graphene materials with desirable properties.

6.3.3 *Type of Solvent*

The choice of electrolyte is a key factor in the exfoliation method's effectiveness. During the exfoliation process, a good electrolyte should be able to perform two roles: it should possess the capacity to facilitate the intercalation of ions into the graphite layers and promote the separation of graphene sheets. Additionally, it must also be capable of effectively dispersing the resulting graphene flakes and minimizing their re-aggregation. The flake tendency for re-aggregation in solution results in poor flake exfoliation yields if the solvent/electrolyte combination is not carefully studied. The most effective solvents for the dispersibility of graphene flakes have been reported to be *N*-methyl pyrrolidone (NMP) and dimethylformamide (DMF). Two key factors determine effective graphene dispersions. The first depends on surface energy, choosing solvents with surface energies that are comparable to those of graphene. Another factor is the solvent's capability to stabilize the graphene sheets, which can be achieved through steric hindrance or electrostatic interactions, which are crucial for the efficient dispersion of graphene materials. Through steric hindrance, steric stabilization prevents the aggregation of graphene sheets when functionalization is present (Abdelkader et al. 2015; Danial et al. 2021).

6.3.4 Voltage Bias

Factors such as the voltage used during exfoliation can directly affect the conductivity and yield of the resulting graphene, making them important considerations for evaluating the effectiveness of large-scale graphite sheet exfoliation processes. At a lower voltage bias, the current density was insufficient to trigger the exfoliation process. The application of higher voltage during the electrochemical exfoliation process resulted in a decreased exfoliation yield, as the pre-treated graphite sheets were prone to breaking into large fragments. This was due to the strong electrochemical forces induced by the increased voltage Electron transport in EG (exfoliated graphene) is hindered by greater oxygen content. A longer duration of the electrochemical process may result in higher oxygen functionality formation due to the increased exposure of graphite electrodes to electrolytes at a lower voltage bias. The development of higher oxygen functionality is promoted by the higher voltage bias, which also accelerates nucleophilic reactions (Abdillah et al. 2021).

6.3.5 Role of Electrolyte Acid Concentration

Lowe et al. investigated the impact of electrolyte acid concentration on produced graphene flakes during exfoliation. They prepared electrochemical graphene oxide products with sulphuric acid in seven different concentration ranges. They observed that the oxygen evolution reaction occurring at the graphite surface during the electrochemical exfoliation process affected the structural changes, oxidation processes, and electrochemical corrosion of the resulting exfoliated graphene oxide (EGO). The conductivity of EGO was found to be highly dependent on the electrolyte's acidity.

The concentration, reactivity, and diffusion capability of the oxidant in the bulk solution impact the degree of oxidation during the exfoliation process of graphite. Oxygen evolution on the graphite electrode is protracted and enhanced under the low acid regime. Evolved oxygen likely contributes to CO_2 evolution and oxidizes the carbon backbone, leading to mechanical strain on the carbon lattice and a more amorphous graphite structure. Higher acid concentrations promote complete intercalation, water oxidation, and crystallinity retention. The structural and compositional variations across samples can be partially explained by the equilibrium between intercalation and water electrolysis. These findings could facilitate future alterations of EGO's chemical and structural properties, including its oxidation level, solution stability, and conductivity (Lowe et al. 2019).

6.3.6 *PH*

The pH of the electrolyte is a crucial factor in determining the quality of the electrode produced through electrochemical exfoliation of graphite. Lower pH values result in a slower exfoliation rate, which promotes the formation of high surface area multilayer van der Waals structures and low resistance paths for the electrolyte. Among sulfate salts, sulfuric acid is the optimal electrolyte for this process. These findings have significant implications for enhancing the electrochemical exfoliation process and producing more efficient electric double-layer capacitor (EDLC) electrodes for energy storage applications (Tamgadge and Shukla 2019). Biranje et al. investigated the kinetics of electrochemical exfoliation of graphite for synthesizing graphene oxide (GO) and found that electrolyte pH plays a vital role in the rate of GO exfoliation. The researchers proposed that the pH influences the dissociation of electrolytes into anions, which controls the exfoliation rate. Their study provides important insights into the electrochemical exfoliation process for GO synthesis (Biranje et al. 2023). Park et al. produced highly water-dispersible graphene by electrochemically exfoliating graphite and inducing mild in situ sulfate functionalization using $S_2O_8^{2-}$ ions. They found that under basic pH conditions, the electrochemical exfoliation led to sulfate functionalization of graphene, resulting in high-quality graphene with low oxygen content and high crystallinity. The sulfate group attached to the graphene sheet increased the surface charge, which stabilized the dispersion without affecting its original characteristics. This study highlights the potential of understanding the persulfate reaction to introduce sulfate functional groups in various materials, thereby increasing dispersion stability (Park et al. 2021).

Overall, these studies demonstrate the significant impact of electrolyte pH on the exfoliation process and properties of graphene materials.

6.4 Synthesis of Graphene Materials by Electrochemical Exfoliation (EE)

The development of cost-effective and efficient large-scale methods is necessary to produce high-quality 2D materials. To date, several top-down fabrication techniques have been developed, including liquid phase exfoliation, chemical exfoliation, and scotch tape method. It has been shown that top-down approaches have disadvantages such as being time-consuming, requiring hazardous chemical usage, and producing defects. Bottom-up techniques like epitaxial growth and chemical vapour deposition are suitable for manufacturing large-area ultrathin 2D materials. Epitaxial growth and CVD are costly and complicated processes that require high temperatures and vacuums. They also require an additional transfer step that may introduce impurities into 2D materials, making production difficult to control. Compared to other manufacturing methods, electrochemical exfoliation offers greater convenience and control since it is conducted under mild conditions (Yang 2019). The electrochemical exfoliation process for graphene and graphene materials is described below.

6.4.1 Graphene

Graphene is a flat, single layer of tightly arranged hexagonal carbon atoms bonded together in a sp^2 configuration, as in Fig. 6.2. It is considered as the foundational unit for other graphitic materials, such as one-dimensional nanotubes, three-dimensional graphite, and zero-dimensional fullerene. The unique properties of graphene, such as high specific surface area, exceptional electronic properties and electron transport capabilities, strong mechanical strength, and excellent thermal and electrical conductivity, have made it highly influential in various fields of science and technology (Campbell and Manning 2018; Chen et al. 2012; Zhen 2018).

Pingale et al. reported a simple method for producing graphene using electrochemical exfoliation assisted by ultrasonication in an acidic bath. The anode and cathode are made of pure graphite rods and platinum-coated titanium electrodes, respectively, and they are dipped vertically into the electrolyte (H_2SO_4 and H_2O). The electrochemical exfoliation process was aided by an external ultrasonic bath that had a 40 kHz ultrasonic frequency and a 180 W ultrasonic output power. The continuous exfoliation of the graphite electrode began when a +10 V DC voltage was supplied between the cathode and anode electrodes. The SO_4^{2-} and OH– ions begin to intercalate into the layers of graphite as soon as the 10 V power supply is supplied. Following that, as a result of SO_2 and O_2 bubbles developing within the graphite layers, graphite exfoliates. The aggregation of graphene in the electrolyte was reduced as a result of the external ultrasonic force and helped in the separation of graphene layers from the graphite electrode (Pingale et al. 2021). Petroleum

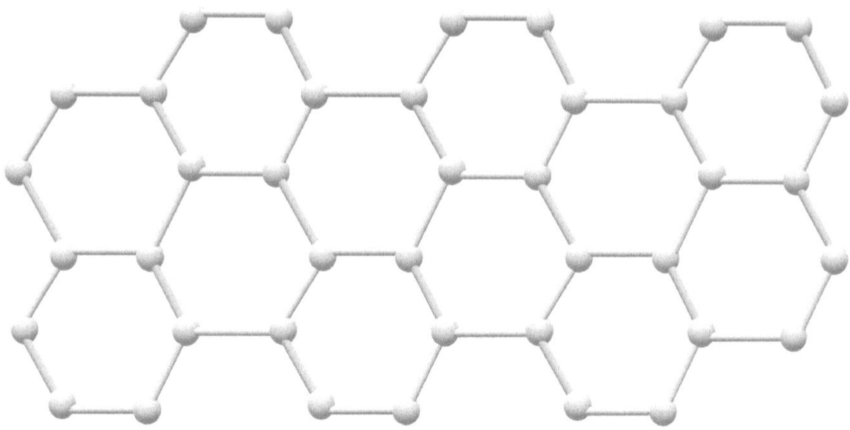

Fig. 6.2 Pristine graphene

coke offers another alternative route, as a precursor for the synthesis of graphene via electrochemical exfoliation (ECE). Petroleum coke, a solid carbonaceous byproduct of oil refining, is frequently used for heating or as an anode in the production of aluminium and steel. Saha et al. recently demonstrated that the electrochemical exfoliation (ECE) process in a compressed expandable chamber can produce high-quality graphene nanosheets on a large scale. The compressed reactor keeps electrical contact throughout the mixture while allowing for the expansion of the parent material during exfoliation. Nanosheets formed, as a result, are known as electrochemically exfoliated graphene (EEG), contain four steps as follows: (a) A starting material that exhibits higher water affinity and wettability is made possible by the nitric acid pre-treatment, consequently, the electrolyte solution is better exposed and utilized throughout the electrochemical exfoliation process (ECE). (b) Each parent material had significantly different particle sizes. Centrifuging was used to eliminate small particles that were less suitable for the process. The pre-reaction separation procedure involves 10 min of centrifugation at 4000 rpm. The sediment served as the ECE process's starting material. (c) Electrochemical exfoliation process, ECE process, was run at 12 V for two hours using the separated parent material (sediment) as the starting material. The material was compacted into a dialysis bag and clipped on both ends after being pre-treated and separated. Platinum wire, the working electrode, was inserted into the material, and copper mesh, the counter electrode, was wrapped around the dialysis bag. Both the working electrode and the counter electrode are immersed in the electrolyte. In order to exfoliate samples for EEG, a positive bias is applied to the working electrode. (d) A two-step centrifugation procedure was used to obtain the end product following the reaction. The first one was to remove the starting material that remains unreacted and the second one was for the separation of the non-graphitic compound. The sediment collected after the second centrifugation was the final product. This centrifugation process is required to separate the final product from the parent material and impurities, because of this the yield of graphene made using these methods is much lower than that of standard methods suggesting that coke is more challenging to exfoliate (Saha et al. 2021). Wu et al. synthesized high-quality graphene using a two-step dual-electrode approach, in that the first phase involved sufficiently intercalating ions into the graphite on both the cathode and anode at once, after that, microwave radiation was used to rapidly expand and exfoliate the intercalated products. By utilizing microwave irradiation, the volume of graphite was rapidly expanded hundreds of times in just a few seconds, saving significant time and energy compared to traditional heating techniques. More than 80% of the graphene flakes present in the end product were larger in size, had fewer defects, had less than four layers, and had low oxygen concentration. The proposed method achieved high yields (81% for anode and 76% for cathode) and allowed for the recycling of electrolytes, implying that this procedure has a considerable possibility of producing at a large scale, and serves as a significant standard for the industrial preparation of environmentally friendly and inexpensive graphene (Wu et al. 2022). The type of electrolyte, composition of different electrolytes, and their concentration affects the exfoliation process. As a result, the yield, morphology, and structure of prepared graphene change as we alter these parameters. Suhaila Hani Ilias et al. investigated the impact of different electrolyte acidity concentrations like H_2SO_4:KOH with different volume ratios (90:10, 80:20, and 70:30) on the graphene prepared by electrochemical exfoliation method. The exfoliation process involved applying a low voltage of +2.5 V for 5 min, followed by a high voltage of +10 V for 10 min. The graphite lattices are helped to expand by the starting low voltage given to the graphite rod so that OH– can intercalate. The graphite exfoliates into two different forms of fragments when a higher voltage is applied. Thin graphene was present on the surface of the electrolyte, while a thick graphite sheet settled at the bottom. Higher electrolyte concentration (H_2SO_4:KOH volume ratio 90:10) results in better graphene yield, reduced oxygen content, and fewer defects during electrochemical exfoliation (Ilias et al. 2021). While studying the exfoliation of graphene in inorganic electrolytes with various concentrations (such as $(NH_4)_2HPO_4$ and $(NH_4)_2SO_4$), Sharif et al. studied the effects of various electrolytes. The EEG produced with electrolytes containing $(NH_4)_2HPO_4$ was less defective than the other EEGs. This shows that electrolytes that contain $(NH_4)_2HPO_4$ are more effective than other electrolytes at protecting exfoliated graphene's basal plane. The surface of graphene sheets was treated with an aqueous electrolyte of $(NH_4)_2HPO_4$, either on its own or combined with sulfate or nitrate salts, resulting in the development of phosphorous functional groups. This treatment helped to inhibit the surface's further oxidation. The EEGs created with the combined electrolyte showed high-temperature stability comparable to that of EEGs created with pure $(NH_4)_2HPO_4$, while also having the largest percentage of monolayer (50%) and bilayer (40%) sheets (Sharif et al. 2020).

6.4.2 Graphene Oxide (GO)

Graphene oxide is one particular area of graphene research. Even now, there is no ambiguous model for GO because its specific molecular structure is still unknown. The nature and distribution of oxygen-containing functional groups, non-stoichiometric atomic composition, and various synthesis techniques all have a significant role in the variation in GO structure. Based on past research, several crucial structural characteristics of graphene oxide have been defined. However, a distinct picture of the fine structure is required. The main difference is that the GO sheet has a hexagonal ring-based carbon structure with predominantly sp^2 hybrid carbon and partially sp^3 hybrid carbon atoms, with functional groups that contain oxygen, in contrast to graphene, which is composed of trigonally bonded sp^2 carbon atoms. The basal plane of sp^3 hybridized carbon contains –OH and –C–O–C groups, while the edge portion is composed of –COOH and –COO groups as represented in Fig. 6.3. Above and below the graphene plane, the hybridized sp^3 carbons are distributed. According to the opinions of various researchers, graphene oxide has a 2D honeycomb lattice structure, a high oxygen content, and a C/O atomic ratio that is often less than three and usually closer to two (Chen et al. 2012, 2021; Anwar et al. 2022).

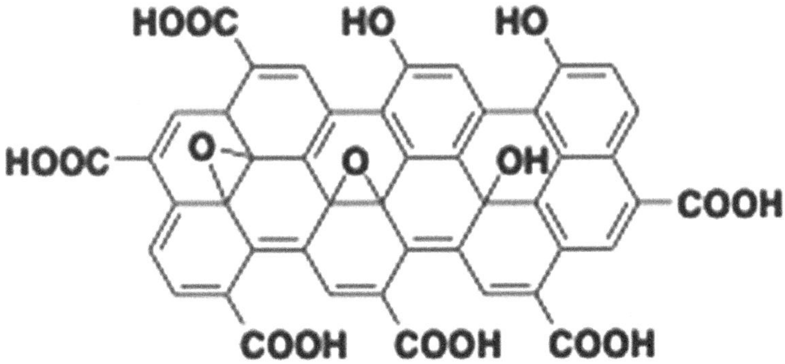

Fig. 6.3 Graphene oxide

The difference between GO and pristine graphene is due to the oxygenated functional group, which has a significant impact on the material's electrical, mechanical, and electrochemical properties. The remarkable hydrophilicity of GO can be attributed to its polar oxygen functional group. This makes GO highly dispersible in a variety of solvents, including water. Additionally, this functional group provides a site for chemical changes or functionalization using well-known chemical techniques. A typical GO is an insulator, and when it is rigorously reduced, the resistance of the material gradually decreases until it transforms into a semiconductor, where it can eventually reach a conductivity of 1250 S/m. The effect of sp3 carbon atoms on electrical conductivity depends on the degree of disorder in GO (Grajek et al. 2020). Biranje et al. developed a method for producing self-rolling GO sheets by adjusting the electrolyte composition through electrochemical exfoliation. The constant current method was used throughout the experiment, and the innovative electrode design and arrangement enabled simultaneous exfoliation and re-exfoliation of the graphite anode. The electrolyte successfully exfoliated the graphite anode with a variable yield of GO using varied molar concentrations of an aqueous sulfuric acid (H_2SO_4) solution. NaOH addition to the solution affected both the exfoliation rate and morphology of the GO sheets. Initially, a high exfoliation rate was observed in the acidic range due to increased anion intercalation with an increase in sulfuric acid concentration, excess sodium ions created a hindrance to the mobility of solvated sulfate ions, in acidic regions, which reduces the intercalation rate due to which the exfoliation rate declined. As hydroxyl ions concentration in basic regions increased, it helped in a slight increase in the rate of exfoliation. The addition of NaOH assisted the boundary functionalization and subsequently increased exfoliation. After the application of the bias, the hydroxyl ions eventually reach the graphite anode due to their great mobility and oxidize it. Oxidation increased with the concentration of the oxidizing agent, leading to the appearance of edge groups such as carbonyl, epoxy, or carboxyl on GO sheets. Although to avoid an additional one step, during exfoliation itself pH was adjusted simultaneously. This plot serves as a starting point for developing an interesting GO morphological framework (Biranje et al. 2021).

The modified Hummers method, CVD, epitaxial growth, liquid phase exfoliation, and electrochemical exfoliation are some of the techniques used to produce GO. The GO production process can be modified according to the intended application. One of these methods, which is electrochemical exfoliation of graphite, is simpler and less dangerous. Surfactants play a crucial role in Electrochemical exfoliation (EE) by promoting intercalation, graphite exfoliation, and oxidized GO formation. The affinity of surfactant chains towards graphite surfaces leads to increased inter-sheet separation and stability of exfoliated sheets. Due to their abundance of sites for dye adsorption, the resultant sEGO (surfactant-assisted exfoliated graphene oxide) composites can easily be utilized to remove the dye from aqueous solutions. The ideal situation would be to develop surfactants that facilitate effective exfoliation and dye removal at the same time. Jamaluddin et al. proposed an exfoliation mechanism for their electrochemical exfoliation process, which involves surfactant intercalation between graphite rod electrode layers using a 7 V power supply for 24 h. The prepared electrolyte (surfactant solution) and applied bias voltage allowed the graphite edge sheets to open up, allowing the triple-chain surfactant to enter the interlayer gap. Ultrasonication induces more sEGO enlargement. The type of surfactant used affects the capacity and efficacy of sEGOs as adsorbents for removing model pollutant dyes surfactant molecules adsorbed on the surface can act as stabilizers and weaken the π–π stacking interactions, which prevents the aggregation of sEGO. Therefore, these sEGO materials have more available adsorbent sites and higher specific surface areas. It was observed that composites made with a typical linear single-chain surfactant had a substantially lower adsorption capacity than those made with a highly branched tri-chain surfactant. This demonstrates that the properties of exfoliated sEGOs can be effectively influenced by surfactant design and open the door for further advancements for being used as an adsorbent for aqueous pollutants (Biranje et al. 2021). Singh et al. employed high voltage to electrochemically synthesize thermally reduced graphene oxide in an aqueous solution containing 0.5 M sulfuric acid and 3% hydrogen peroxide as the electrolyte. By choosing a high anode voltage, the graphite was able to gradually exfoliate from the edge. The ideal temperature range for optimal GO exfoliation yield throughout the experiment was found to be between 60 and 70 °C. The vacuum-filtered exfoliated graphene sheets were collected and dried at 70 °C for 12 h. After that, previously gathered exfoliated graphene sheets were combined with a 3% H_2O_2 solution and mechanically stirred (at a speed of 1000 rpm) for 30 min. elimination of the oxygen-containing functional groups associated with the graphitic layers after the thermal reduction. With the addition of H_2O_2 and re-heating, the amount of oxygen functional group and the number of defects were controlled. As the temperature of the electrolyte's exfoliation, sonication, and H_2O_2 addition increases, the proportion of oxygen functional groups decreases (Singh et al. 2022a, b). A scalable and affordable electrochemical exfoliation technique was created by Yu et al. to produce GO from exfoliated graphite at room temperature (RTEG). A container with holes was utilized in the preparation technique to confine graphite particles during electrochemical oxidation and ensure a highly entangled and continuous structure for the RTEG in the anode region. In this study, to obtain exfoliated graphite, graphite flakes were first exfoliated with concentrated H_2SO_4 and H_2O_2 at R.T. RTEG was added to a container with holes for electrochemical oxidation, confining graphite particles and preventing shedding from the reactive region. Refilling the RTEG in the container allows for the continued preparation of EGO (exfoliated Graphene oxide), resulting in a reliable, affordable method of preparing GO (Yu et al. 2022).

6.4.3 Heteroatom-Doped Graphene

Nitrogen, fluorine, sulphur, and chlorine are some of the most commonly used heteroatoms and halogens used for doping graphene as represented in Figs. 6.4 and 6.5. In fluorographene, the graphene is chemically modified by the addition of fluoride ions which bind covalently or ionically with certain carbon atoms in the graphene exoskeleton. The highly electronegative fluorine atom alters the hybridization of the carbon atom from sp^2 to sp^3, thereby introducing defects in the lattice structure. The higher lattice defect produces magnetic moment, thus giving tuneable magnetic property to fluorographene (Padamata et al. 2021). It gives paramagnetic properties to the otherwise diamagnetic material. The addition of a fluorine atom not just improved the electron dispersion, but also enhanced the dielectric nature of the material. Also, the heterogeneous electron transfer (HET) rate has been shown to increase with a decrease in atomic number and is thereby high in fluorographene (Narayanan et al. 2017). The fluorine atoms disrupt the electron distribution of the graphene by altering the original 2D conjugated structure of graphene thereby changing the bandgap with F/C ratios in the range of 0–3.8 eV. The change in the bandgap thereby has led to enhancement in optical, thermal, mechanical, electrical, and magnetic properties of fluorographene and have found several applications in energy storage devices, photocatalyst, bio-application, etc. (Padamata et al. 2021; Aghamohammadi et al. 2020).

Fluorographene can be synthesized by simultaneously fluorinating and exfoliating the graphene using the electrochemical method. It can be carried out in an acidic environment with hydrofluoric acid (HF) or HF and H_2SO_4 in a 1:1 ratio as electrolyte and graphite electrode acting as both anode and cathode. Here, initially, the kinetics of exfoliation increased with voltage and HF concentration. Further increase in HF concentration beyond a certain limit resulted in more graphite remaining in the fluoride compound and increased defect density in the exfoliated powder. Similarly, even though the increase in voltage improved the exfoliation rate, the fluorographene synthesized at a lower voltage showed a higher degree of fluorination and defect density (Padamata et al. 2021).

Electrochemical exfoliation can also be achieved under mild neutral conditions by using platinum foil as a counter electrode and graphite as the working electrode in fluorine-based inorganic salts like $NaBF_4$, NH_4BF_4, KPF_6, etc., as electrolyte during the synthesis. Here when a bias voltage is applied, strong nucleophilic hydroxyl ions (OH^-) are produced due to the reduction of water. These OH^- ions generated initially attack the edges of the graphite. Subsequently, the oxidation is accompanied

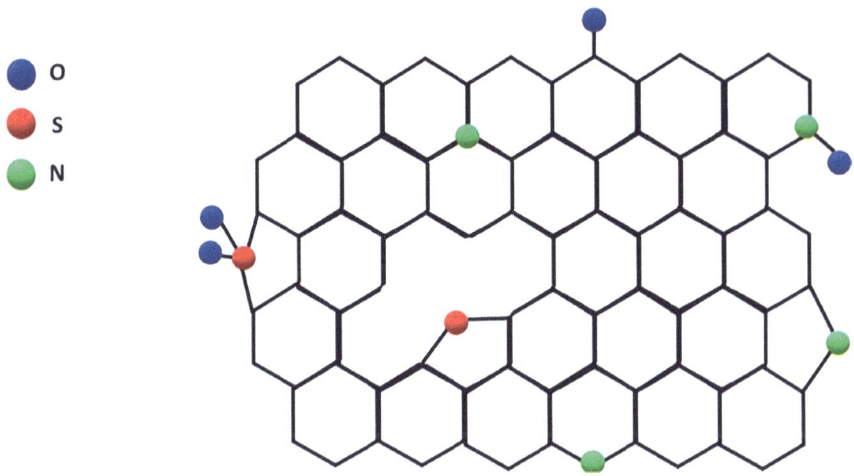

Fig. 6.4 Hetero atom doped graphene

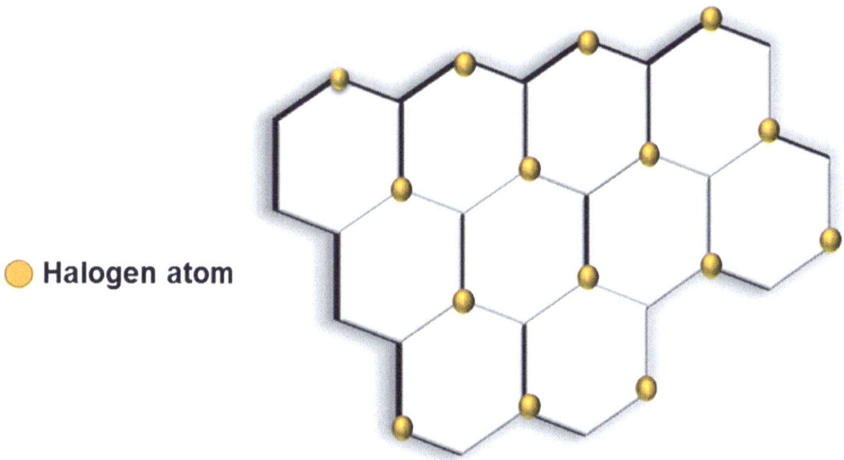

Fig. 6.5 Halogen doped graphene

by fluorination by BF_4^- penetrating into the interior of the inter-layer spacing. The oxidation and fluorination produce gaseous by-products like O_2 which forces the weakly bounded layers of graphite apart, thereby exfoliating the fluorographene into the electrolyte solution. This exfoliated fluorographene can be then collected after filtration and washing (Zhou et al. 2018).

Similarly, doping graphene with Nitrogen atoms has been shown to improve the selectivity of graphene for electro-Fenton systems. The N-doped graphene favours the

2-electron transfer mechanism in an acidic medium, i.e., H_2O_2 generation over the 4-electron reduction mechanism. The electrochemical synthesis of N-doped graphene can be achieved in one step by dipping the graphite rod in an ammonium nitrate solution. At higher potential, there is an acceleration in the rate of formation of 2D graphene sheets as well as in the reduction of nitrate ions at the cathode (Kamaraj et al. 2021). Apart from ammonium nitrate, using urea-ammonium sulfate (Kakaei and Ghadimi 2021) and ammonium sulfate-ammonium phosphate (Momodu et al. 2021) electrolytes also can be used for the synthesis of N-doped graphene. Likewise, Sulphur-doped graphene can also be prepared by dipping a graphite rod in an electrolyte containing sodium thiosulphate. On applying current, initially, the electrode becomes rough and then the graphite exfoliation starts which reacts with the electrolyte solution to form the heteroatom-doped graphene (Magerusan et al. 2022).

6.4.4 Other 2D Graphene Derivatives

Apart from heteroatoms, graphene sheets have also been electrochemically doped with nanomaterials, where the resultant composite structure shows enhanced properties. For the synthesis of Ag nanoparticles doped graphene, the AgNPs and graphene were dissolved in water in the presence of a surfactant as the AgNPs and graphene were insoluble in deionized water. When the current was passed through this solution, AgNPs-doped graphene nanosheets were obtained (Htwe et al. 2022). Apart from this, the graphene sheets obtained by electrochemical exfoliation can also be doped with nanomaterial like $TiNb_2O_7$ by mixing the graphene nanosheets with the precursor $NbCl_5$ and tertbutyl titanate of $TiNb_2O_7$. The resultant mixture was then given solvothermal treatment to obtain the nanocomposite of $TiNb_2O_7$/Graphene (Aghamohammadi et al. 2020).

6.4.5 Conclusion and Future Perspectives

Exfoliated graphene from electrochemical methods shows promising possibilities for high-performance applications. Electrochemical exfoliation is more convenient and controllable than other synthetic methods. The van der Waals forces or electrostatic interactions that hold 2D layers together can be weakened by electrochemical exfoliations including anodic exfoliation and cathodic exfoliation, which can be used to exfoliate the bulk layered materials into single-layer or few-layer 2D materials like graphene. The electrochemical exfoliation technique has undergone significant improvement in an effort to achieve low production costs, high product purity, and high output yields of graphene materials. The structure and chemical and physical properties of the resulting graphene materials can be altered by adjusting the voltage, concentration and type of electrolyte, and even the shape of the bulk materials. To continue this cutting-edge research, it is necessary to be able to produce a large quantity of graphene material. It is possible to advance the development of graphene materials with a wide range of potential applications in the fields of electronics, energy storage, energy conversions, sensors, corrosion prevention, and catalysis by comprehending the mechanism of electrochemical exfoliation and controlling parameters in the synthesis process. In summary, this chapter has focused on electrochemical exfoliation (EE) as a cutting-edge technique for producing nanomaterials, with a specific emphasis on graphene and its derivatives. We have delved into the mechanism of EE, its types, and how it outperforms other synthetic approaches. EE offers a milder, more convenient, and easier to control means for large-scale nanomaterial synthesis. Thus, this chapter underscores the significance of electrochemical exfoliation as a cost-effective and high-quality method for creating a range of nanomaterials; particularly graphene and its derivatives.

References

Abdelkader, A.M., Cooper, A.J., Dryfe, R.A.W., Kinloch, I.A.: How to get between the sheets: a review of recent works on the electrochemical exfoliation of graphene materials from bulk graphite. Nanoscale **7**(16), 6944–6956 (2015)

Abdillah, O.B., Floweri, O., Iskandar, F.: The effect of voltage bias on the yield and electrical properties of Exfoliated Graphene synthesized via electrochemical method. IOP Conf. Ser.: Mater. Sci. Eng. **1098**, 062069 (2021)

Aghamohammadi, H., Eslami-Farsani, R.: Synthesis and electrochemical performance of $TiNb_2O_7$ nanoparticles grown on electrochemically prepared graphene as anode materials for Li-ion batteries. J. Power. Sources **535**, 231418 (2022)

Aghamohammadi, H., Heidarpour, A., Ghasemi, S.: Electrochemical synthesis of fluorinated graphene nanoplatelets in electrolytes containing hydrofluoric acid and TiO_2 nanoparticles. FlatChem **22**, 100172 (2020)

Ambrosi, A., Chua, C.K., Bonanni, A., Pumera, M.: Electrochemistry of graphene and related materials. Chem. Rev. **114**(14), 7150–7188 (2014)

Anwar, A., Chang, T.P., Chen, C.T.: Graphene oxide synthesis using a top–down approach and discrete characterization techniques: a holistic review. Carbon Lett. **32**(1), 1–38 (2022)

Bandara, T.M.W.J., Thennakoon, T.M.A.A.B., Gamachchi, G.G.D.M.G., Bandara, L.R.A.K., Pemasiri, B.M.K., Dahanayake, U.: An electrochemical route to exfoliate vein graphite into graphene with black tea. Mater. Chem. Phys. **289**, 126450 (2022)

Biranje, P.M., Prakash, J., Srivastava, A.P., Biswas, S., Patwardhan, A.W., Joshi, J.B., Dasgupta, K.: In situ tuning of graphene oxide morphology by electrochemical exfoliation. J. Mater. Sci. **56**(35), 19383–19402 (2021)

Biranje, P.M., Patwardhan, A.W., Joshi, J.B., Prakash, J., Dasgupta, K.: Kinetic study of graphene oxide synthesis by electrochemical exfoliation of graphite. J. Ind. Eng. Chem. **119**, 335–345 (2023)

Campbell, B., Manning, J.: The rise of victimhood culture: Microaggressions, safe spaces, and the new culture wars. Rise Vict. Cult. Microaggressions, Safe Spaces, New Cult. Wars 1–265 (2018)

Chen, D., Feng, H., Li, J.: Graphene oxide: preparation, functionalization, and electrochemical applications. Chem. Rev. **112**(11), 6027–6053 (2012)

Chen, J., Perez-Page, M., Ji, Z., Zhang, Z., Guo, Z., Holmes, S.: One step electrochemical exfoliation of natural graphite flakes into graphene oxide for polybenzimidazole composite membranes

giving enhanced performance in high temperature fuel cells. J. Power. Sources **491**, 229550 (2021)
Dalal, M.H., Lee, C.Y., Wallace, G.G.: Cathodic exfoliation of graphite into graphene nanoplatelets in aqueous solution of alkali metal salts. J. Mater. Sci. **56**(4), 3612–3622 (2021)
Danial, W.H., Norhisham, N.A., Ahmad Noorden, A.F., Abdul Majid, Z., Matsumura, K., Iqbal, A.: A short review on electrochemical exfoliation of graphene and graphene quantum dots. Carbon Lett. **31**(3), 371–388 (2021)
Gao, X., Liu, H., Wang, D., Zhang, J.: Graphdiyne: synthesis, properties, and applications. Chem. Soc. Rev. **48**(3), 908–936 (2019)
García-Dalí, S., Paredes, J.I., Munuera, J.M., Villar-Rodil, S., Martínez-Alonso, A., Tascón, J.M.D.: An aqueous cathodic delamination route towards high quality graphene flakes for oil sorption and electrochemical charge storage applications. Chem. Eng. J. **372**, 1226–1239 (2019)
Grajek, H., Jonik, J., Witkiewicz, Z., Wawer, T., Purchała, M.: Applications of graphene and its derivatives in chemical analysis. Crit. Rev. Anal. Chem. **50**(5), 445–471 (2020)
Hossain, S.T., Wang, R.: Electrochemical exfoliation of graphite: effect of temperature and hydrogen peroxide addition. Electrochim. Acta. Acta **216**, 253–260 (2016)
Htwe, Y.Z.N., Abdullah, M.K., Mariatti, M.: Water-based graphene/AgNPs hybrid conductive inks for flexible electronic applications. J. Mater. Res. Technol. **16**, 59–73 (2022)
Ilias, S.H., Murshidi, J.A., Ying, K.K.: Effect of electrolyte concentration on the synthesis of graphene by electrochemical exfoliation process. IOP Conf. Ser.: Mater. Sci. Eng. **1106**, 012013 (2021)
Ivanovskii, A.L.: Graphynes and graphdyines. Prog. Solid State Chem. **41**(1–2), 1–19 (2013)
Kakaei, K., Ghadimi, G.: A green method for Nitrogen-doped graphene and its application for oxygen reduction reaction in alkaline media. Mater. Technol. **36**(1), 46–53 (2021)
Kamaraj, R., Aravind, P., Vasudevan, S.: Nitrogen doped Graphene Nano sheets (N−Gns) as electrocatalyst for electro-fenton process for the degradation of highly toxic chlorophenoxy acid herbicides from water. Chem. Select **6**(11), 2804–2810 (2021)
Kumar, N.: Top-down synthesis of graphene: a comprehensive review. Flat Chem **27**, 100224 (2021)
Liu, F., Wei, L., Chen, Y.: Synthesis of graphene materials by electrochemical exfoliation : recent progress and future potential. Carbon Energy **1**, 173–199 (2019)
Liu, W.W., Aziz, A.: Review on the effects of electrochemical exfoliation parameters on the yield of graphene oxide. ACS Omega **7**(38), 33719–33731 (2022)
Lowe, S.E., Ge, S., Zhang, Y., Qin, J., Jiang, L., Jiang, S., Mamun, M.A., Liu, P., Zhong, Y.L., Zhao, H.: The role of electrolyte acid concentration in the electrochemical exfoliation of graphite: mechanism and synthesis of electrochemical graphene oxide. Nano Mater. Sci. **1**(3), 215–223 (2019)
Magerusan, L., Pogacean, F., Pruneanu, S.: Eco-friendly synthesis of sulphur-doped graphenes with applicability in caffeic acid electrochemical assay. Bioelectrochemistry **148**(67), 108228 (2022)
Momodu, D., Zeraati, A.S., Pablos, F.L., Sundararaj, U., Roberts, E.P.: Hybrid energy storage using nitrogen-doped graphene and layered-MXene (Ti_3C_2) for stable high-rate supercapacitors. Electrochim. Acta. Acta **388**, 138664 (2021)
Munuera, J.M., Paredes, J.I., Villar-Rodil, S., Martínez-Alonso, A., Tascón, J.M.D.: A simple strategy to improve the yield of graphene nanosheets in the anodic exfoliation of graphite foil. Carbon **115**, 625–628 (2017)
Narayanan, T.N., Biroju, R.K., Renugopalakrishnan, V.: Fluorographene: synthesis and sensing applications. J. Mater. Res. **32**(15), 2848–2859 (2017)
Padamata, S.K., Yasinskiy, A., Stopic, S., Friedrich, B.: Fluorination of two-dimensional graphene: a review. J. Fluor. Chem. **255–256**, 109964 (2021)
Park, S.W., Jang, B., Kim, H., Lee, J., Park, J.Y., Kang, S.O., Choa, Y.H.: Highly water-dispersible graphene nanosheets from electrochemical exfoliation of graphite. Front. Chem. **9**, 1–8 (2021)
Pingale, A.D., Owhal, A., Katarkar, A.S., Belgamwar, S.U., Rathore, J.S.: Facile synthesis of graphene by ultrasonic-assisted electrochemical exfoliation of graphite. Mater. Today Proc. **44**(1), 467–472 (2021)

Prekodravac, J.R., Kepić, D.P., Colmenares, J.C., Giannakoudakis, D.A., Jovanović, S.P.: A comprehensive review on selected graphene synthesis methods: from electrochemical exfoliation through rapid thermal annealing towards biomass pyrolysis. J. Mater. Chem. C **9**(21), 6722–6748 (2021)

Saha, S., Lakhe, P., Mason, M.J., Coleman, B.J., Arole, K., Zhao, X., Yakovlev, S., Uppili, S., Green, M.J., Hule, R.A.: Sustainable production of graphene from petroleum coke using electrochemical exfoliation. NPJ 2D Mater. Appl. **5**(1), 1–8 (2021)

Sharif, F., Zeraati, A.S., Ganjeh-Anzabi, P., Yasri, N., Perez-Page, M., Holmes, S.M., Sundararaj, U., Trifkovic, M., Roberts, E.P.L.: Synthesis of a high-temperature stable electrochemically exfoliated graphene. Carbon **157**, 681–692 (2020)

Singh, S., Hasan, M.R., Sharma, P., Narang, J.: Graphene nanomaterials: the wondering material from synthesis to applications. Sensors Int. **3**, 100190 (2022a)

Singh, P.K., Sharma, K., Singh, P.K.: A low cost, bulk synthesis of the thermally reduced graphene oxide in an aqueous solution of sulphuric acid and hydrogen peroxide via electrochemical method. Inorg. Chem. Commun.commun. **140**, 109378 (2022b)

Su, C.Y., Lu, A.Y., Xu, Y., Chen, F.R., Khlobystov, A.N., Li, L.J.: High-quality thin graphene films from fast electrochemical exfoliation. ACS Nano **5**(3), 2332–2339 (2011)

Tamgadge, R.M., Shukla, A.: A pH-dependent partial electrochemical exfoliation of highly oriented pyrolytic graphite for high areal capacitance electric double layer capacitor electrode. Electrochim. Acta. Acta **325**, 134933 (2019)

Wu, W., Zhang, C., Hou, S.: Electrochemical exfoliation of graphene and graphene-analogous 2D nanosheets. J. Mater. Sci. **52**(18), 10649–10660 (2017)

Wu, J., Wang, H., Qiu, J., Zhang, K., Shao, J., Yan, L.: Efficient preparation of high-quality graphene via anodic and cathodic simultaneous electrochemical exfoliation under the assistance of microwave. J. Colloid Interface Sci. **608**(2), 1422–1431 (2022)

Yang, Y.: Electrochemical exfoliation of graphene-like two-dimensional nanomaterials. Nanoscale **11**(1), 16–33 (2019)

Yang, S., Lohe, M.R., Müllen, K., Feng, X.: New-generation graphene from electrochemical approaches: production and applications. Adv. Mater. **28**(29), 6213–6221 (2016)

Yu, P., Lowe, S.E., Simon, G.P., Zhong, Y.L.: Electrochemical exfoliation of graphite and production of functional graphene. Curr. Opin. Colloid Interface Sci.opin. Colloid Interface Sci. **20**(5–6), 329–338 (2015)

Yu, Q., Wei, L., Yang, X., Wang, C., Chen, J., Du, H., Shen, W., Kang, F., Huang, Z.H.: Electrochemical synthesis of graphene oxide from graphite flakes exfoliated at room temperature. Appl. Surf. Sci. **598**, 153788 (2022)

Zhang, Y., Xu, Y., Liu, R.: Regulating cations and solvents of the electrolyte for ultra-efficient electrochemical production of high-quality graphene. Carbon **176**, 157–167 (2021)

Zhen, Z., Zhu, H.: Structure and properties of graphene. In: Graphene, pp. 1–12 (2018)

Zhou, F., Huang, H., Xiao, C., Zheng, S., Shi, X., Qin, J., Fu, Q., Bao, X., Feng, X., Mullen, K., Wu, Z.S.: Electrochemically scalable production of fluorine-modified graphene for flexible and high-energy ionogel-based microsupercapacitors. J. Am. Chem. Soc. **140**(26), 8198–8205 (2018)

Chapter 7
Unique Characteristics of Electrochemically Exfoliated Multidimensional Graphene and Its Derivatives

Aishik Das, Saneeya Vichare, and Soumya Basu

Abstract Graphene, a single atomic layer of graphite, is considered superior to other carbon-based graphitic allotropes due to its distinctive topologies, intriguing electrical, optical, and mechanical properties, and promising nanoelectronics. It has numerous applications in nanotechnology, including electrochemical exfoliation, which involves the use of natural graphite foils, sheets, or rods. Ion intercalations increase the interlayer gap in the electrochemical exfoliation of graphite, producing pure graphene in monolayer or multilayer forms using chemical vapor deposition and pulsed laser deposition techniques. Electrochemical exfoliation techniques dope graphene materials, incorporate covalent functional groups, integrate nanoparticles, and assemble graphene structures. Compared to other processes, electrochemical exfoliation offers benefits such as simplicity, cost-effectiveness, environmental friendliness, high production rate, and high-quality few-layered graphene sheets. EG also provides a large sheet, less oxygen content, a high C/O ratio, and outstanding electrical characteristics. This chapter's goal is to inform readers about the characteristics and applications of currently recognized electrochemically exfoliated multidimensional graphene structures, including graphene quantum dots (GQDs), one-dimensional (1D) graphene nanoribbons (GNRs) and graphene fibers (GFs), two-dimensional (2D) graphene nanomeshes (GNMs), rippled/wrinkled graphene, and 2D graphene membranes (GMs) and three-dimensional graphene (3DG) architectures made of graphene oxide (GO) and reduced GO (rGo) and highlight their unique properties which are based on the configuration of carbon atoms.

Keywords Electrochemical · Exfoliated · Multilayered graphene · Character

---

A. Das · S. Vichare · S. Basu (✉)
Cancer and Translational Research Centre, Dr. D. Y. Patil Biotechnology and Bioinformatics Institute, Dr. D. Y. Patil Vidyapeeth, Tathawade, Pune 411033, Maharashtra, India
e-mail: soumya.bs@gmail.com

R. Khan et al. (eds.), *Electrochemical Exfoliation of Graphene and its Derivatives*, Engineering Materials, https://doi.org/10.1007/978-981-97-2128-3_7

7.1 Introduction

Graphene, a single-atom-thick, 2D sheet of sp^2-hybridized carbon atoms, can be floated freely or attached to another substrate (Bianco et al. 2013). It has impressive carrier mobility (200 000 cm^2 V^{-1} s^{-1}), extraordinary thermal conductivity (5000 W m^{-1} K^{-1}), a wide surface area (2630 m^2 g^{-1}), ambipolar electric field effect, room-temperature quantum Hall effect, and high transparency toward visible light (97.7%) (Zhu et al. 2010), as shown in Fig. 7.1. The lattice structure of graphene is a single carbon sheet with no edge or basal plane variations and a zero bandgap (Meyer et al. 2007). Engineering such an infinite planar structure to a limited size or good design will result in diverse electrical configurations, allowing graphene's bandgap to be opened and employed for target-specific purposes (Ritter and Lyding 2009; Yan et al. 2012). The spatial confinement of the conduction electrons and valence holes caused by the reduction of graphene to GQDs or the coupling of organic molecules to GQDs causes the bandgap to open. It causes unusual phenomena such as multi-photon excitation, Coulomb blockade, and chaotic Dirac billiards (Ponomarenko et al. 2008; Libisch et al. 2009). GQDs have exceptional and tunable electrical, optical, chemical, and structural capabilities due to their numerous edges and size impact (Zhang et al. 2008). GQDs are exciting prospects for bioimaging, biosensing, and photovoltaic devices due to their photostability, biocompatibility, and adjustable photoluminescence features resulting from quantum confinement phenomena (Bak et al. 2016). GNRs' limited width and chiral angle control electrical and magnetic features like carrier density and ferromagnetism, making them ideal for electronics (Barone et al. 2006). Because of the atomic thickness and strong electrical conductivity for biosensing, particularly for quick DNA sequencing, the GNM or GNR with holes offers a better platform (Merchant et al. 2010; Schneider et al. 2010). The one-atom-thick mesh may serve as a very effective and permeable filtering membrane, opening up possibilities for water desalination and gas separation (Suk and Aluru 2010). The graphene scaffold is combined with a range of oxygen-containing entities in GO, which may be produced in large quantities (Park and Ruoff 2009). This increases the range of interactions that are possible, including hydrogen bonds, Csp^2 interactions, electrostatic and van der Waals interactions, and enables customization of the properties of the graphene surface (e.g., wettability) (Compton and Nguyen 2010). One-of-a-kind advantages in lightweight, flexible, and stretchy electronics may be seen in 1D GFs and 2D GMs, which exhibit excellent mechanical flexibility and rigidity, are simple to functionalize, and are highly conductive (Wang and Wallace 2015). High surface areas, hierarchical pore architectures, solid conductivities, low densities, and mechanical characteristics are typical features of 3DG materials. They could be utilized in a variety of applications, including ecological sustainability, energy conversion, and storage. The types of graphene and their prospective applications have been described in Table 7.1.

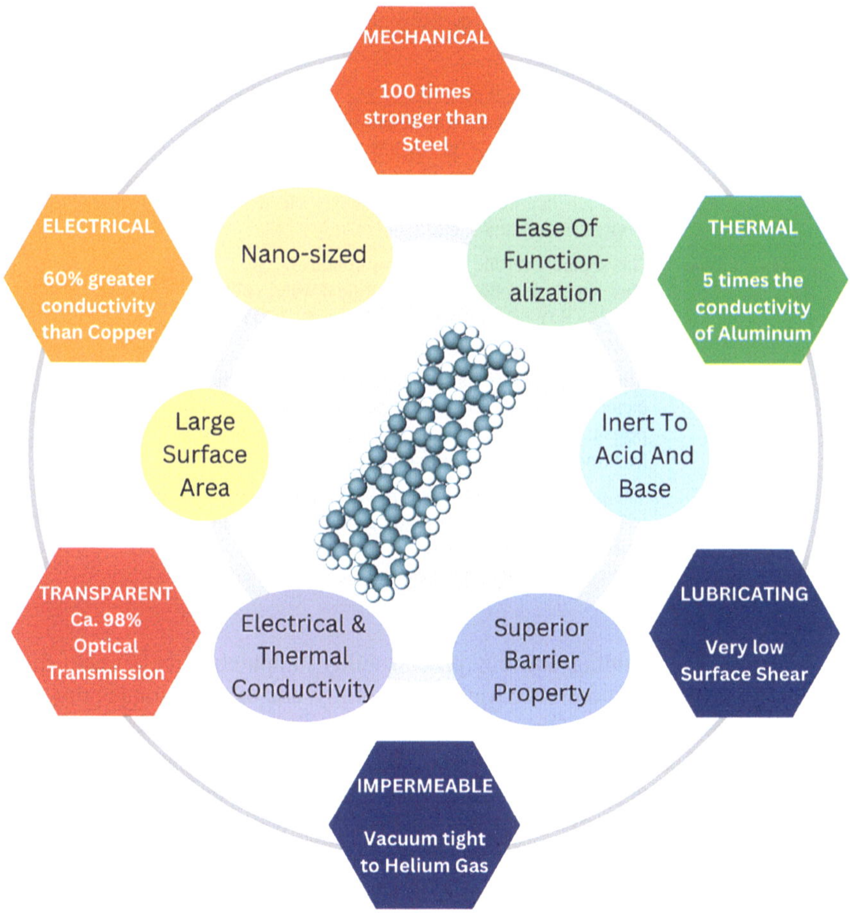

Fig. 7.1 Unique characteristics of graphene

7.2 Atomic Interactions in Diverse Types of Graphene

7.2.1 Graphene Oxide

Graphene oxide (GO) is graphene in its oxidized state, as illustrated in Fig. 7.2. It is a single-atomic-layered substance created by oxidizing inexpensive and widely accessible graphite. Graphene oxide (GO) has demonstrated various unusual physicochemical qualities, including tiny size, vast area, extraordinary tenacity in 2D structure, and intriguing electrical and optical capabilities, among others (Voiry et al. 2016; Miao et al. 2007). Exfoliated graphene's hydrophobic nature prevents it from readily dispersing in water, which makes functionalization more challenging, whereas GO, when oxidized, becomes hydrophilic and hence water-soluble. As a

Table 7.1 Types of graphene and their prospective applications

Sr. no.	Types of graphene	Applications	References
1	Graphene oxide	GO is generally used for the production of reduced graphene oxide (rGO), which is in turn used for energy-storing devices, biosensors, supercapacitors, water purification processes, catalysts, electronic devices, membranes, etc.	Jiříčková et al. (2022)
2	Graphdiyne	Gas separation (due to its porous nature), photocatalysis, photoelectrochemical water splitting (PEC), GDY-based lithium-ion batteries, GDY-based electrolyzer, humidity sensors, etc.	Gao et al. (2019)
3	Graphane	Since graphene exists only in theory, its applications cannot be pinpointed. Its charge carriers can potentially serve as a source of a 2D electron gas with high mobility and varying concentration	Sofo et al. (2007)
4	Graphyne	The properties of graphyne make it a potent substance to be used in making nanofillers, desalinators, sensors, anisotropic conductors, transistors, and semiconductor metal hybrids	Peng et al. (2014)

result, GO has outstanding aqueous processability, amphiphilicity, surface functionalization capacity, fluorescence quenching ability, and other desirable properties for prospective biotechnological applications (Gomes et al. 2012). Additionally, the combination of GO with gold, polymers, and other magnetic nanoparticles has led to a variety of biomedical applications, including light therapy, bioimaging, drug, and genetic transfer, biosensing, and antibacterial activity, among others. The pi–pi* transition, in particular, promotes reduced electron transport, which enhances GO's bioimaging and biosensing capabilities.

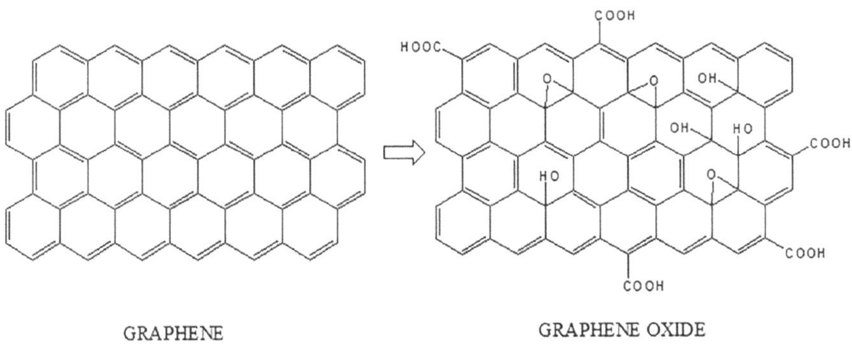

Fig. 7.2 Synthesis of graphene oxide from graphene

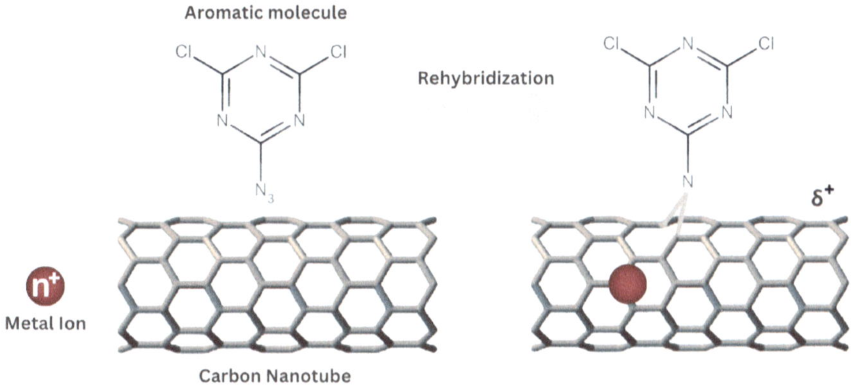

Fig. 7.3 Metal ion adsorption on the surface of graphene through charge transfer and rehybridization of the CNT with an aromatic compound via $\pi - \pi$ (Van der Waals) stacking

7.2.1.1 *Van Der Waals ($\pi - \pi$) Interaction*

Aromatic compounds are frequently used to alter the electrical characteristics of carbon nanotubes (CNTs). To produce a stable hybrid structure and increase their electrical characteristics, the molecules engage with the sidewalls of CNTs via $\pi - \pi$ stacking, a type of van der Waals force related to their π-electron backbone system (see Fig. 7.3). Surface stacking is generally highly durable against washing or other solution processing due to the strong $\pi - \pi$ stacking between the graphene surface and the aromatic molecules (Zhang et al. 2011a, b, c). Because π-π interactions between graphene and aromatic molecules rely heavily on dispersion and electrostatic forces, London dispersion forces must be the most attractive component in the binding. When extremely electronegative fluorine atoms are substituted into aromatic compounds, π–π interactions become stronger and more dispersive (less electrostatic) (Wang et al. 2014). The density functional theory with empirical dispersion correction (DFT-D) is currently the most practicable tool for studying the π–π interactions between graphene and aromatic compounds (Wang et al. 2014).

7.2.1.2 Ionic Interaction

The ion effect is significant for attaching organic molecules to the surface of graphene oxide (GO), as in environmental remediation and water treatment procedures. Because ions differ in charge, size, and complexing capacity, the adherence of organic compounds to the GO surface can be substantially influenced, particularly for organics ending with polar groups of –COO(H). By drop-casting a metal chloride and GO solution, metal cations (Mn^{2+}, Co^{2+}, Cu^{2+}, and Fe^{3+}) can self-assemble into a hydroxide/GO lattice structure due to the positively charged hydroxide interacting electrostatically with the negatively charged GO nanosheets. The cation species can control the interlayer spacing of the superlattice. The GO/cation systems' development process and structural evolution may be divided into four stages: the first step includes the GO interlayer being intercalated with hydrated cations in an aqueous solution, then when the GO layers are annealed, metal hydroxide grows between them, and at about 250 °C, a superlattice structure forms. In the third stage, metal oxide nucleation occurs between the reduced GO (rGO) interlayers as the temperature rises. And finally, the complete disintegration of the graphene layer at 600 °C and production of metal oxide nanosheets take place. This study provides a novel viewpoint on metal cations' interactions and growth behavior with GO (Li et al. 2021). Ion bridging is crucial in attaching –COO(H) tips to the GO surface. The –COOH tip has an enormous adhesion reaction at high pH, whereas the $-CH_3$ tip exhibits a relatively modest but substantial response. These provide essential insights into how solutions and mineral surfaces interact with adsorbed organic molecules. They also offer suggestions for enhancing remediation techniques like environmental cleanup and water treatment with GO, which can be used as catalysts, membranes, adsorbents, and coating materials (Liu et al. 2019).

7.2.1.3 Electrostatic Interaction

Electrostatic interaction bonds stacked GO nanosheets. Due to the ionization of carboxylate groups in water (Li et al. 2008), GO nanosheets can act as polyanions because they are negatively charged and can bind to positively charged polyelectrolytes like polyallylamine hydrochloride (PAH) electrostatically (Kovtyukhova et al. 1999). Compared to covalent bonding, electrostatic bonding may result in GO membrane production with the advantages of a more straightforward reaction, more flexible chemistry control, less use of organic solvent, and fewer by-products. Additionally, an electrostatically bonded GO membrane may be regenerable, meaning that by changing the solution chemistry to an extremely high pH and surfactant concentrations, the existing GO layers on the membrane may be released, allowing the deposition of new GO layers to restore membrane features (Hu and Mi 2014). Ubiquitin, a globular protein, has been discovered to bind electrostatically with graphene oxide with variable degrees of oxidation and to undergo a rapid and reversible exchange (quick exchange regime) from the GO surface (Malik et al. 2020).

7.2.1.4 Hydrogen Bonding

Graphene oxide (GO) platelets randomly stacked to form a multilayered material are promising candidates for novel uses in nanoelectromechanical systems. The large-scale properties of these composites are controlled by hydrogen bond networks involving functional groups on specific GO platelets and molecules of water in the interlayer voids. Water content impacts the interlayer spacing and elastic moduli of the composite, which controls the size and overall strength of these interlayer hydrogen bond networks. Additionally, the mechanical properties of the composite are greatly influenced by the chemical makeup of the individual GO platelets. A higher functional group density results in more hydrogen bonds and, as a result, more interlayer adhesion, which enhances stiffness (Medhekar et al. 2010). In a ternary system, GO primarily interacts with the metals, and the GO-metal complexes subsequently attach to SSLBs via hydrogen bonding as opposed to cation bridging across the metals (Li et al. 2020).

7.2.1.5 Lewis Acid–Base Interaction

The active sites for H2 activation are the frustrated Lewis acid–base pairs on graphene sheets (Navalón et al. 2020). Because of the Lewis acid–base interaction, the adsorption energy, as well as the total charge transfer value of N-doped graphene, are significantly higher than those of pure graphene. Higher negative adsorption energy results in stronger Lewis acid–base interaction in N-doped graphene due to the interstice of Lewis bases and acids of nitrogen and boron, respectively (Rad et al. 2015). Lewis-acid sites and Bronsted acidic edges of GO have been found to aid in the synthesis of triazoloquinazolinones. Because the number of edge carboxylic groups is limited, the increasing number of epoxide Lewis acid sites becomes the decisive factor in creating highly active GO promoters (Ebajo et al. 2019).

7.2.2 Graphdiyne

Graphdiyne (GDY), a novel two-dimensional (2D) carbon allotrope, is gaining popularity. Its unusual sp–sp^2 carbon atoms, homogeneous pores, and highly conjugated structure, as shown in Fig. 7.4, make it a viable candidate for use in practical fields such as energy-related businesses, water treatment, gas separation, catalysis, and humidity sensors. A 2D planar structure, known as GDY, is created by adding a diacetylenic link between the two benzene rings in a GR structure. With a direct bandgap of 0.46 eV and high carrier mobility at room temperature, it is the most durable diacetylene-bonded quasi-carbon allotrope, making it attractive for upcoming nanoelectronics (Gao et al. 2019).

7.2.2.1 Two-Hybrid State (sp–sp^2) and a Highly π-conjugated Structure

GDY is regarded as the most stable sp and sp^2 carbon compound that contains diacetylene. The C6 hexagons in graphene that connect the graphyne (GY)-GDY networks via acetylene connections have an unusual atomic configuration and electrical structure. These flat carbon networks should function similarly to graphene due to their high π-conjugation levels, evenly dispersed pores, and tunable electrical properties. They could be used for a variety of things, such as energy storage, anode materials for

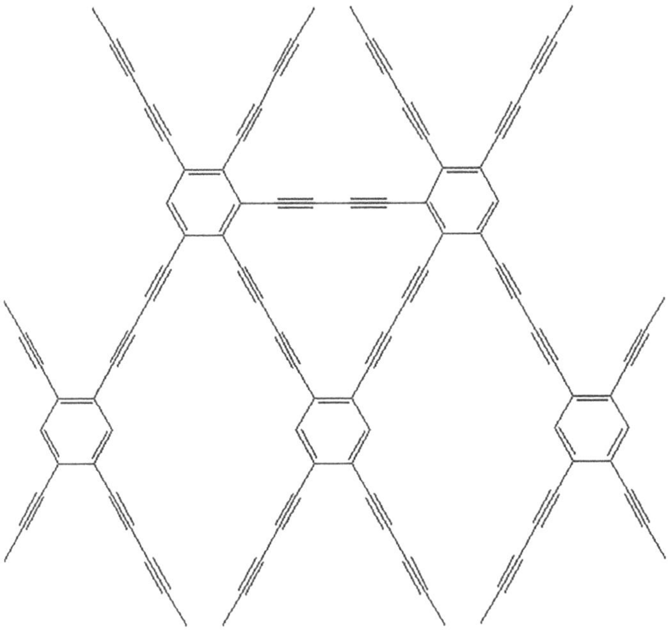

Fig. 7.4 Chemical structure of a graphdiyne plane with Sp- and Sp^2-hybridized carbon atoms

batteries, and biological purposes. With its multiple big triangle-like pores, GDY's unusual structure provides many Li storage sites and allows for quick electron and ion movement. GDY, a new highly conjugated carbon material, possesses charge transport and semiconductive properties (Jia et al. 2017). A material's ultimate response can theoretically be tuned by correct molecular design by manipulating topology, connectivity, or the sp–sp^2 carbon ratio to create an endless variety of alternative structures, each of which may be characterized by different electronic/vibrational properties (Serafini et al. 2021).

7.2.3 *Graphane*

7.2.3.1 Hydrogen, Covalent Bonding, and Modification from sp^2 to sp^3

The chemical symbol for graphane is CH, a completely hydrogenated derivative of graphene. Carbon atoms in graphene hybridize with hydrogen, changing from flat sp^2 to tetrahedral sp^3 structure. Changing conductive graphene to dielectric graphane or opening up the zero bandgaps of graphene to create a bandgap in graphane makes it possible to create nanoelectronic devices based on graphane and develop hydrogen-fuel technologies. With a stoichiometric formula unit of CH, graphane has the most

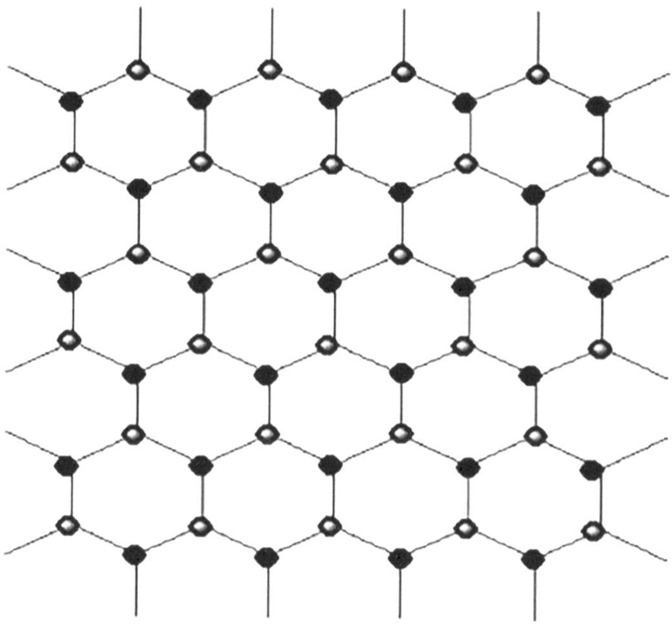

Fig. 7.5 Representation of graphane in the chair conformation

durable extended 2D form of hydrocarbon because it has a higher BE than other types of hydrocarbons, like benzene and acetylene (Banerjee 2018).

This hydrocarbon is expected to have two possible conformers, "chair" and "boat," both of which have a covalently attached hydrogen atom (sp^3) to each carbon atom in the sheet and have a 2D wrinkled honeycomb-like structure. The H atoms bonded to the C atoms of the sheet switch on both sides (1up-1down) in the chair conformer, while in the boat conformer, they do so in pairs (2up-2down) (Bhattacharya et al. 2011). The strong hybridization between C–s/p and H–s states at the valence band and the conduction band in chair-like graphane (see Fig. 7.5) results in very strong covalent bonds, which might be one factor in the great stability of this structure, as opposed to boat-like, which only exhibits substantial C–s/p and H–s hybridization about -11.0 eV, suggesting that covalent bonds are only concentrated in this energy range (Reshak and Auluck 2014).

7.2.4 Graphyne

7.2.4.1 Covalent Bonding and Acetylene Bonds

Graphyne is a graphene derivative, a 2D sheet of carbon atoms with sp and sp^2 links that is one atom thick and has acetylenic linkages (sp hybridization) linking

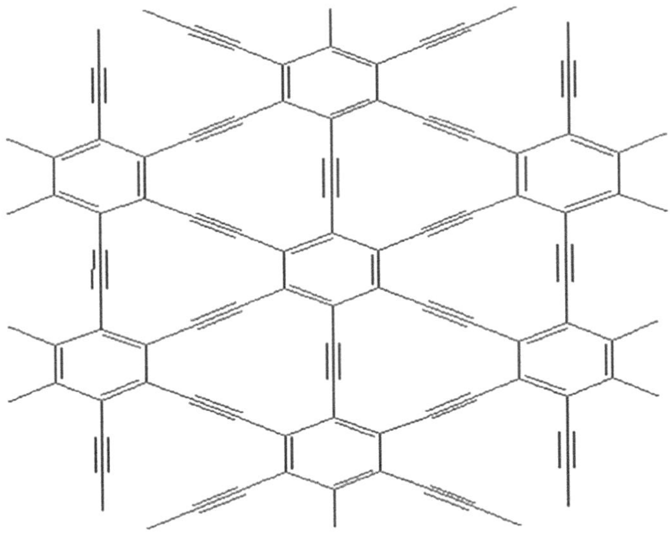

Fig. 7.6 Chemical structure of a graphyne plane

graphene's hexagons (benzene rings with sp^2 hybridization; see Fig. 7.6). One linear acetylenic chain connects carbon hexagons in the graphyne structure, whereas graphdiyne contains two acetylenic (di-acetylenic) chains between carbon hexagons (Banerjee 2018). Graphynes are a kind of structure in which two neighboring sp^2-hybridized carbon atoms are linked by n "$-C \equiv C-$" connections. The peculiar electrical structure, mechanical strength, and optical features of graphyne pique the scientific community's curiosity. Graphynes contain well-distributed pore architectures and large p-conjugated systems, making them suitable for gas separation, catalysis, and energy-related sectors (Gao et al. 2019). Bonds formed by sp-hybridized carbon atoms have recently been used as a dynamic covalent synthetic method to create a crystalline γ-graphyne network (Hu et al. 2022). γ-Graphyne is the most symmetric sp^2/sp^1 carbon allotrope, and it can operate as a defect-free catalyst in the oxygen reduction process (Desyatkin et al. 2022; Yuan et al. 2020).

7.3 Properties of Graphene and Its Derivatives

7.3.1 Optical Properties

Understanding the basic optical characteristics of graphene is critical for developing any graphene-based optoelectronic device. Interband carrier transitions contribute the most to optical conductivity in the visible region. The optical conductivity of inherent or mildly doped graphene is virtually constant. In contrast, intraband scattering dominates the optical conductivity of multilayer graphene in the THz region. In photothermal therapy (PTT), graphene-based nanomaterials are used to target cancer cells, followed by NIR radiation to produce enough heat to trigger cancer cell death (Banerjee 2018). This is possible because of graphene's excellent optical adsorption in the near-infrared (NIR) range. Graphene's optical absorption (A) follows a straightforward relationship $A = \pi.\alpha = 2.3\%$, where $\alpha = 1/137$ is the fine structure constant. We can determine the amount of light that is absorbed by each layer of graphene by measuring the light intensity. On a silicon device, graphene is visible due to the strong nuclear force of just a single layer of atoms with light, which enables us to perform mechanical resonance measurements. Because of this feature, graphene may theoretically be utilized to construct incredibly effective solar panels.

7.3.2 *Mechanical Properties*

Because of its outstanding mechanical, physical, and thermal qualities, graphene has achieved widespread recognition and attention in the scientific community. Graphene's exceptional mechanical and other beneficial features demonstrate a significant potential for creating dominant electronics and have earned the moniker of "wonder substance" (Iqbal et al. 2020). Electromechanical devices, field-effect transistors (FET), strain sensing and electronics devices, supercapacitors, H_2 storage, and power-driven solar cells extensively use graphene materials. It is commonly transformed into quantum dots and nanoribbons for use in hydrogels, fizzes, and semiconductor devices for a range of energy applications (Sahu et al. 2021). Recent research has shown that inventive modifications of various graphene-supported polymers and metal-based grid nanocomposites have produced the most promising results. Since then, the number of new works associated with graphene-based nanocomposite has increased significantly during the past few decades. These investigations show that alternative graphene components could be added to the polymer framework to enhance the composite materials' mechanical, electrical, thermal, and other functional capabilities, thereby increasing the overall efficacy of the substances manufactured with these composites (Dorri Moghadam et al. 2015). Because of its unique sheet-like form, graphene has intriguing qualities such as high mechanical strength (Young's modulus: 2.4 0.4 TPa-single layer and 2 0.5 TPa-bilayered). The tensile strength of graphene is 130 GPa. It is also exceptionally stiff, has exceptional heat conductivity (5000 Wm^1K^1) (Balandin et al. 2008), has a high specific surface area (2630 m^2g^1), exhibits the quantum hall effect, which is necessary for applications such as vicinity sensing and pace detection, has an extraordinary optical transmission (97.7%), and is excellent for electronic transportation (Becerril et al. 2008). With zero bandgaps, it also efficiently stops the spread of flames.

7.3.3 *Electronic Properties*

The fact that graphene is a semimetal with both holes and electrons as charge carriers, as well as its outstanding electrical conductivity, is one of its most essential properties (Cooper et al. 2012). In graphene, each carbon atom is connected to three additional carbon atoms in a 2D plane (the electrical configuration of a carbon atom contains six electrons; two in the inner shell and four in the outermost shell). As a result of these four outer shell electrons participating in chemical bonding, in graphene, one electron from each carbon atom is free in the third dimension for electronic conduction. These highly mobile electrons are known as pi (π), which may be found both above and below the graphene sheet. The orbitals overlap and contribute to the enhancement of carbon–carbon bonding in graphene. Fundamentally, the electronic characteristics of graphene are determined by the bonding and anti-bonding of these orbitals (the valence and conduction bands). These π electrons contribute to the enhancement of c–c bonds in graphene. Graphene's electronic mobility has been demonstrated in tests. It is claimed that due to their lack of mass, graphene electrons behave similarly to photons in their mobility. Ballistic transport describes the ability of these charge carriers to travel sub-micrometer distances without dispersion. However, the substrate used and the graphene's purity will be the limiting factors. Mobility may be limited to 40,000 $cm^2V^1s^1$ when silicon dioxide is employed as the substrate (Radadiya 2015).

7.3.4 *Photochemistry*

Photogenerated free radicals have recently been shown to successfully trigger graphene's chemical processes. This graphene photochemistry creates a novel path for zero-gap graphene's covalent functionalization and band structure alteration. Because of its infinite 2D geometry, graphene photochemistry is particularly appealing for studying geometrical covalent chemical reactions, such as single and double-sided covalent addition reactions, asymmetric interactions on two surfaces of monolayer graphene, edge-selective sciences, as well as interlayer coupling-dependent graphene chemistry (Zhang and Liu 2015). Free radicals are highly reactive chemical entities easily produced by photochemical or thermochemical processes. It is possible to break the inert –C = C– bonds in graphene chemistry despite the high energy barriers because of the significant reactivity of free radicals. Fluoropolymer-covered graphene has recently been laser irradiated in order to create fluorographene using a straightforward photochemical process, in which fluoropolymer molecules release F, functionalizing graphene in the process (Lee et al. 2012). Photocatalytic oxidation based on TiO_2 is a common photochemical process widely used to break down various organic contaminants for environmental protection. Under UV irradiation and appropriate humidity, TiO_2 may be used to create highly reactive OH. As a result of these photogenerated radicals' stepwise reactions with the graphenic carbons, oxy groups (such as epoxy, hydroxyl, carbonyl, and carboxyl groups) are integrated within the graphene basal plane. The photogenerated OH can behave as precise chemical scissors that cut graphene into random forms by adjusting the UV light intensity and irradiation time, enabling the varied design and manufacture of graphene-based devices and circuits (Zhang et al. 2011a, b, c). Understanding graphene's photochemical reactivity is critical for assessing its potential as a transparent electronic material. Furthermore, photochemistry may enable unique techniques to covalently change graphene, allowing it to be tailored in terms of chemical, electrical, and mechanical characteristics (Zhang et al. 2011a, b, c).

7.3.5 *Large Surface Area*

Studies on thin graphene nanoparticle sheets with a big surface area but a small thickness have indicated that they provide improved morphological, flexural, tensile, and other thermal characteristics, increasing the final composites' effective performance (Sahu et al. 2021). A high surface area with corrugated surfaces allows huge amounts of molecules/ions to be absorbed and faster charge buildup, enhancing sensitivity and response time (Yol Jeong et al. 2015). Because of their huge surface area and high conductivity, the proposed 3D graphene materials may be employed in EDLCs. They can also support pseudocapacitors due to their high conductivity and superior mechanical and electrochemical stability (Zhao et al. 2013). Supercapacitor electrodes made of 3D hierarchical graphene/polypyrrole aerogels exhibit high specific capacitance of 253 Fg^{-1}, strong rate performance, and exceptional cycle stability (Ye and Feng 2014). Graphene's huge surface area is ideal for lithium-ion storage, while its high electrical conductivity facilitates lithium ion and electron movement (Gong et al. 2013). In addition to these two main advantages of 3D graphene materials for electrochemical energy storage, they also offer space to accommodate the significant volume shift that high-capacity electrode materials like Sn, Si, or transition metal oxides/sulfides encounter during charge–discharge cycles (Ma et al. 2020). For instance, 3D porous graphene networks with Sn nanoparticle anchors contained in graphene shells performed incredibly well as a superior LIB anode.

7.3.6 *Easy Bio-Functionalization*

The lateral aspect, surface chemistry, grade, density of defect, and pureness of graphene oxide (GO) sheets can all vary. Because it has epoxy, hydroxyl, and carboxyl groups on its surface, GO may be covalently linked with various materials, allowing it to be functionalized and endowed with the required activity. Nanotube toxicity appears to be impacted by their degree of dispersion, amount of functionalization, and length (Nanda et al. 2015). Because pure graphene is largely inert with little chemical reactivity and solubility, surface functionalization is required to permit its usage in biological applications. By modifying the degree of ionization, the functionalization of graphene alters its surface charge and hydrophobicity, which determines its physical effects. For example, negatively charged surfaces are produced by GO, which is high in carboxyl groups (COOH). The surface becomes positive when COOH groups are functionalized with amino groups such as NH2 and PEG through coupling mechanisms (Guo et al. 2021). In contrast, functionalization with poly-m-aminobenzene sulfonic acid (NH_2/SO_3H) or methoxyl (OCH_3) ended functional groups makes the surface neutral (Tu et al. 2014). Stable dispersion in organic solvents can be achieved by covalently functionalizing graphene materials with isocyanates, aliphatic amines, and aromatic chemicals (Su et al. 2009). Furthermore, surface functionalization of GBMs with big polymers (like PEG, poly-L-lysine, dextran, polycaprolactone, or polyvinyl alcohol), proteins, or DNA significantly reduces toxicity and improves biocompatibility.

7.3.7 Availability of Delocalized π-Electrons

A conjugated system, which may lower the overall energy of the molecule and improve its stability, is a web of interconnected p-orbitals containing delocalized electrons in complexes with varying single or multiple bonds. The overlap of one p-orbital with another over an intervening sigma bond is known as conjugation. Overlapping p-orbitals that span the adjacent single bonds are seen in conjugated systems. They make it possible for π-electron delocalization to occur between all nearby aligned p-orbitals. The π electrons are connected to a collection of atoms rather than a single bond or atom. Graphene, graphite, conductive polymers, and carbon nanotubes contain the most conjugated systems. Great electrical conductivity is detected due to the delocalized out-of-plane p bonds formed by the sp^2 hybridization of carbon atoms.

7.3.8 High Drug Loading Capacity

Graphene has a greater drug loading capacity than other nano-drug carrier systems, such as single-walled carbon nanotubes (i.e., up to a 200% loading ratio of loaded drug weight to vehicle). Ibuprofen, DOX, heparin, ellagic acid, 5-fluorouracil, and camptothecin have all been employed as nanocarriers for bioactive chemicals and pharmaceuticals (Zainal-Abidin et al. 2020). Because of noncovalent forces, GO, the carrier, has a high loading capacity for the anticancer medication doxorubicin hydrochloride (DOX), and drug release may be initiated by a lower pH value (Chang et al. 2018). Several papers on in vitro controlled loading of the drug and its targeted delivery using graphene, as well as its derivatives as bio-delivery carriers for cancer and other non-cancer therapies, have been published. A superparamagnetic GO-Fe_3O_4 nanohybrid, for example, is employed as a drug delivery vehicle for directing drugs toward tumor cells, with the release of drugs regulated by pH levels (Banerjee 2018). Similar to this, exfoliated graphene with surface modifications is used as a nanocarrier to deliver the breast cancer drug tamoxifen citrate in a time-dependent manner, preferentially increasing the mortality of transformed cancer cells over normal cells (Misra et al. 2012).

7.3.9 Affinity Toward Water

For many years, graphene was thought to have a hydrophobic surface with wetting qualities comparable to graphite (Shin et al. 2010). According to recent research, graphene on copper is originally more hydrophilic, and its observed hydrophobicity is caused by airborne pollutants or water adsorption. Graphene's intrinsic hydrophilicity is connected to its Fermi level state rather than charge trapping at the interface between graphene and water. Doping can be used to modify graphene's hydrophilicity (Hong et al. 2016). Anti-corrosive protective coatings are critical for increasing material sustainability without affecting bulk qualities. Furthermore, an anti-corrosive coating may successfully perform its function because it has toughness and adhesion to endure impacts, appropriate flexibility, inherent durability, weathering or mechanical abuse, and appearance and cracking when stressed. The super-hydrophobic characteristic of graphene will be an extra benefit for anti-wear applications. Graphene incorporation might benefit devices with anti-corrosion, anti-sticking, self-cleaning, and reduced friction (Ashok Kumar et al. 2022).

7.4 Unique Characteristics of Multidimensional Graphene Structures and Their Derivatives

7.4.1 Graphene Quantum Dots

Graphene Quantum dots (also called graphene quantum disks or graphene quantum islands) can be defined as nanoparticles of 0D transverse size lesser than 100 nm. Usually, GQDs have a single atomic layer of carbon. Still, some GQDs can be synthesized with attached functional groups (oxygen and hydrogen) and more than one (usually 5) atomic layer smaller than 10 nm. Such functional groups at the edges of a GQD enable it to be dispersed in water easily. GQDs show unique properties like structural and chemical stability, robustness, multi-photon excitation, biocompatibility, catalytic activity, and photoluminescence. Its structural and chemical properties can be easily modulated.

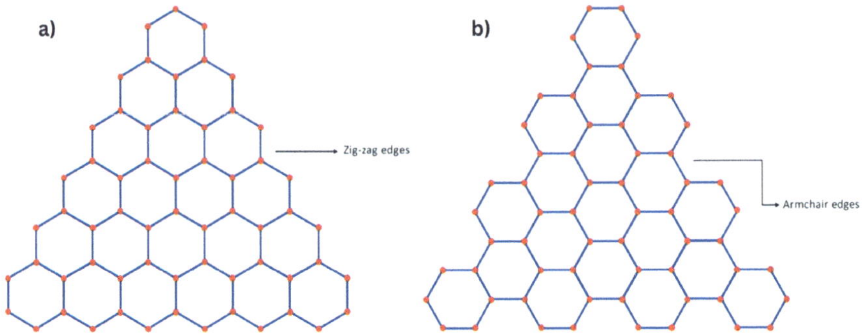

Fig. 7.7 A significant increase is observed in the differential conductance (influenced by rzig value) in (**a**) GQD with zig-zag edges, unlike in (**b**) GQD with armchair edges

7.4.1.1 Structural Features

a. They possess diameters of 1–5 nm and topological height of 0.5–1.5 nm, as represented in an Atomic Force Microscopy image (AFM).
b. Their crystalline structure has a lattice spacing of about 2.11 Angstrom, derived with the help of a High-Resolution Transmission Electron Microscope (HRTEM).
c. The Quantum confinement effect refers to changes in a macroscopic material's properties when its size is reduced to below 100 nm. The material is now considered nanomaterial and has different properties than bulk materials. This confinement of the particles turns continuous energy levels into discrete energy levels. In graphene quantum dots, quantum confinement induces changes in the magnetic, optical, thermal, and electronic properties that are drastic compared to those of their bulk counterpart.

Edge effect: The fraction of zig-zag edges is a crucial factor that affects the electronic properties of GQDs. The fraction (rzig value) is defined as the ratio of the zig-zag edges to the complete length of the resolved zig-zag and armchair edges. This ratio does not include the armchair and unassigned edges—the fraction of zig-zag influences the measured energy gap. An increase in the ratio leads to an increase in conductance. Adding some functional groups at the edges produces fluctuating variations in the positions of the zig-zag energy state. Thus, edges that are modified by adding various functional groups or moieties tune the electronic properties of graphene quantum dots and have a specified effect on their optical properties (see Fig. 7.7).

7.4.1.2 Photoluminescence

Most of the GQDs show luminescence, with their absorption spectrum peaking at 230 nm. This activity is credited to the pi–pi* excitatory transition of the p bonds. It is noted that the larger GQDs exhibit higher photoluminescence energy than smaller

GQDs, contradicting the quantum effect. This is because the edge microstructure and the corresponding strong bound exciton (pair of an electron and positive hole that moves together as a unit) override the anomalous photoluminescence behavior of larger GQDs. The photoluminescence of GQDs can be tuned to obtain emission spectra of near-infrared, blue, green, red, and yellow by modulating the number of oxygen atoms. The amount of oxygen is directly proportional to the bandgap. The PL is sometimes pH dependent. For example, a blue shift is observed in the exciton absorption/emission with an increasing concentration of H.

7.4.1.3 Cell Membrane Permeability

A 2014 study explored graphene quantum dots' permeability and transport mechanism across the biological membrane (Wang et al. 2015). GQDs were tested as potential drug carriers by investigating their membrane permeability and transport mechanisms on a Madin Darby Canine Kidney (MDCK) cell monolayer. According to the research, MDCK cells are capable of transporting and uptaking GQDs in a time and concentration-dependent fashion. The membrane transport was facilitated by incubation as it assisted lipid raft formation. The experimentally derived Papp values for two different sizes of GQDs demonstrated that smaller GQDs are well-transported species while the larger ones show mediocre membrane permeability.

7.4.1.4 Applications of GQDs

a. They are extensively exploited as probes for environmental, electrochemical, biological, and optoelectronic operations.
b. Owing to the dimensional resemblance of GQDs to molecules like small proteins, nucleic acids and antibodies can be easily grafted on these molecules.
c. They heighten the surface of biosensors.
d. They are widely harnessed as fluorescent probes by virtue of their broad excitation wavelength, photon stability, and narrow emission range. Besides, their background in the carbon family makes them the ideal fluorescent materials due to their non-toxicity and biocompatibility.
e. Horseradish peroxidase (HRP) is a metalloenzyme that can be conjugated to a labeled molecule. It yields a colorful, luminescent, or fluorometric derivative of that labeled molecule upon interaction with the compatible substrate for detection and quantification. GQDs dispose of peroxidase (POD)-like catalytic properties that carry out a redox reaction between peroxide and an electron-donating substrate. The GQDs are now replacing HRP-bound secondary bioreceptors as they are faster and more cost-effective (Mansuriya and Altintas 2020; Zhao et al. 2020).

7.4.2 *Graphene Nanoribbons*

Graphene nanoribbons are narrow strips of graphene with a width (W) less than 100 nm. This owes to change a lot of properties of the micromaterial. GNRs possess tunable bandgaps, making them distinct compared to 2D graphene. They have a high amount of charge-carrier mobilities which makes them favorable materials for various applications like radio-frequency transmitters or field-effect transistors (FETs).

7.4.2.1 Structure

GNRs are quasi-1D graphene. They are categorized based on the structure of their edges, which can be Armchair or Zig-zag. These edge structures determine the electronic and magnetic properties of the GNRs (see Fig. 7.8). A few differences between the two are.

a. GNRs with zig-zag edges possess edge states energies close to the Fermi level, while the GNRs with armchair edges possess bandgaps that are sensitive to the width of the ribbons.
b. GNRs with zig-zag edges show greater reactivity toward halogenation than those with armchair edges due to the larger number of electrons.
c. Raman spectroscopy visualized a relatively less intense D band of GNRs, which possess zig-zag edges than that of GNRs with armchair edges (Kim et al. 2018; Bronner 2018).

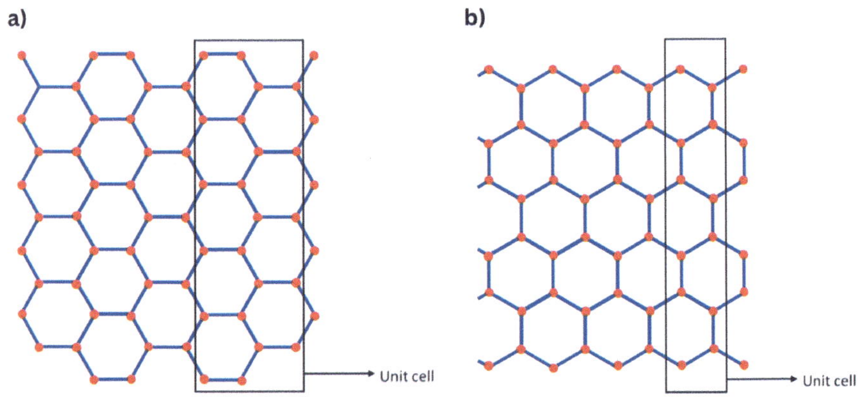

Fig. 7.8 Lattice structure of **a** an armchair-ended graphene nanoribbon, **b** a zig-zag ended graphene nanoribbon

7.4.2.2 Electronic Properties

It was found that the bandgaps of zig-zag GNRs are smaller than that of armchair GNRs. The size of the bandgaps increases with the corresponding increase in the size of GNRs. In an experiment, two GNRs differing only in the type of edge structures were analyzed. Upon performing X-ray Photoelectron spectroscopy, it was determined that the difference between the binding energies of GNRs with zig-zag edges and armchair edges was around 0.8 eV.

7.4.2.3 Mechanical Properties

The mechanical properties of the two types of GNRs were analyzed by computational modeling with the help of the fine element method (it is a method used to solve differential equations numerically), molecular dynamics (a digital simulation to analyze the movements of atoms and molecules), and density functional theory (properties of a many-electron system using functionals of spatially dependent electron density). Poisson's ratio, Young's modulus, and the shear modulus of GNRs vary in size (length and width) and shape. Young's modulus of armchair GNRs was found to be 1.24 TPa using the molecular dynamics method, while its tensile strength was found to be 175 GPa. Using the finite element method, it was determined that Young's modulus, tensile strength, and ductility of armchair GNRs are higher than those of zig-zag GNRs. Graphene nanoribbons are used in Field Effect Transistors (FETs), heterojunctions like p-n junctions, fast DNA sequencing (GNRs are suitable for making highly sensitive sensors due to their opening of bandgaps, thus allowing detection of translocation of DNA molecules), gas sensors, etc.

7.4.3 Graphene Fibers

Graphene fibers are 1D macroscopically assembled fibrous homo/heteropolymers of graphene and its derivatives. They are made by regularly aligning graphene sheets as building blocks continuously uniaxially. These fibers have the same properties as graphene-like mechanical and electrical properties, high flexibility, less weight, and ease of functionalization (Xu and Gao 2015).

7.4.3.1 Mechanical Properties

GFs have excellent mechanical properties like high tensile strength, lightweight, high elastic modulus, good electrical and thermal conductivities, etc. GFs have a tensile strength of about 130 GPa, an elastic modulus of 1.1 TPa, and high flexibility.

7.4.3.2 Electronic Properties

GFs possess superior electrical and thermal conductivities. Their high electrical conductivity reaches the mark of about 10^8 S/m, with carrier mobility of 200,000 cm^2/V s and ampacity (maximum current in amperes that a conductor can carry continuously when under use without exceeding its temperature rating) of 1–2GA/cm^2. It has a thermal conductivity of 5000W/m K. It was recently discovered that doping of GFs with silver ions leads to a remarkable increase in electrical conductivity.

7.4.3.3 Wet-Spinning Method and Neat GFs

Wet-spinning is a method of preparing fibers by solution spinning. In this, the chosen polymer powder is dissolved in a suitable solvent. Then this solution is further extruded through a spinneret into a coagulant (a solvent and non-solvent mixture). Here the diffusion of solvent and non-solvent leads to the coagulation of polymer solution, thus forming fibers. The polymer solution is forced through spinning pipes with a continuous uniaxial flow to allow fiber formation. The first continuous neat graphene fibers (graphene sheets aligned properly along the fiber axis) were produced by Gao et al. by bypassing graphene oxide liquid crystals through a coagulation bath which was then followed by chemical reduction. These manufactured fibers showed high flexibility and resistance to torsion.

The process of wet-spinning GFs from GO liquid crystals (LCs) involved the following steps (see Fig. 7.9):

a. Extrusion of GO LCs into coagulation baths by passing them through spinning pipes with a uniaxial flow. This process unified the previously random configurations and alignment of the GO liquid crystals along the flow direction.
b. Next, the LCs are passed through coagulation baths that contain solutions compatible with the solvents of the LCs but incompatible with the GO sheets taken. This allowed solvent exchange, leading to GO liquid crystalline gel fibers forming.

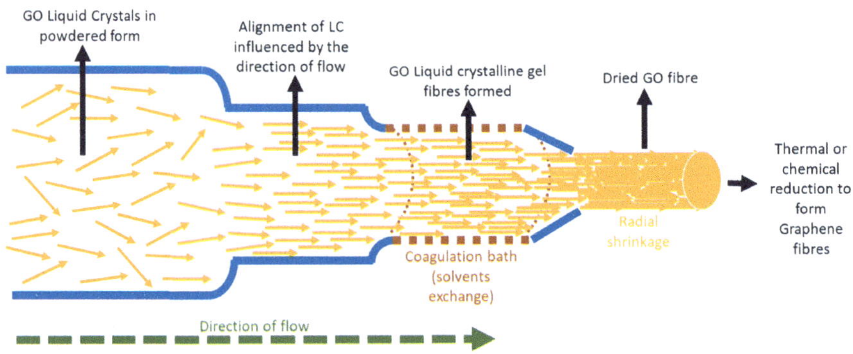

Fig. 7.9 Steps of fabrication of graphene fibers through the process of wet-spinning

c. These gel fibers are then dried, which causes them to shrink in a radial direction.
d. Then the graphene oxide fibers are converted to graphene fibers using chemical or thermal reduction.

7.4.3.4 Biomimetic Graphene Composite Fibers

These enhanced fibers are produced by adding graphene to polymeric composites. These composites comprise an extremely high level of graphene, ranging from 50% to more than 90%. This gives the composites their laminated structure. This structure resembles the "brick-and-mortar" configuration of the nacre-like fibers (called "biomimetic"); the graphene behaves as the bricks while the interlayer macromolecules act as the mortar.

7.4.4 Graphene Nanomeshes

We know graphene sheets generally do not possess any energy bandgaps, thus making them unfavorable for semiconductor applications. But energy bandgaps can be introduced by cutting the graphene sheets into graphene nanoribbons. The use of tensile or shear stress leads to the formation of graphene nanomeshes with periodic nanoholes, where the bandgap can be changed or regulated with the geometry and dimensions of these nanoholes. GNMs are prepared using block polymer lithography (Bai et al. 2010; Zhang et al. 2018).

7.4.4.1 Photocatalytic Cutting

This method is a progressive oxidation process in which oxygenic groups are introduced, and sp^2 bond breaking in the carbon network is done with the aid of photocatalysts like TiO_2 and ZnO. In this technique, the nanoholes are produced on the graphene sheet that is to be tailored into a nanomesh by regulating the photocatalytic reaction rather than the actual cutting of the graphene sheet. This process involves covering graphene with a patterned mask made of a UV-sensitive photocatalyst. Upon irradiation, UV generates electrons and holes that are injected into the graphene sheet. They, in turn, lead to the formation of highly reactive OH radicals that act as photo scissors and cut the sp^2 bonds in the carbon network. The OH radicals are formed by the reaction of the photocatalyst with the surrounding H_2O and O_2 molecules under UV.

7.4.4.2 Field Effect Transistors

Graphene, due to its 2D structure and high conductivity, is an ideal choice for manufacturing FET devices. Still, the absence of bandgap leads to Ion/Ioff ratios being lesser than ten which is too small to be used in digital applications. Thus, GNM possesses tunable electronic properties that allow their use in the fabrication of FETs. FETs made of graphene nanomeshes can support currents about 100 times larger than those made of individual graphene nanoribbons. Also, the on–off ratio can be regulated by changing the neck width.

7.4.4.3 Thermal Transition Properties

It is established that nanomeshes with circular pores have a thermal conductivity that is an exponential function of the nanomesh's neck width and a little on the pore lattice structure. The thermal conductivity of graphene nanomeshes with elliptical pores becomes anisotropic, which increases with decreasing density.

7.4.5 Rippled/Wrinkled Graphene

Transmission electron microscopy and scanning-probe microscopy experiments revealed that 2D graphene sheets exhibited nanosized, out-of-plane ripple-like deformations. Graphene is not always planar. In some cases, ripples are observed on its surface. These ripples and the differences in their height and range cause significant changes in the electronic and mechanical properties of the graphene sheet. The carbon–carbon bonds cause these ripples to attempt to minimize free energy during thermal vibrations and interatomic interactions by temporal and spatial modulation of the bond lengths to occupy space in the third dimension. Rippled graphene possesses bandgaps which makes it a good semiconductor material. The increase in the bandgap is directly proportional to the increase in the amplitude of the ripple, which is the distance between the graphene sheet and the substrate. This can also be induced by the p-cloud of delocalized electrons, which causes the asymmetric distribution of bond lengths. Wrinkles and crumples heavily influence electronic properties, including the suppression of weak localization electron–hole puddles, bandgap opening, the pseudo-magnetic field in bilayers and carrier scattering, enhancement of chemical activities, bio-interfacing, etc. (Deng and Berry 2016).

7.4.5.1 Electronic Properties

Ripples violate the electronic homogeneity of graphene. They suppress electron transport and mobility and lead to weak localization. Ripples have two main effects:

a. p–s rehybridization occurs between the neighboring atoms and causes a significant shift in the p-orbital energy, creating a notable variation in the electrochemical potential.
b. A change in the surrounding atom's hopping integral induces an effective vector potential.

7.4.5.2 Surface Properties

As reported by Zang et al., high wettability and transparency are two surface properties of crumpled graphene. It was demonstrated that the wrinkled graphene produced by the relaxation of highly pre-strained substance could be considered a potent strain sensor in which the change in the applied strain stimulates a proportional change in the resistance of the device.

7.4.6 Graphene Membranes

Graphene films are highly crystalline, one-atom-layer thick films of sp^2 hybridized carbon atoms. They are used to make thicker films by layer-by-layer stacking. Monolayers or more than one layer but less than ten layers constitute thin films, while the rest are considered thick films. Graphene-based films are composed of graphene oxide or reduced graphene oxide by virtue of their easy functionalization and low production costs. The produced graphene membranes are lightweight, stretchable, stiff, and flexible. The sample graphene is exfoliated from the given graphite by peeling a thin layer off of it with the help of Scotch tape. This technique is called the Peel-off or the "Scotch tape" method. This forms graphene flakes that have lateral dimensions of about 10–100 mm.

Also, there is a precipitation technique known as chemical vapor deposition. Here carbon atoms are dissolved in a metal catalyst substrate under high temperatures and hydrocarbon gas exposure. The metal substrates include nickel, copper, platinum, iridium, etc. When the substrate is cooled off, the carbon atoms precipitate out in the form of graphene sheets.

7.4.6.1 Heteroatom-Doped Graphene

Graphene is usually doped with various heteroatoms to enhance structural, mechanical, electrical, and chemical properties as required. This is done by single-element doping or dual-element doping, or metal doping. The differences in the size and electronegativity of the replacing heteroatoms and that of the carbon atoms induce variations in graphene's structural, mechanical, chemical, electrical, optical, and magnetic aspects. This substitution by the heteroatom is stable because of the covalent bond linkage of the dopant in the graphene lattice. But this disturbs graphene's unique honeycomb structure, leading to defects and disorders.

7.4.6.2 Antimicrobial Application

Graphene membranes can be used as antimicrobial coats. They do not degrade over time or release any biocides that might harm the environment. Besides, graphene shows antibacterial activity, the exact mechanism of which is still elusive. But its high mechanical and structural strength, water dispersibility, and facile surface modification are associated with the antibacterial properties it possesses. Graphene films show sheet adsorption on the surface of the bacterial cell membrane, followed by puncturing and penetration through the lipid bilayer membrane, lipid extraction, and oxidative stress.

7.4.7 3D Graphene Architectures

These include graphene sponges, foams, aerogels, hydrogels, and vertically aligned graphene nanosheet arrays (VAGNAs). Usually, restacking of graphene sheets by strong p-p stacking and van der Waals interactions causes suppression of the unique properties of graphene-like high mechanical strength, high conductivity, etc. Thus, 3DGs are constructed by bonding individual graphene sheets together, which will make these sheets repel, stop restacking, and help maintain the intrinsic properties of the 2D graphene sheets. They possess large surface areas and pore volumes, excellent electrical conductivities, low densities, and desirable mechanical properties. They exhibit low contact resistance, robustness, and easy functionalization. Their high surface area, conductivity, and mechanical and electrochemical stability make them suitable for electrical double-layer supercapacitors. Pollution caused by spills of petroleum products, toxic organic solvents, and crude oil causes extreme ecological crises. The conventional absorbers usually have less oil loading capacity and end up just absorbing large quantities of water, making further extraction and recycling difficult and unfeasible. Large pore sizes, high capacity to absorb both organic as well as inorganic contaminants, pore size and structure that can be regulated, surface exhibiting a high level of oleophilic and hydrophobic properties, and chemical and mechanical stability are some characteristic features possessed by 3D architectures derived from graphene nanosheets.

Multidimensional graphene structures along with their wide range of applications have been illustrated above in a tabular format (Table 7.2). In order to distinguish graphene-based materials from conventional catalysts, their characterization methods must consider both their distinctive surface structure and exceptional bulk properties (see Fig. 7.10).

Table 7.2 Multidimensional graphene structures and their applications

Sr. No	Multidimensional Graphene Structures	Applications	References
1	Graphene quantum dots	Biological therapies, Drug delivery systems, Bioimaging, photothermal therapy, Sensors, Fluorescent probes, etc.	Zhao et al. (2020)
2	Graphene nanoribbons	Molecular sensors, Acoustic detectors, Drug delivery vehicles, Photodetectors, thermodetectors, Sequencers, and tissue engineering constructs	Zakharova et al. (2021)
3	Graphene fibers	Electronic textiles, GF motors, GF actuators, biomimetic graphene composite fibers, and dye-sensitized photovoltaic wires	Cheng et al. (2014)
4	Graphene nanomeshes	Potential applications include field effect transistors, Supercapacitors, Chemical sensing, DNA sequencing, photothermal therapy, etc.	Yang et al. (2014)
5	Rippled/Wrinkled graphene	Supercapacitors, Lithium-ion batteries, Artificial muscle actuators, and Field effect transistors	Ma et al. (2020), Deng and Berry (2016)
6	Graphene membranes	Gas separation, Water purification, Electrodes of energy storage devices like lithium-ion batteries, and water desalination	Ma et al. (2020), Homaeigohar and Elbahri (2017)
7	3D graphene architectures	Supercapacitors, Fuel cells, solar cells, Hydrogen production and storage, and absorbent for organic and inorganic pollutants	Ma et al. (2020)

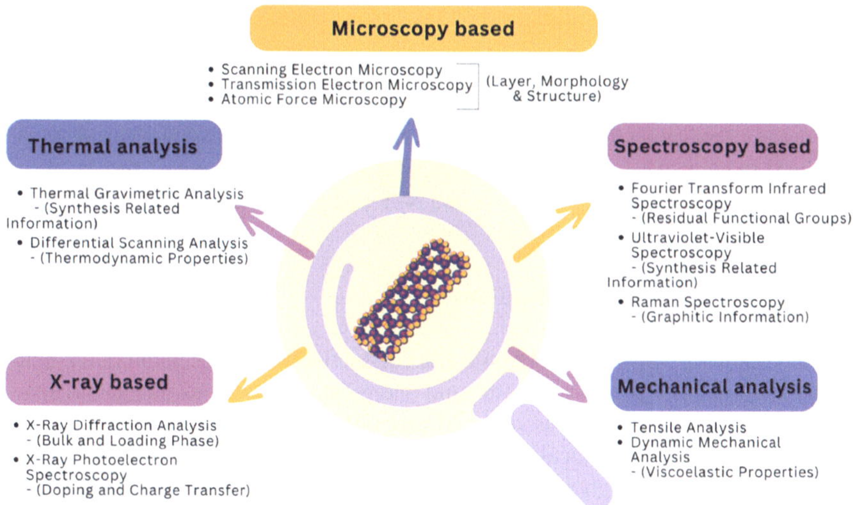

Fig. 7.10 Schematic diagram of characterization procedures used for validating the characteristics of graphene-based materials

7.5 Properties of Electrochemically Exfoliated Graphene Doped with Heteroatoms

7.5.1 Nitrogen

Graphene is doped with metals and other heteroatoms to enhance the mechanical and electrochemical properties like conductivity and carrier mobility of the pure form of graphene. Nitrogen is one of the most common heteroatoms used to dope graphene because of its small size, similar to that of a carbon atom. It also forms stable covalent bonds with surrounding carbon atoms by virtue of the valence electrons. The nitrogen atoms in the doped graphene have the potential to induce pseudo-capacitance, which in turn increases the specific capacitance of the material. The conventional method to produce N-doped graphene included heat processing. But this process yielded a product with low electrical conductivity and poor hydrophilic properties. Thus, N-doped graphene is made through the electrochemical exfoliation of graphite in an inorganic electrolyte (Jing et al. 2020).

7.5.1.1 Electrical Conductivity

It was reported that electrochemically exfoliated N-doped graphene in aqueous solutions of glycine and ammonia had high levels of nitrogen (6.05%) and oxygen (22.28%). It shows excellent electrical conductivity of up to 19 S/cm (Jing et al. 2020; Chao et al. 2017; Yang et al. 2016).

7.5.1.2 Wettability

Incorporating pyrrolic nitrogen (nitrogen atom that contributes two p electrons to the p system) induces the formation of active sites on the basal plane of the graphene sheet. This improves the wettability of the N-doped graphene.

7.5.1.3 Supercapacitor Applications

Graphene shows exceptional electronic and conductive properties, making it a noteworthy candidate for use in electronic and energy storage devices like supercapacitors. Upon examination, it was reported that n-doped graphene showed excellent performance in electrochemical aspects like high columbic efficiency, good electrochemical reversibility, high specific capacity (34.2 at a current density of 0.5 A/g), etc. It was also shown that N–Gh/N–Gh symmetric supercapacitors showed 92.3% retention of their initial specific capacity even after 5000 cycles, indicating their high cycling stability.

7.5.2 Boron

Boron-doped graphene shows excellent properties like high capacitance and electrocatalytic activity for oxygen reduction.

7.5.2.1 B-Doped Graphene in Energy Storage

B-doped graphene is used in the fabrication of Lithium-ion batteries. Boron stabilizes the lithium ions, and the 2 s electrons on Li atoms tend to migrate to the empty spaces associated with the dopant. It also enables an increase in the lithium ions' binding energy, which prevents the clustering of the lithium ions. Li atoms diffuse easily, even under strong interactions with Boron atoms.

7.5.2.2 Gas Sensing

Boron doping has a significant impact on the sensing of small gaseous molecules. It improves interactions with small molecules like NO, NH_3, NO_2, etc., that previously only weakly interacted with CO. This property allows B-doped nanoribbons, prepared by on-surface Ullmann reaction, to make efficient sensors (Shi et al. 2018a, b; Agnoli and Favaro 2016).

7.5.3 Sulfur

7.5.3.1 Supercapacitor Applications

Evaluating the electrochemical capacitance performance of S-graphene showed that those containing about 3.37% sulfur exhibited superior capacitive performance. The capacitive properties of S-graphene decrease with an increase in the amount of sulfur content by virtue of poor ionic transportation and insufficient contact between the electrode and electrolyte. The highest specific capacitance obtained for S-graphene at a current density of 10 was 160 F/g.

7.5.3.2 Electron Transfer

An analysis using the Electrochemical Impedance Spectroscopy (EIS) technique concluded that sulfur-doped graphene shows low electron transfer resistance due to its high electron-conductive properties (Parveen et al. 2016).

Table 7.3 Applications of electrochemically exfoliated graphene doped with elements

Sr. No.	Doping agent used	Applications	References
1	Nitrogen	Electrocatalytic activities, Liquid fuel cells, and supercapacitors	Shao et al. (2010)
2	Boron	They are used in the fields of biology, electrochemistry, Photovoltaics, sensing (gas sensors), and catalysis	Agnoli and Favaro (2016)
3	Sulfur	Electrocatalysts for oxygen reduction, supercapacitors, and biosensing probes	Shi et al. (2018a, b), Ganganboina et al. (2021)
4	Phosphorous	Adsorptions and catalytic/electrocatalytic degradation of organic or inorganic pollutants	Feng et al. (2020)

7.5.4 Phosphorous

7.5.4.1 Enhanced Capacitive Performance

It was reported that electrochemically exfoliated phosphorus-doped graphene had a capacitance value of 367 Fg^{-1} and an energy density of 59 Wh kg^{-1}, which is significantly higher than the phosphorus-free counterparts of graphene. Graphene doped with phosphorus shows enhanced capacitive properties in an aqueous electrolyte, owing to the increased pseudo-capacitance and good electrical conducting network in the electrode structure. Thus, supercapacitors manufactured using P-doped graphene exhibited high specific capacitance of 367 Fg^{-1}, superior frequency response, and charge–discharge profiles of 5000 cycles, with energy and power densities of 9 kW kg^{-1} and 59 Wh kg^{-1} (Karthika et al. 2013).

For applications related to the above doping agent, refer to Table 7.3 given below.

7.6 Conclusion

It can thus be concluded that electrochemical exfoliation of graphite is a highly potent approach of yielding graphene sheets in high quality and amount. This is attributed to the reduced oxygen content, greater sheet size, and a high carbon/oxygen ratio, coupled with excellent electrical and chemical properties equivalent to chemical vapor-formed graphene. In addition, the remarkable quality and processability of these sheets offer huge prospects for flexible as well as rigid organic electronic devices. Future research can be done in GO, rGO (pristine and biofunctionalized), graphene, graphone, graphdiyne, graphyne, fluorographene, and their metallo-doped derivatives (as biomedical materials). Graphene's outstanding mechanical robustness, adjustable morphology, high carrier mobility, and overall stability make it and its derivatives favorable candidates that can be used over a wide range of applications. Investigations must be done to develop a proper, scalable synthesis protocol, understanding and evaluating the toxicity and biocompatibility, electronic, chemical, structural, mechanical, optical, electrochemical, and surface properties of these nanomaterials.

References

Agnoli, S., Favaro, M.: Doping graphene with boron: a review of synthesis methods, physico-chemical characterization, and emerging applications. J. Mater. Chem. A **4**, 5002–5025 (2016). https://doi.org/10.1039/C5TA10599D

Ashok Kumar, S.S., Bashir, S., Ramesh, K., Ramesh, S.: A comprehensive review: super hydrophobic graphene nanocomposite coatings for underwater and wet applications to enhance corrosion resistance. FlatChem. **31**, 100326 (2022). https://doi.org/10.1016/j.flatc.2021.100326

Bai, J., Zhong, X., Jiang, S., Huang, Y., Duan, X.: Graphene nanomesh. Nat. Nanotechnol.nanotechnol. **5**, 190–194 (2010). https://doi.org/10.1038/nnano.2010.8

Bak, S., Kim, D., Lee, H.: Graphene quantum dots and their possible energy applications: a review. Curr. Appl. Phys. **16**, 1192–1201 (2016). https://doi.org/10.1016/j.cap.2016.03.026

Balandin, A.A., Ghosh, S., Bao, W., Calizo, I., Teweldebrhan, D., Miao, F., Lau, C.N.: Superior thermal conductivity of single-layer graphene. Nano Lett. **8**, 902–907 (2008). https://doi.org/10.1021/nl0731872

Banerjee, A.N.: Graphene and its derivatives as biomedical materials: future prospects and challenges. Interface Focus. **8**, 20170056 (2018). https://doi.org/10.1098/rsfs.2017.0056

Barone, V., Hod, O., Scuseria, G.E.: Electronic structure and stability of semiconducting graphene nanoribbons. Nano Lett. **6**, 2748–2754 (2006). https://doi.org/10.1021/nl0617033

Becerril, H.A., Mao, J., Liu, Z., Stoltenberg, R.M., Bao, Z., Chen, Y.: Evaluation of solution-processed reduced graphene oxide films as transparent conductors. ACS Nano **2**, 463–470 (2008). https://doi.org/10.1021/nn700375n

Bhattacharya, A., Bhattacharya, S., Majumder, C., Das, G.P.: Third conformer of graphane: a first-principles density functional theory study. Phys. Rev. B **83**, 033404 (2011). https://doi.org/10.1103/PhysRevB.83.033404

Bianco, A., Cheng, H.-M., Enoki, T., Gogotsi, Y., Hurt, R.H., Koratkar, N., Kyotani, T., Monthioux, M., Park, C.R., Tascon, J.M.D., Zhang, J.: All in the graphene family–a recommended nomenclature for two-dimensional carbon materials. Carbon N. Y. **65**, 1–6 (2013). https://doi.org/10.1016/j.carbon.2013.08.038

Bronner, C.: Bottom-up synthesis and electronic structure of graphene nanoribbons on surfaces. In: Encyclopedia of Interfacial Chemistry, pp. 210–225. Elsevier (2018)

Chang, X., Zhang, Y., Xu, P., Zhang, M., Wu, H., Yang, S.: Graphene oxide/$MnWO_4$ nanocomposite for magnetic resonance/photoacoustic dual-model imaging and tumor photothermo-chemotherapy. Carbon N. Y. **138**, 397–409 (2018). https://doi.org/10.1016/j.carbon.2018.07.058

Chao L, Qin Y, Liu Y, Kong Y, Chu F (2017) Electrochemically exfoliating graphite into N-doped graphene and its use as a high efficient electrocatalyst for oxygen reduction reaction. J. Solid State Electrochem. **21**, 1287–1295 (2017). https://doi.org/10.1007/s10008-016-3480-4

Cheng, H., Hu, C., Zhao, Y., Qu, L.: Graphene fiber: a new material platform for unique applications. NPG Asia Mater. **6**, e113–e113 (2014). https://doi.org/10.1038/am.2014.48

Compton, O.C., Nguyen, S.T.: Graphene oxide, highly reduced graphene oxide, and graphene: versatile building blocks for carbon-based materials. Small **6**, 711–723 (2010). https://doi.org/10.1002/smll.200901934

Cooper, D.R., D'Anjou, B., Ghattamaneni, N., Harack, B., Hilke, M., Horth, A., Majlis, N., Massicotte, M., Vandsburger, L., Whiteway, E., Yu, V.: Experimental review of graphene. ISRN Condens. Matter Phys. **2012**, 1–56 (2012). https://doi.org/10.5402/2012/501686

Deng, S., Berry, V.: Wrinkled, rippled and crumpled graphene: an overview of formation mechanism, electronic properties, and applications. Mater. Today **19**, 197–212 (2016). https://doi.org/10.1016/j.mattod.2015.10.002

Desyatkin, V.G., Martin, W.B., Aliev, A.E., Chapman, N.E., Fonseca, A.F., Galvão, D.S., Miller, E.R., Stone, K.H., Wang, Z., Zakhidov, D., Limpoco, F.T., Almahdali, S.R., Parker, S.M., Baughman, R.H., Rodionov, V.O.: Scalable synthesis and characterization of multilayer γ-graphyne, new carbon crystals with a small direct band gap. J. Am. Chem. Soc. **144**, 17999–18008 (2022). https://doi.org/10.1021/jacs.2c06583

Dorri Moghadam, A., Omrani, E., Menezes, P.L., Rohatgi, P.K.: Mechanical and tribological properties of self-lubricating metal matrix nanocomposites reinforced by carbon nanotubes (CNTs) and graphene—a review. Compos. Part B Eng. **77**, 402–420 (2015). https://doi.org/10.1016/j.compositesb.2015.03.014

Ebajo, V.D., Santos, C.R.L., Alea, G.V., Lin, Y.A., Chen, C.-H.: Regenerable acidity of graphene oxide in promoting multicomponent organic synthesis. Sci. Rep. **9**, 15579 (2019). https://doi.org/10.1038/s41598-019-51833-2

Feng, L., Qin, Z., Huang, Y., Peng, K., Wang, F., Yan, Y., Chen, Y.: Boron-, sulfur-, and phosphorus-doped graphene for environmental applications. Sci. Total. Environ. **698**, 134239 (2020). https://doi.org/10.1016/j.scitotenv.2019.134239

Ganganboina, A.B., Dega, N.K., Tran, H.L., Darmonto, W., Doong, R.-A.: Application of sulfur-doped graphene quantum dots@gold-carbon nanosphere for electrical pulse-induced impedimetric detection of glioma cells. Biosens. Bioelectron.. Bioelectron. **181**, 113151 (2021). https://doi.org/10.1016/j.bios.2021.113151

Gao, X., Liu, H., Wang, D., Zhang, J.: Graphdiyne: synthesis, properties, and applications. Chem. Soc. Rev. **48**, 908–936 (2019). https://doi.org/10.1039/C8CS00773J

Gomes, K.K., Mar, W., Ko, W., Guinea, F., Manoharan, H.C.: Designer dirac fermions and topological phases in molecular graphene. Nature **483**, 306–310 (2012). https://doi.org/10.1038/nature10941

Gong, Y., Yang, S., Liu, Z., Ma, L., Vajtai, R., Ajayan, P.M.: Graphene-network- backboned architectures for high-performance lithium storage. Adv. Mater. **25**, 3979–3984 (2013). https://doi.org/10.1002/adma.201301051

Guo, Z., Chakraborty, S., Monikh, F.A., Varsou, D., Chetwynd, A.J., Afantitis, A., Lynch, I., Zhang, P.: Surface functionalization of graphene-based materials: biological behavior, toxicology, and safe-by-design aspects. Adv. Biol. **5**, 2100637 (2021). https://doi.org/10.1002/adbi.202100637

Homaeigohar, S., Elbahri, M.: Graphene membranes for water desalination. NPG Asia Mater. **9**, e427–e427 (2017). https://doi.org/10.1038/am.2017.135

Hong, G., Han, Y., Schutzius, T.M., Wang, Y., Pan, Y., Hu, M., Jie, J., Sharma, C.S., Müller, U., Poulikakos, D.: On the mechanism of hydrophilicity of graphene. Nano Lett. **16**, 4447–4453 (2016). https://doi.org/10.1021/acs.nanolett.6b01594

Hu, Y., Wu, C., Pan, Q., Jin, Y., Lyu, R., Martinez, V., Huang, S., Wu, J., Wayment, L.J., Clark, N.A., Raschke, M.B., Zhao, Y., Zhang, W.: Synthesis of γ-graphyne using dynamic covalent chemistry. Nat. Synth. **1**, 449–454 (2022). https://doi.org/10.1038/s44160-022-00068-7

Hu, M., Mi, B.: Layer-by-layer assembly of graphene oxide membranes via electrostatic interaction. J. Memb. Sci. **469**, 80–87 (2014). https://doi.org/10.1016/j.memsci.2014.06.036

Iqbal, A.A., Sakib, N., Iqbal, A.K.M.P., Nuruzzaman, D.M.: Graphene-based nanocomposites and their fabrication, mechanical properties and applications. Materialia **12**, 100815 (2020). https://doi.org/10.1016/j.mtla.2020.100815

Jia, Z., Li, Y., Zuo, Z., Liu, H., Huang, C., Li, Y.: Synthesis and properties of 2D carbon—graphdiyne. Acc. Chem. Res. **50**, 2470–2478 (2017). https://doi.org/10.1021/acs.accounts.7b00205

Jing, M., Wu, T., Zhou, Y., Li, X., Liu, Y.: Nitrogen-doped graphene via in-situ alternating voltage electrochemical exfoliation for supercapacitor application. Front. Chem. **8** (2020). https://doi.org/10.3389/fchem.2020.00428

Jiříčková, A., Jankovský, O., Sofer, Z., Sedmidubský, D.: Synthesis and applications of graphene oxide. Materials (Basel) **15**, 920 (2022). https://doi.org/10.3390/ma15030920

Karthika, P., Rajalakshmi, N., Dhathathreyan, K.S.: Phosphorus-doped exfoliated graphene for supercapacitor electrodes. J. Nanosci. Nanotechnol.nanosci. Nanotechnol. **13**, 1746–1751 (2013). https://doi.org/10.1166/jnn.2013.7112

Kim, J., Lee, N., Min, Y.H., Noh, S., Kim, N.-K., Jung, S., Joo, M., Yamada, Y.: Distinguishing zigzag and armchair edges on graphene nanoribbons by X-ray photoelectron and raman spectroscopies. ACS Omega **3**, 17789–17796 (2018). https://doi.org/10.1021/acsomega.8b02744

Kovtyukhova, N.I., Ollivier, P.J., Martin, B.R., Mallouk, T.E., Chizhik, S.A., Buzaneva, E.V., Gorchinskiy, A.D.: Layer-by-layer assembly of ultrathin composite films from micron-sized graphite oxide sheets and polycations. Chem. Mater. **11**, 771–778 (1999). https://doi.org/10.1021/cm981085u

Lee, W.H., Suk, J.W., Chou, H., Lee, J., Hao, Y., Wu, Y., Piner, R., Akinwande, D., Kim, K.S., Ruoff, R.S.: Selective-area fluorination of graphene with fluoropolymer and laser irradiation. Nano Lett. **12**, 2374–2378 (2012). https://doi.org/10.1021/nl300346j

Li, D., Müller, M.B., Gilje, S., Kaner, R.B., Wallace, G.G.: Processable aqueous dispersions of graphene nanosheets. Nat. Nanotechnol.nanotechnol. **3**, 101–105 (2008). https://doi.org/10.1038/nnano.2007.451

Li, H., Ji, H., Zhang, R., Zhang, W., Pan, B., Liu, W., Sun, W.: Hydrogen bonding rather than cation bridging promotes graphene oxide attachment to lipid membranes in the presence of heavy metals. Environ. Sci. Nano **7**, 2240–2251 (2020). https://doi.org/10.1039/D0EN00366B

Li, Z., Wang, R., Wu, S., Xue, Z., Zhu, D., Zou, J., Li, X.: In situ observation of metal ion interactions with graphene oxide layers: From the growth of metal hydroxide to metal oxide formation. Carbon N. Y. **184**, 721–727 (2021). https://doi.org/10.1016/j.carbon.2021.08.073

Libisch, F., Stampfer, C., Burgdörfer, J.: Graphene quantum dots: Beyond a Dirac billiard. Phys. Rev. B **79**, 115423 (2009). https://doi.org/10.1103/PhysRevB.79.115423

Liu, Z., Rios-Carvajal, T., Andersson, M.P., Ceccato, M., Stipp, S.L., Hassenkam, T.: Ion effects on molecular interaction between graphene oxide and organic molecules. Environ. Sci. Nano. **6**, 2281–2291 (2019). https://doi.org/10.1039/C9EN00274J

Ma, R., Zhou, Y., Bi, H., Yang, M., Wang, J., Liu, Q., Huang, F.: Multidimensional graphene structures and beyond: unique properties, syntheses and applications. Prog. Mater. Sci. **113**, 100665 (2020). https://doi.org/10.1016/j.pmatsci.2020.100665

Malik, S.A., Mohanta, Z., Srivastava, C., Atreya, H.S.: Modulation of protein–graphene oxide interactions with varying degrees of oxidation. Nanoscale Adv. **2**, 1904–1912 (2020). https://doi.org/10.1039/C9NA00807A

Mansuriya, B., Altintas, Z.: Applications of graphene quantum dots in biomedical sensors. Sensors **20**, 1072 (2020). https://doi.org/10.3390/s20041072

Medhekar, N.V., Ramasubramaniam, A., Ruoff, R.S., Shenoy, V.B.: Hydrogen bond networks in graphene oxide composite paper: structure and mechanical properties. ACS Nano **4**, 2300–2306 (2010). https://doi.org/10.1021/nn901934u

Merchant, C.A., Healy, K., Wanunu, M., Ray, V., Peterman, N., Bartel, J., Fischbein, M.D., Venta, K., Luo, Z., Johnson, A.T.C., Drndić, M.: DNA translocation through graphene nanopores. Nano Lett. **10**, 2915–2921 (2010). https://doi.org/10.1021/nl101046t

Meyer, J.C., Geim, A.K., Katsnelson, M.I., Novoselov, K.S., Booth, T.J., Roth, S.: The structure of suspended graphene sheets. Nature **446**, 60–63 (2007). https://doi.org/10.1038/nature05545

Miao, F., Wijeratne, S., Zhang, Y., Coskun, U.C., Bao, W., Lau, C.N.: Phase-Coherent Transport in Graphene Quantum Billiards. Science (80-.). **317**, 1530–1533 (2007). https://doi.org/10.1126/science.1144359

Misra, S.K., Kondaiah, P., Bhattacharya, S., Rao, C.N.R.: Graphene as a nanocarrier for tamoxifen induces apoptosis in transformed cancer cell lines of different origins. Small **8**, 131–143 (2012). https://doi.org/10.1002/smll.201101640

Nanda, S.S., Papaefthymiou, G.C., Yi, D.K.: Functionalization of graphene oxide and its biomedical applications. Crit. Rev. Solid State Mater. Sci. **40**, 291–315 (2015). https://doi.org/10.1080/10408436.2014.1002604

Navalón, S., Ong, W.-J., Duan, X.: Sustainable catalytic processes driven by graphene-based materials. Processes **8**, 672 (2020). https://doi.org/10.3390/pr8060672

Park, S., Ruoff, R.S.: Chemical methods for the production of graphenes. Nat. Nanotechnol.nanotechnol. **4**, 217–224 (2009). https://doi.org/10.1038/nnano.2009.58

Parveen, N., Ansari, M.O., Ansari, S.A., Cho, M.H.: Simultaneous sulfur doping and exfoliation of graphene from graphite using an electrochemical method for supercapacitor electrode materials. J. Mater. Chem. A **4**, 233–240 (2016). https://doi.org/10.1039/C5TA07963B

Peng, Q., Crean, J., Han, L., Liu, S., Wen, X., De, S., Dearden, A.: New materials graphyne, graphdiyne, graphone, and graphane: review of properties, synthesis, and application in nanotechnology. Nanotechnol. Sci. Appl. **1** (2014). https://doi.org/10.2147/NSA.S40324

Ponomarenko, L.A., Schedin, F., Katsnelson, M.I., Yang, R., Hill, E.W., Novoselov, K.S., Geim, A.K.: Chaotic Dirac Billiard in graphene quantum dots. Science (80-.). **320**, 356– 358 (2008). https://doi.org/10.1126/science.1154663

Rad, A.S., Shadravan, A., Soleymani, A.A., Motaghedi, N.: Lewis acid-base surface interaction of some boron compounds with N-doped graphene; first principles study. Curr. Appl. Phys. **15**, 1271–1277 (2015). https://doi.org/10.1016/j.cap.2015.07.018

Radadiya, T.: A properties of graphene. Eur. J. Mater. Sci. **2**, 6–18 (2015)

Reshak, A.H., Auluck, S.: Electronic and optical properties of chair-like and boat-like graphane. RSC Adv. **4**, 37411–37418 (2014). https://doi.org/10.1039/C4RA05124F

Ritter, K.A., Lyding, J.W.: The influence of edge structure on the electronic properties of graphene quantum dots and nanoribbons. Nat. Mater. **8**, 235–242 (2009). https://doi.org/10.1038/nmat2378

Sahu, D., Sutar, H., Senapati, P., Murmu, R., Roy, D.: Graphene, graphene-derivatives and composites: fundamentals, synthesis approaches to applications. J. Compos. Sci. **5**, 181 (2021). https://doi.org/10.3390/jcs5070181

Schneider, G.F., Kowalczyk, S.W., Calado, V.E., Pandraud, G., Zandbergen, H.W., Vandersypen, L.M.K., Dekker, C.: DNA translocation through graphene nanopores. Nano Lett. **10**, 3163–3167 (2010). https://doi.org/10.1021/nl102069z

Serafini, P., Milani, A., Tommasini, M., Bottani, C.E., Casari, C.S.: Topology-dependent conjugation effects in graphdiyne molecular fragments. Carbon N. Y. **180**, 265–273 (2021). https://doi.org/10.1016/j.carbon.2021.04.094

Shao, Y., Zhang, S., Engelhard, M.H., Li, G., Shao, G., Wang, Y., Liu, J., Aksay, I.A., Lin, Y.: Nitrogen-doped graphene and its electrochemical applications. J. Mater. Chem. **20**, 7491 (2010). https://doi.org/10.1039/c0jm00782j

Shi, P., Wang, Y., Liang, X., Sun, Y., Cheng, S., Chen, C., Xiang, H.: Simultaneously exfoliated boron-doped graphene sheets to encapsulate sulfur for applications in lithium-sulfur batteries. ACS Sustain. Chem. Eng. **6**, 9661–9670 (2018a). https://doi.org/10.1021/acssuschemeng.8b00378

Shi, X., Zhang, J., Huang, T.: Significant advantages of sulfur-doped graphene in neutral media as electrocatalyst for oxygen reduction comparing with Pt/C. IOP Conf. Ser. Earth Environ. Sci. **121**, 022011 (2018b). https://doi.org/10.1088/1755-1315/121/2/022011

Shin, Y.J., Wang, Y., Huang, H., Kalon, G., Wee, A.T.S., Shen, Z., Bhatia, C.S., Yang, H.: Surface-energy engineering of Graphene. Langmuir **26**, 3798–3802 (2010). https://doi.org/10.1021/la100231u

Sofo, J.O., Chaudhari, A.S., Barber, G.D.: Graphane: a two-dimensional hydrocarbon. Phys. Rev. B **75**, 153401 (2007). https://doi.org/10.1103/PhysRevB.75.153401

Su, Q., Pang, S., Alijani, V., Li, C., Feng, X., Müllen, K.: Composites of graphene with large aromatic molecules. Adv. Mater. **21**, 3191–3195 (2009). https://doi.org/10.1002/adma.200803808

Suk, M.E., Aluru, N.R.: Water transport through ultrathin graphene. J. Phys. Chem. Lett. **1**, 1590–1594 (2010). https://doi.org/10.1021/jz100240r

Tu, Q., Pang, L., Chen, Y., Zhang, Y., Zhang, R., Lu, B., Wang, J.: Effects of surface charges of graphene oxide on neuronal outgrowth and branching. Analyst **139**, 105–115 (2014). https://doi.org/10.1039/C3AN01796F

Voiry, D., Yang, J., Kupferberg, J., Fullon, R., Lee, C., Jeong, H.Y., Shin, H.S., Chhowalla, M.: High-quality graphene via microwave reduction of solution-exfoliated graphene oxide. Science (80-.). **353**, 1413–1416 (2016). https://doi.org/10.1126/science.aah3398

Wang, C., Wallace, G.G.: Flexible Electrodes and electrolytes for energy storage. Electrochim. Acta. Acta **175**, 87–95 (2015). https://doi.org/10.1016/j.electacta.2015.04.067

Wang, W., Zhang, Y., Wang, Y.-B.: Noncovalent π–π interaction between graphene and aromatic molecule: Structure, energy, and nature. J. Chem. Phys. **140**, 094302 (2014). https://doi.org/10.1063/1.4867071

Wang, X., Lei, R., Huang, H., Wang, N., Yuan, L., Xiao, R., Bai, L., Li, X., Li, L., Yang, X.: The permeability and transport mechanism of graphene quantum dots (GQDs) across the biological barrier. Nanoscale **7**, 2034–2041 (2015). https://doi.org/10.1039/C4NR04136D

Xu, Z., Gao, C.: Graphene fiber: a new trend in carbon fibers. Mater. Today **18**, 480–492 (2015). https://doi.org/10.1016/j.mattod.2015.06.009

Yan, L., Zheng, Y.B., Zhao, F., Li, S., Gao, X., Xu, B., Weiss, P.S., Zhao, Y.: Chemistry and physics of a single atomic layer: strategies and challenges for functionalization of graphene and graphene-based materials. Chem. Soc. Rev. **41**, 97–114 (2012). https://doi.org/10.1039/C1CS15193B

Yang, Y., Shi, W., Zhang, R., Luan, C., Zeng, Q., Wang, C., Li, S., Huang, Z., Liao, H., Ji, X.: Electrochemical exfoliation of graphite into nitrogen-doped graphene in glycine solution and its energy storage properties. Electrochim. Acta. Acta **204**, 100–107 (2016). https://doi.org/10.1016/j.electacta.2016.04.063

Yang, J., Ma, M., Li, L., Zhang, Y., Huang, W., Dong, X.: Graphene nanomesh: new versatile materials. Nanoscale **6**, 13301–13313 (2014). https://doi.org/10.1039/C4NR04584J

Ye, S., Feng, J.: Self-assembled three-dimensional hierarchical graphene/polypyrrole nanotube hybrid aerogel and its application for supercapacitors. ACS Appl. Mater. Interfaces **6**, 9671–9679 (2014). https://doi.org/10.1021/am502077p

Yol Jeong, S., Jeong, S., Won Lee, S., Tae Kim, S., Kim, D., Jin Jeong, H., Tark Han, J., Baeg, K.-J., Yang, S., Seok Jeong, M., Lee, G.-W.: Enhanced response and sensitivity of self-corrugated graphene sensors with anisotropic charge distribution. Sci. Rep. **5**, 11216 (2015). https://doi.org/10.1038/srep11216

Yuan, Y., Wu, S., Ai, H., Lee, J.Y., Kang, B.: γ-Graphyne nanotubes as defect-free catalysts of the oxygen reduction reaction: a DFT investigation. Phys. Chem. Chem. Phys. **22**, 8633–8638 (2020). https://doi.org/10.1039/D0CP00644K

Zainal-Abidin, M.H., Hayyan, M., Ngoh, G.C., Wong, W.F.: Doxorubicin loading on functional graphene as a promising nanocarrier using ternary deep eutectic solvent systems. ACS Omega **5**, 1656–1668 (2020). https://doi.org/10.1021/acsomega.9b03709

Zakharova, O.V., Mastalygina, E.E., Golokhvast, K.S., Gusev, A.A.: Graphene nanoribbons: prospects of application in biomedicine and toxicity. Nanomaterials **11**, 2425 (2021). https://doi.org/10.3390/nano11092425

Zhang, Z.Z., Chang, K., Peeters, F.M.: Tuning of energy levels and optical properties of graphene quantum dots. Phys. Rev. B **77**, 235411 (2008). https://doi.org/10.1103/PhysRevB.77.235411

Zhang, Z., Huang, H., Yang, X., Zang, L.: Tailoring electronic properties of graphene by π–π stacking with aromatic molecules. J. Phys. Chem. Lett. **2**, 2897–2905 (2011a). https://doi.org/10.1021/jz201273r

Zhang, L., Diao, S., Nie, Y., Yan, K., Liu, N., Dai, B., Xie, Q., Reina, A., Kong, J., Liu, Z.: Photocatalytic patterning and modification of graphene. J. Am. Chem. Soc. **133**, 2706–2713 (2011b). https://doi.org/10.1021/ja109934b

Zhang, H., Zhao, M., He, X., Wang, Z., Zhang, X., Liu, X.: High mobility and high storage capacity of lithium in sp–sp^2 hybridized carbon network: the case of graphyne. J. Phys. Chem. C **115**, 8845–8850 (2011c). https://doi.org/10.1021/jp201062m

Zhang, J., Zhang, W., Ragab, T., Basaran, C.: Mechanical and electronic properties of graphene nanomesh heterojunctions. Comput. Mater. Sci.. Mater. Sci. **153**, 64–72 (2018). https://doi.org/10.1016/j.commatsci.2018.06.026

Zhang, L., Liu, Z.: Photochemistry of Graphene. Presented at the (2015)

Zhao, Y., Liu, J., Hu, Y., Cheng, H., Hu, C., Jiang, C., Jiang, L., Cao, A., Qu, L.: Highly compression-tolerant supercapacitor based on polypyrrole-mediated graphene foam electrodes. Adv. Mater. **25**, 591–595 (2013). https://doi.org/10.1002/adma.201203578

Zhao, C., Song, X., Liu, Y., Fu, Y., Ye, L., Wang, N., Wang, F., Li, L., Mohammadniaei, M., Zhang, M., Zhang, Q., Liu, J.: Synthesis of graphene quantum dots and their applications in drug delivery. J. Nanobiotechnology. **18**, 142 (2020). https://doi.org/10.1186/s12951-020-00698-z

Zhu, Y., Murali, S., Cai, W., Li, X., Suk, J.W., Potts, J.R., Ruoff, R.S.: Graphene and graphene oxide: synthesis, properties, and applications. Adv. Mater. **22**, 3906–3924 (2010). https://doi.org/10.1002/adma.201001068

Chapter 8
Electrochemistry and Energy Storage Applications of Graphene and Its Derivatives

Gaurav Tatrari, Mayank Pathak, Diksha Bhatt, Kamal Garwal, Faiz Ullah Shah, and Nanda Gopal Sahoo

Abstract The energy demand cannot be fully accomplished as the rate of increasing worldwide population is larger than the production of energy. The increasing population also leads to the development of more electrical components, which need a lot of energy to store, so energy storing devices are the need of the hour. Batteries and supercapacitors are the main class of such energy storage devices. Graphene is a 2D nanomaterial suitable for energy storage devices as electrode material due to its remarkable properties like high theoretical specific surface area and high electrical conductivity. Still, scientific works are underway to optimize the synthesis and applicability of graphene and its derivative materials in energy storage systems. This chapter discusses graphene and its derivatives for supercapacitor applications. Further, the electrochemistry behind storing energy in storage devices is discussed. This in-depth examination of graphene-based materials for energy storage may help researchers better comprehend the advantages and the most promising outcomes of such materials.

Keywords Graphene · Electrochemistry · Energy storage · Supercapacitors · Batteries

8.1 Introduction

The current global scenario necessitates a consistent supply of sustainable energy, renewable energy storage, and convertion of that energy to different forms for usage in other sectors (Edenhofer et al. 2014). When fossil fuels are burnt, the greenhouse

G. Tatrari (✉) · F. U. Shah
Chemistry of Interfaces, Luleå Technology University, Luleå, Sweden
e-mail: gaurav.tatrari@associated.ltu.se

M. Pathak · D. Bhatt · K. Garwal · N. G. Sahoo
Rajendra Singh Nanoscience and Nanotechnology Lab, Department of Chemistry, Kumaun University, DSB Campus, Nainital, Uttrakhand, India

R. Khan et al. (eds.), *Electrochemical Exfoliation of Graphene and its Derivatives*, Engineering Materials, https://doi.org/10.1007/978-981-97-2128-3_8

effect is enhanced, and the release of various hazardous gases produces environmental damage (Chaudhuri et al. 2022). As a result of rapid industrialization, exploding populations, and widespread illiteracy, modern society has become critically reliant on fossil fuels. Efficient utilization of fossil fuels may forestall disastrous pollution and energy shortages (Tatrari et al. 2024). Therefore, providing clean energy to solve these issues is vital. Since the yearly growth rate of energy consumption has grown by more than 50% in recent years, and the needs will double in the coming decades (Yang et al. 2018), exploring the beginnings of clean energy is crucial.

Graphene has gained promise as a carbon nanomaterial since Andre Geim and Konstantin Novoselov isolated it from graphite in 2004 (Ahmad and Zhang 2020). Graphene is a nanomaterial with advanced physical, chemical, and electrical properties (Lelieveld et al. 2019), such as high thermal and electrical conductivity, intrinsic carrier mobility, theoretical specific surface area, optical transmittance, resilience, low weight, and mechanical strength (Tewari et al. 2019; Tatrari et al. 2021a, b, c, d). In addition, Van der Waals forces and covalent interactions between neighboring carbon atoms keep monolayer graphene sheets in the same plane (Karakoti et al. 2022a). Therefore, this single-atom thick layer has a honeycomb structure, with two equal sub-lattices of carbon atoms connected by a carbon–carbon sigma connection spanning just 0.142 nm (Tewari et al. 2022). Although graphene is mainly composed of pure carbon atoms, the electrochemical properties of its edge and basal plane are distinct. In the meantime, the graphene analog GO has several oxygen functional groups. Its layer edge contains carbonyl groups, whereas its basal plane contains hydroxyl groups (Pandey et al. 2021).

Because of its reactive functional groups, free electrons, and aromatic rings, graphene and its analogs, especially GO and rGO, have been explored in various fields where graphene-based materials have shown their promising performance (Tatrari et al. 2021a; Karakoti et al. 2022b). For example, (Zhao et al. 2015) created sulfur-nobbled graphene, which is 40 times more efficient than carbon. Furthermore, an outstanding electrochemical performance for energy storage devices was revealed by Li et al. (2018) in a well-controlled pyrene-decorated graphene composite.

This graphene composite had a specific capacitance of 310.7 Fg^{-1} and retained 99% of its capacitance after 15,000 cycles at the current density of 1.5 Ag^{-1}. Scientists all around the globe are trying to develop more efficient batteries and supercapacitors. The effective conversion and storage of energy may also need the development of new materials. Most electrochemical applications often use carbon nanomaterials such as graphene, graphene oxide (GO), and reduced graphene oxide (rGO) due to their accessibility, processing simplicity, stability, and eco-friendliness. Graphene material-based electrodes perform better than other electrode materials, notably in energy storage applications, due to their chemical stability in acidic or basic solutions across a wide temperature range (Tewari et al. 2019; Tatrari et al. 2021a, b, c, d). This chapter summarizes top-to-down and bottom-to-up techniques for synthesizing graphene and its derivatives. Furthermore, a thorough analysis of the production and characteristics of hybrid composites based on graphene for use in supercapacitors has been described. Finally, a comprehensive review of the various kinds of energy storage devices and the electrochemistry involved in supercapacitors is offered.

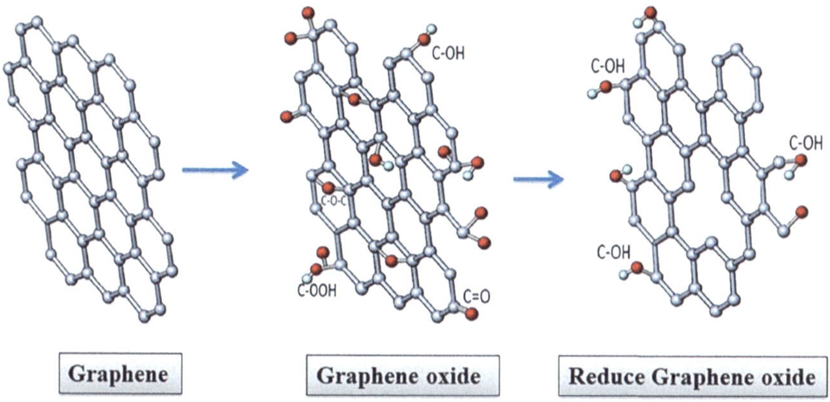

Fig. 8.1 Graphene and its derivatives

8.2 Graphene and Its Derivatives

Graphene and its analogs constitute a 2D structure, although tiny changes affect their physico-chemical characteristics. Graphene's leading derivatives are GO and reduced GO (rGO). However, the ratio of oxygen-containing groups to surface defects in the rGO may be regulated by the degree of the reduction reaction. Because of its functional groups (such as hydroxyl, carboxyl, and epoxy groups), GO may be attached to many compounds, making GO a promising material for different applications (Karakoti 2021; Pandey et al. 2023). Further, the synthesis of graphene can be done by reducing graphite to rGO or GO (Fig. 8.1).

8.3 Graphene

In graphene, a single layer of carbon atoms forms a hexagonal lattice, giving the substance a total thickness of only one atomic layer. The synthesis is usually done by using graphite as a precursor, which is a naturally occurring mineral; sometimes carbon nanotubes and fullerenes are also used. Since its isolation in 2004, graphene has been the focus of a great deal of study owing to its unusual and appealing qualities. Graphene possesses outstanding mechanical strength, flexibility, and electrical solid and thermal conductivities (Bohra et al. 2022). Because of its transparency and resistance to gas and liquid penetration, it has the potential to be used in a variety of fields, including electronics, energy storage, composites, sensors, biomedical devices, and more. The top-down method involves converting a bulk substance, often graphite, into graphene, whereas the bottom-up process includes generating graphene from individual atomic or molecule precursors. Both approaches have their benefits and drawbacks, and researchers are constantly working to enhance the quality and scalability of graphene manufacturing for a wide range of uses (Tatrari et al. 2021b).

8.4 Graphene Oxide

Graphene called GO has added oxygen-containing functional groups, including hydroxyl, carboxyl, and epoxide. The synthesis is done by oxidizing graphite to produce GO, usually with chemical oxidants like sulfuric acid or potassium permanganate. GO is hydrophilic, or water-soluble, due to its surface's oxygen-containing functional groups, making it simpler to handle and treat than graphene. Due to its hydrophilic properties, it is beneficial in water-based applications such as synthesizing graphene nanocomposites and manufacturing composites based on graphene (Tatrari et al. 2023a, b, c). In addition, the manufacture of various graphene-based compounds uses GO as a precursor for reducing graphene oxide back to graphene. However, unlike graphene, GO has functional groups that affect its electrical and mechanical characteristics, making it less appropriate for various uses like biological applications, electronics applications, and energy storage devices (Oli et al. 2023; Tewari et al. 2023a, b). Despite this, the unique qualities of GO are valuable and recognized as well-known materials for multiple applications, such as energy storage, catalysis, water purification, and biological imaging (Pandey et al. 2021; Tatrari et al. 2021a).

8.5 Reduced Graphene Oxide

Reduced graphene oxide (rGO) may be produced by reducing graphene oxide with various oxygen groups. Chemical, microwave-assisted, photo-assisted, thermal, and solvothermal reduction may be used for this strategy. Reducing graphene oxide is an exciting way to get properties similar to graphene. It is also possible to reduce graphene oxide chemically, thermally, or photothermally. However, the pure graphene structure is inaccessible to rGO (Tatrari et al. 2022). Even after considerable reduction, the remaining oxygen and structural defects from the chemical oxidation that produces GO are still present in rGO.

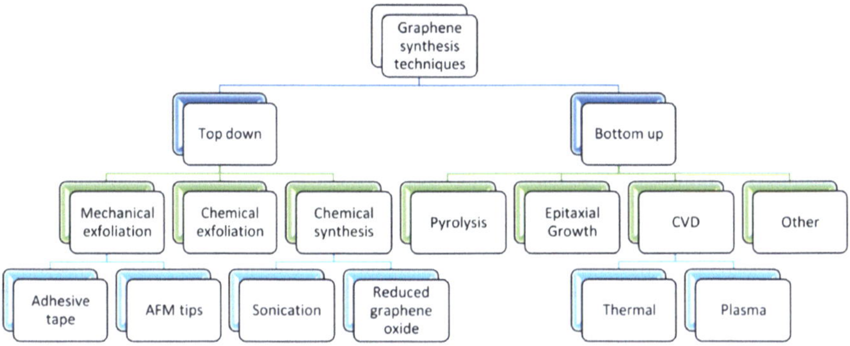

Fig. 8.2 A flowchart synthesis of graphene

8.6 Synthesis of Graphene and Its Derivatives

After discovering graphene, many methods have been developed to synthesize graphene and its derivatives. Graphene can be synthesized in two ways: top-down and bottom-up. Top-down mainly consists of mechanical exfoliation, chemical exfoliation, and chemical synthesis, while bottom-up consists of epitaxial growth and chemical vapor deposition (CVD). An overview of these synthesis processes can be seen in Fig. 8.2.

8.7 Top-Down Approach

The top-down method for synthesizing graphene entails the conversion of bulk material, often graphite, into graphene. The mechanical exfoliation approach involves obtaining graphene by carefully peeling off thin layers from a graphite flake using tape or a sharp blade. This technique is widely recognized as the most popular method for producing graphene from top-down. This method may create high-quality graphene sheets with sharp edges and fewer layers. Exfoliating single-layer graphene flakes mechanically on appropriate surfaces may be the most efficient and novel way. Materials on the surface of layered structures are subjected to axial or bending stresses, thanks to top-down technology. As a result of intercalation and colloidal graphite, graphene undergoes chemical modifications; however, graphene may be manufactured by the chemical reduction of graphite oxide (Tatrari et al. 2023a, b, c; Tewari et al. 2023a, b). The Hummer method transforms graphite into graphite oxide (GO) and uses concentrated sulfuric acid, nitric acid, and potassium permanganate. Chemical etching is another top-down approach for synthesizing graphene. A chemical solution dissolves the material between graphene layers, leaving a single, atomically thin graphene sheet behind. However, these approaches have drawbacks, such as the small size of the graphene sheets they create and the difficulty in regulating the number of layers. Nevertheless, the top-down technique is advantageous for producing high-quality graphene sheets, which may be used in electronics and energy storage devices (Toh et al. 2014). The top-down approach to graphene synthesis has been an active field of study, with various studies targeting enhancing graphene production's quality and scalability using this technique. Some current top-down research includes the following aspects:

Improving mechanical exfoliation: Researchers have been attempting to improve the process to boost the quantity and quality of graphene generated. This involves developing innovative procedures for peeling graphene off graphite, such as utilizing ultrasonic or laser treatments to improve the separation of the graphene layers.

Scaling up production: Researchers also focus on scaling graphene manufacturing using top-down approaches. This involves the development of large-scale exfoliation methods as well as the utilization of roll-to-roll processing to generate graphene sheets on a big scale.

Controlling the number of layers: Another study area is controlling the number of layers in graphene sheets created via top-down approaches. This involves the development of innovative strategies for separating graphene into single or few-layer sheets, such as employing a mix of mechanical exfoliation and chemical treatment.

Improving the quality of graphene: Researchers are also looking at ways to increase graphene produced using top-down approaches. This entails researching the impact of different process factors on graphene characteristics, such as growth temperature and pressure, and the kind of catalyst utilized.

8.8 Bottom-Up Approach

The bottom-up method includes generating graphene from its individual atomic or molecule precursors. Chemical vapor deposition (CVD), the most prevalent technique for bottom-up graphene synthesis, is the process of forming graphene. This approach decomposes a carbon-containing gas on a metal catalyst to produce graphene. During this procedure, the metal catalyst creates a surface on which a sheet of graphene may form from the deposition and growth of individual carbon atoms. The quality and uniformity of the graphene made via CVD may be manipulated by changing the circumstances of the process. These factors include the kind of catalyst, the growth temperature, and the gas composition. Solution-based techniques can synthesize graphene from the bottom up, such as liquid phase exfoliation, solvent-assisted graphite intercalation, and molecular beam epitaxy. The bottom-up technique provides a highly controlled method for synthesizing graphene, allowing for the creation of large-area graphene sheets with precise characteristics.

Graphene made via bottom-up methods has low defect rates and large surface areas, but its production is time-consuming and costly. Regarding graphene sheets, the bottom-up fabrication methods provide a wide range of variations in terms of kind, average size, and thickness, as well as benefits and downsides. Many years ago, valuable fractions were extracted from heavy petroleum fuel via pyrolysis, a thermochemical process that destroys materials without oxygen at high temperatures. This innovation has recently acquired prominence due to its ability to recycle conventional garbage into high-tech products. Studies have shown that feed materials produce carbon, bio-oil, and fuel-grade pyrolysis gas, including old tires, wheat straw, sugarcane bagasse, old wood sawdust, and rice husk. Sometimes, diesel engines may be converted to bio-oil refined through pyrolysis (Tatrari et al. 2021a, b, c, d; Sahoo et al. 2022). As a result, the utilization of pyrolysis grew to produce cutting-edge application materials like CNTs (Zahid et al. 2018).

Breaking down silicon carbide (SiC) at 120–160 °C in a vacuum or inert atmosphere produces graphene. Silicon (Si) sublimates at 1100 °C, enabling carbon (C) atoms to cluster and form an sp^2 hybridized network, creating graphene. This CVD process thermally decomposes the substrate between two precursor layers and coats transition metal substrates (Ni, Pd, Ru, Ir, and Cu) with high-quality graphene. However, the CVD on copper and nickel surfaces produces good-quality graphene. The plasma-enhanced chemical vapor deposition uses plasma generated in a void chamber to deposit a thin layer on a substrate, and gases are the primary medium for chemical reactions. However, large-scale industries use catalyst-free graphene production at low temperatures, which is expensive and requires precursor materials in the gas phase.

8.9 Electrochemistry and Electrochemical Properties

The vast quantity of research on graphite (mainly highly orientated pyrolytic graphite) and carbon nanomaterials that have been published may be used to study the electrochemistry of graphene (Pumera 2009). Graphene sheets provide a 2D environment for electron transport. On the margins of a graphene sheet, heterogeneous electron transfer (HET) occurs from or to the sheet. However, this transfer from the sheet's plane is exceptionally near zero (Davies et al. 2005). The margins of the graphene sheets with oxygen-containing groups affect the electrochemistry of the material.

But whether this influence is favorable (i.e., increases the rate of heterogeneous electron transfer) or negative is debatable (slows the heterogeneous electron transfer rate). For instance, the abundance of carbon–oxygen bonding in graphene oxide results in various oxygen functional groups, such as carboxyl, epoxide, and peroxide groups, as well as carbonyl (aldehyde, ketone, and quinone) groups. These oxygen functional groups are arranged in various ways depending on how graphene oxide is made. Because they are already electrochemically active, certain oxygen functional groups may be oxidized or reduced at low electrochemical potentials. These groups include quinones, hydroxyls, aldehydes, epoxides, and peroxides. Reducing carboxyl groups is possible at harsher potentials and pH (typically at possibilities lower than 2 V) (Grimshaw 2000). Additionally, the electrochemical reduction of solid-state graphene oxide enables the detection of oxygen functional groups on graphene surfaces.

The number of oxygen functional groups that can be electrochemically reduced is determined by measuring the amount of transferred charge during the reduction of solid-state graphene oxide material (Chng and Pumera 2011). However, the impact of oxygen-containing groups on graphene's electrochemistry is not to be underestimated. In addition to influencing the adsorption and desorption of molecules before and after an electrochemical reaction, they also affect the rate at which heterogeneous electrons may be transferred between atoms. The electrochemical reduction approach may alter the electrochemical characteristics of graphene oxide sheets since it is easy to regulate and monitor. It has been shown that changing the cathodic potential of a graphene oxide layer over a certain length of time may alter the C/O ratio, which, in turn, affects the electrochemical characteristics (Ambrosi and Pumera 2013).

Due to graphene oxide's sensitivity to electrochemical reduction, several applications, such as energy storage and sensing, may suffer. Graphite and pure graphene have similar HET rates. The electrochemical probes used determine this. Electrochemistry is slower at folded graphene edges than at opened edges. Pristine graphene lacks electrochemistry. Graphene oxide (GO) has low conductivity and k0 values (Ambrosi et al. 2014). Chemical, electrochemical, or thermal reduction of oxygen-containing groups enhances the HET rate. Graphene oxide electrochemical reduction can modify the HET rate accurately. Reduced graphene has many flaws, raising HET rates.

GO and chemically rGO have significant inherent electrochemistry in the cathodic or cathodic regions depending on whether they are synthesized using chlorate or permanganate. Chemical, electrochemical, or thermal reduction of oxygen-containing groups increases the HET rate (Ambrosi et al. 2014).

Further, the electrochemical properties of graphene and its derivatives must be understood to know the electrochemistry of such nanomaterials properly. Specific surface area is the primary concern when talking about energy storage applications. Theoretically, graphene has a surface area of 2630 $m^2\ g^{-1}$, which is twice as big as that of CNTs (1315 $m^2\ g^{-1}$) and surpasses that of graphite (10 $m^2\ g^{-1}$) (Brownson and Banks 2010). According to calculations, graphene has an electrical conductivity of 64 mS cm^{-1}. As a result, suspended graphene has seen exceptionally high electron mobility, with mobilities above 200,000 $cm^2\ V^{-1}\ s^{-1}$ recorded at room temperature (Geim and Novoselov 2007; Sato et al. 2010).

In contrast, the maximal electron mobility in silicon is roughly 1,000 $cm^2\ V^{-1}\ s^{-1}$, making the electron mobility in Graphene 200 times greater (Bolotin et al. 2008). It has been hypothesized that a GNS may carry a super-current due to graphene's unique characteristics (Heersche et al. 2007), and it is obvious that its theoretical electron transfer rates are higher when compared to graphite and CNTs. In addition to being continuous, the carrier qualities of graphene (and other 2D materials) were discovered to have exceptional crystal quality, which is crucial for graphene since it allows charge carriers to cross tens of thousands of interatomic distances without scattering (Sun et al. 2011).

According to the prior discussion, we can come to the conclusion that graphene possesses distinctive qualities that set it apart from other carbon allotropes of varying diameters as well as any other electrode material, "theoretically" claiming that graphene is a perfect electrode material that may be useful in a variety of electrochemical applications; this is what we discussed in greater detail in the following section.

8.10 Composites of Graphene in Energy Storage Systems

As the twenty-first century progresses, concerns about pollution, the depletion of fossil resources, and global warming continue to grow.

Therefore, it is crucial to have access to clean and renewable energy sources and related technologies (Mahmood et al. 2014). Putting renewable energy to use entails two stages. In the first place, renewable resources like the oxygen reduction process (ORR), solar power, and water may be effectively converted into proper forms of energy like electricity or fuel. Most effectively, this may be accomplished by using solar cells, fuel cells, and water splitting. Devices used to store energy must be viable regarding cost, performance, and environmental impact. This is mainly because renewable energy sources are notoriously intermittent. Lithium-based batteries, such as lithium-ion, lithium-air, and lithium-sulfur, are the most practical and effective storage alternatives. One more effective and promising energy storage device is the supercapacitor, which can store and release energy in seconds (Paek et al. 2009).

Graphene possesses exceptional thermal conductivity, mechanical stability, and charge carrier properties at room temperature. Conduction and valence bands in semimetal graphene intersect at Dirac points, making them zero-gap semiconductors. The monolayer structure of graphene gives it a specific surface area of 2630 $m^2 g^{-1}$. Graphene and graphene-based materials (GBMs) exhibit novel electrochemical properties such as low charge transfer, a vast potential window, high electrochemical activity, and quick electron transfer. Because of their characteristics, GBMs, such as graphene oxide (GO), reduced graphene oxide (r-GO), few-layer graphene (FLG), and graphene nanoplatelets (GNP), are suited for energy storage (Zhu et al. 2014).

In graphene, there is sp^2 hybridization of the carbon bonds, where the in-plane c–c bonds are the strongest bonds and the out-of-plane bonds, which lead to a delocalized network of electrons, favor graphene's electronic conduction and provide a weak interaction between graphene layers (Razaq et al. 2022).

Even though graphene has many fantastic qualities, its applicability is severely restricted in particular fields due to its poor water solubility and lack of bandgap. Making graphene derivatives is a practical approach to get around these restrictions and increase the scope of graphene's potential uses. As an illustration, adding epoxy, hydroxyl, and carboxyl groups to the basal plane of graphite layers will produce graphene oxide (GO). GO has high hydrophilicity due to these polar oxygen-containing functional groups. Because of this, GO disperses very well in various solvents, particularly water.

Additionally, the functional groups that include oxygen can act as reactive sites for GO to be chemically modified or functionalized. This can then be used to create materials functional in electrode development. The oxygen-containing groups can eliminate some of graphene's drawbacks but can also lead to issues. For instance, they make GO electrically insulating. However, GO's conductivity can be somewhat restored by chemical reduction. The resultant reduced GO (rGO), which still contains certain functional groups, is well dispersed in various solvents. Most importantly, adjusting the quantity of the remaining functional groups makes altering the electrical performance and solubility of rGO effortless. This chemically reduced graphene's characteristics roughly resemble those of pristine graphene (Huang et al. 2011).

Graphene's synergistic interactions with other nanomaterials, such as metals, metal oxides, sulfides, alloys, and polymeric materials, make it a promising candidate for energy storage and conversion devices.

Accordingly, graphene and related materials are recommended for use as low-cost, eco-friendly, transparent, conductive, and electro-active materials in solar cells, in proton-based oxygen reduction catalysts in fuel cells, and as electrodes in SCs and LIBs (Sun et al. 2011). Figure 8.3 is a schematic depicting the big picture of graphene-based nanomaterials and their use in energy storage systems.

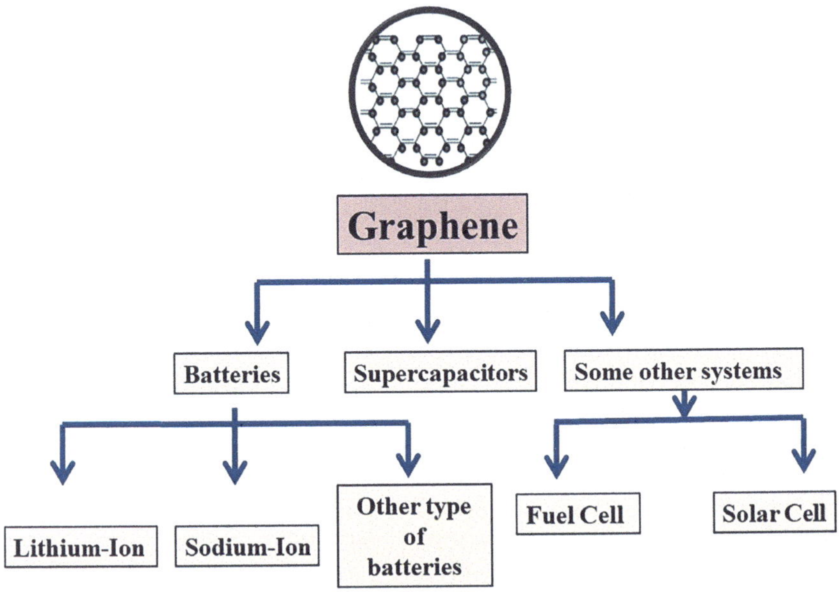

Fig. 8.3 Application of graphene and its derivatives in various energy storage systems

8.11 Batteries

Batteries are devices that store chemical energy and convert it to electrical energy. They comprise one or more electrochemical cells that create a voltage differential between two electrodes, the positive electrode (cathode) and the negative electrode (anode), and an electrolyte that permits the movement of ions between the electrodes. Primary and secondary batteries are distinguished by the number of times they may be charged and drained. Primary batteries are intended for single usage and cannot be recharged; secondary batteries can be restored. The most prevalent secondary battery types are lead-acid, nickel–cadmium, nickel-metal hydride, and lithium-ion. Batteries are vital in contemporary civilization since they provide portable power for various gadgets and equipment. As a result of their elevated surface-to-volume ratio and strong electrochemical characteristics, multilayered carbon nanomaterials (CNMs) are fresh choices for battery electrodes.

8.11.1 Lithium-Ion Batteries

Lithium-ion (Li-ion) batteries are rechargeable batteries that use lithium ions as charge carriers. Batteries have a cathode, an anode, and an electrolyte that allows ions to travel between the electrodes and store energy. Due to their high energy density, Li-ion batteries are often used in portable gadgets and electric vehicles. In addition, they have a low self-discharge rate, allowing them to retain their charge for an extended period and a limited number of charge–discharge cycles. Unfortunately, lithium batteries are sensitive to high temperatures and tend to deteriorate over time, which may shorten their total lifetime.

The development of Li-ion batteries started in the 1970s, intending to generate a rechargeable battery that was more efficient and lightweight.

Sony produced the first commercial Li-ion battery in 1991, and since then, Li-ion batteries have grown prevalent in a broad range of consumer and industrial applications. Due to their high energy density and low self-discharge rate, Li-ion batteries are suitable for portable gadgets such as smartphones, laptops, and cameras. In addition, the expansion of the market for electric vehicles has also increased the need for Li-ion batteries, which are required to power the electric motor of these vehicles. As a result, Li-ion batteries have a long history of research and development focused on enhancing their performance, safety, and durability (Sun et al. 2011).

In the past 20 years, the LIB, initially developed in 1991, has taken the lead in providing power for portable electronics, particularly laptop computers and cordless phones. Due to their high voltage, high energy density, extended cycling life, and favorable environmental compatibility, lithium-ion batteries (LIB) are considered one of the most valuable batteries in portable electronics (Mahmood et al. 2014). High-performance batteries are required for the next generation of always-on and mobile electronics. These batteries must have increased power and energy densities, longer life spans, greater capacities, greater environmental robustness, and additional features like flexibility and self-healing (Brownson and Banks 2010). A LIB's performance strongly depends on its electrodes' structures and properties. Due to its high coulombic efficiency and excellent life cycle performance, graphite is the most widely used anode. However, by generating an intercalation compound of LiC_6, graphite's specific capacity (Cs) is restricted to 372 mA h g^{-1}. Therefore, anode materials with specific capacities (Cs) greater than graphite are needed to increase the performance of LIBs (Sato et al. 2010). For this, a range of carbon nanomaterials has been investigated, including mesoporous carbons, carbon nanotubes, and carbon nanofibers. Lithium ions diffuse across short distances in nanostructured carbons, improving rate performances as alternatives to graphite such as the anode material for LIBs, silicon, metals, and oxides were also investigated. Compared to graphite, these materials have substantially higher Cs values. However, one of these materials' most significant flaws is how much their volume changes when the lithium is inserted or extracted. The electrodes' volume changes may result in complete electrode pulverization. In addition, most metal oxides have low electrical conductivities, another issue. Therefore, as mentioned earlier, the most effective way to address both problems is to incorporate carbon nanotubes into these inorganic materials (Geim and Novoselov 2007). Based on these factors, graphene and its derivatives seem to have excellent potential for use as electrode materials for LIBs.

Modern electronics and communication equipment require batteries with quicker recharge times, longer lives, and reduced costs. The anode, cathode, and electrolyte are the fundamental elements of LIB cells. The development of anode materials for LIBs is ongoing. Because of their high conductivity, carbon materials have been seen as viable options for use in LIBs. Effective Li-ion transport and electron exchange require a conductive carbon network. Graphite was the most popular trade negative electrode for a very long period. However, due to its low theoretical capacity (372 mA h g^{-1}), graphite is not a good material for virtual anodes. In lithium-ion batteries (LIBs), there is a continuous exchange of ions between a Li-emitting +ve electrode, typically a sheet of Li-oxide, and a Li-receiving −ve electrode (graphite). The capacity of the battery, and hence its energy, is proportional to the amount of ions housed for each gm. In the same way, graphite may serve the purpose of an −ve electrode in which Li^+ is hosted. Graphene can also be employed as a matrix of activated carbons used in hybrids with other components that can store Li (Zhang et al. 2018).

One sheet of graphene can accommodate twice as many Li^+ ions as ordinary graphite, as proposed by Dahn et al. in 1995. One lithium-ion is stored on each graphene side, creating a Li_2C_6 stoichiometry with a specific capacity of 744 mAh g^{-1}, double that of graphite (372 mAh g^{-1}). Unlike graphite, which has lithium intercalated between its stacked layers 2–3, one-sheet graphene has the potential to retain Li^+ particles through sorption on its inner walls and vacant nanopores present between the single layers' randomly formed arrangements (Hossain et al. 2006). Due to their exceptional qualities, such as mechanical strength, superior conductivity, large specific area of 2600 m^2 g^{-1}, electrochemical activity, and elastic nature,

Graphene, graphene oxide (GO), and its derivatives have garnered many side spots that might be utilized to link other electrode materials (Ohzuku and Brodd 2007).

Li^+ diffusing over its flat surfaces enhances graphene's reactiveness in LIB. Due to its long-range conjugation, graphene exhibits unprecedented ionic mobility and ballistic electron transport. Graphene has the appropriate characteristics to function as a negative electrode and a conducting component in metal oxide negative electrodes. Compared to graphite electrodes, graphene has double the theoretical capacity at 1,388 mAh g^{-1}. Nanoporous sheets similar to graphene were used by Zhao et al. to create high-efficiency LIB anodes. At 100 milliamperes, porous graphene electrodes discharged at 830.4 mAh g^{-1}. The LIB characteristics of graphene are intriguing. The LIB characteristics of graphene are intriguing.

The LIB characteristics of graphene are fascinating. Despite the high current density, they have a rated capacity of 211.6 mAh g^{-1} (20 Ag^{-1}). After being cycled 10,000 times, the electrode still had the same ability as when it was first used. The electrochemical performance was greatly improved by the nanosheet-like porous structure (Dahn et al. 1995; Zhu et al. 2014).

GO with heterogeneous chemical and electrical structures can be employed as an electrode material for lithium batteries. Numerous groups, including epoxy, carboxyl, and hydroxyl, are found in the GO. A material with tunable electrical characteristics can be created from GO by varying the concentration of these oxygen-containing groups. The functional groups on GO might be used as locations for chemical modification or functionalization, which would immobilize different active species and produce hybrid topologies for electrode materials. Overall, materials based on graphene/GO that have been altered with metal oxides show improved electrochemical performance for LIB applications. With high cycle, performance, and rate capability, GO is the binder-free anode material in LIBs to boost the reversible capacity to 690 mA h g^{-1} at the rate of 0.5 C (1 C = 372 mA g^{-1}). The original graphite in the mixed material keeps its conductivity, while GO contributes to its increased capacity as a lithium accommodator. Another graphene derivative called reduced graphene oxide may be produced by a reduction procedure (rGO). Chemical or physical reactions can make the conversion of GO to rGO. The oxygenated functional groups in the GO are removed during the reduction process to create rGO. The traditional graphite in the anode can be replaced by the conductive rGO itself (Kumar et al. 2019). Reduced GO is the preferred material for lithium-ion storage in most investigations that have been reported. Reduced GO displayed high-capacity values of >2,000 mAh g^{-1} during the initial Li^+ insertion, which is more than the potential of single-layer graphene (Raccichini et al. 2015; Zhang et al. 2016).

8.11.2 Sodium-Ion Batteries

Instead of lithium ions, which are utilized in lithium-ion batteries, sodium ions are employed as the charge carriers in sodium-ion batteries (SIBs). They share the same cathode, anode, and electrolyte as lithium-ion batteries. Because sodium is so plentiful on Earth, sodium-ion batteries may be a cost-effective alternative to lithium-ion batteries. As Li is costly and physically limited, developing a new battery with superior performance and low production cost is necessary.

Na is the second softest and tiny alkali metal adjacent to Li; also, they have similar physical and chemical properties (Al Hassan et al. 2019). Na's richness and low production cost with minimal redox potential (marginally better than Li) support SIBs as a good substitute for LIBs. The E_D for SIBs is theoretically about 200 Wh kg^{-1}, making them similar prospects, and the charge-storing mechanism provides superior reversibility with long cycling stability. However, oxidation occurs when the positive electrode (cathode) is charged with simultaneous Na de-insertion and electron deficiency (Zhang et al. 2019). Research is underway to enhance the qualities of sodium-ion batteries, but they still face hurdles in their overall effectiveness and consistency.

8.11.3 Other Types of Batteries

Graphene and its composites have potential uses in various batteries outside LIBs and SIBs. These include Li–S, Mg, zinc-air/nickel, and Li-air. For example, the components of a standard Li–S battery are lithium at the anode, sulfur at the cathode, and an electrolyte. Lithium is oxidized at the anode when a storm is discharged, producing lithium ions and electrons. At the same time, sulfur is reduced to lithium sulfide at the cathode, and during the charging process, backward reactions occur (Al Hassan et al. 2019; Zhang et al. 2019).

8.12 Supercapacitors

Supercapacitors are among the most extensively researched energy storage technologies, similar to batteries. Supercapacitors offer a substantially larger energy density than traditional electrostatic or electrolytic capacitors, allowing quick and highly reversible energy charging and discharging. Though their lack of energy density still restricts their practical usage, there is a great demand for innovative electrode materials. In general, the accessible specific surface area of the electrode material and the designed pore structure of the supercapacitor are critical factors in its performance. If a larger size is exposed to the electrolyte, ions may move more freely inside the cell. EDLCs operate by charge separation at the electrode/electrolyte interface. In contrast, redox-based electrochemical capacitors use metal oxide/hydroxide or conducting polymer electrodes and exhibit reversible redox reactions at the electrode surface.

A high energy density does not always accompany high conductivity and power capacities in carbonaceous materials since the carbon surface is sometimes only partly accessible (Tatrari et al. 2024). Supercapacitors use graphene-based electrodes with various graphene topologies, including 0D, i.e., dots; 1D, i.e., fiber; 2D, i.e., graphene and based films; and 3D graphene viz. aerogels. Electrochemical capacitors store charge in an electrical double layer formed by ions between large-surface electrodes and an electrolyte, unlike standard capacitors, which store charge on plates with a modest surface area (Tatrari et al. 2021a; Tatrari et al. 2023a, b, c).

A supercapacitor maintains a specific working voltage for its electrodes made of graphene. This voltage drives ions through the supercapacitor by establishing an electric field between the graphene electrodes. Due to the large specific surface area and high porosity of graphene, the migration of electrolytes toward the electrode with the opposite polarity is high. As a result, it becomes more physically adsorbed upon reaching the electrode. An electrode and several adsorbed ions with opposing polarities produce a charge layer at this contact. The interface can effectively be thought of as two layers of charges with opposing polarities, giving rise to the parallel plate capacitor concept, in which the two plates are kept at opposing polarities. However, charges can be transferred across the double layer if redox species are at the electrode–electrolyte surface. This second mechanism is more common for polymer or metal-oxide-based electrodes. Here, charge storage is accomplished by charge transfer in addition to double-layer development.

The third method of charge storage used in pseudocapacitors is if the ions may permeate through the electrode materials and intercalate or de-intercalate between the electrode layers. Such supercapacitors respond rapidly to potential change due to the absence of redox reactions in the EDLC charge transfer process (as opposed to slower faradaic reactions in pseudocapacitors). Still, they have a lower energy density because most of the charged ions are only present on the surface of the electrodes (Mahmood et al. 2014; Tatrari et al. 2021a, b, c, d). For a general overview, the working mechanism of a supercapacitor is shown in Fig. 8.4.

8.13 Some Other Energy Storage Systems

Except for batteries and supercapacitors, some other energy storage systems store the energy and act as energy conversion systems, including fuel cells and solar cells. For instance, the usage of fuel cells, which have shown promising results in various applications, allows for the direct conversion of the chemical energy of the fuel into electricity with a high efficiency level. Graphene is typically used to support catalysts on fuel cell membranes, particularly PEM. Many studies are currently being conducted to determine whether replacing platinum catalysts with other non-precious metals/metal oxides and nitrogen-coordinated metal catalysts would be possible. These catalysts are less stable and active than platinum catalysts, which is a problem. Active carbon can overcome these obstacles, but it frequently has its limitations. Active carbon tends to have a large surface area, but its support is unstable, creating severe concerns unless combined with another material.

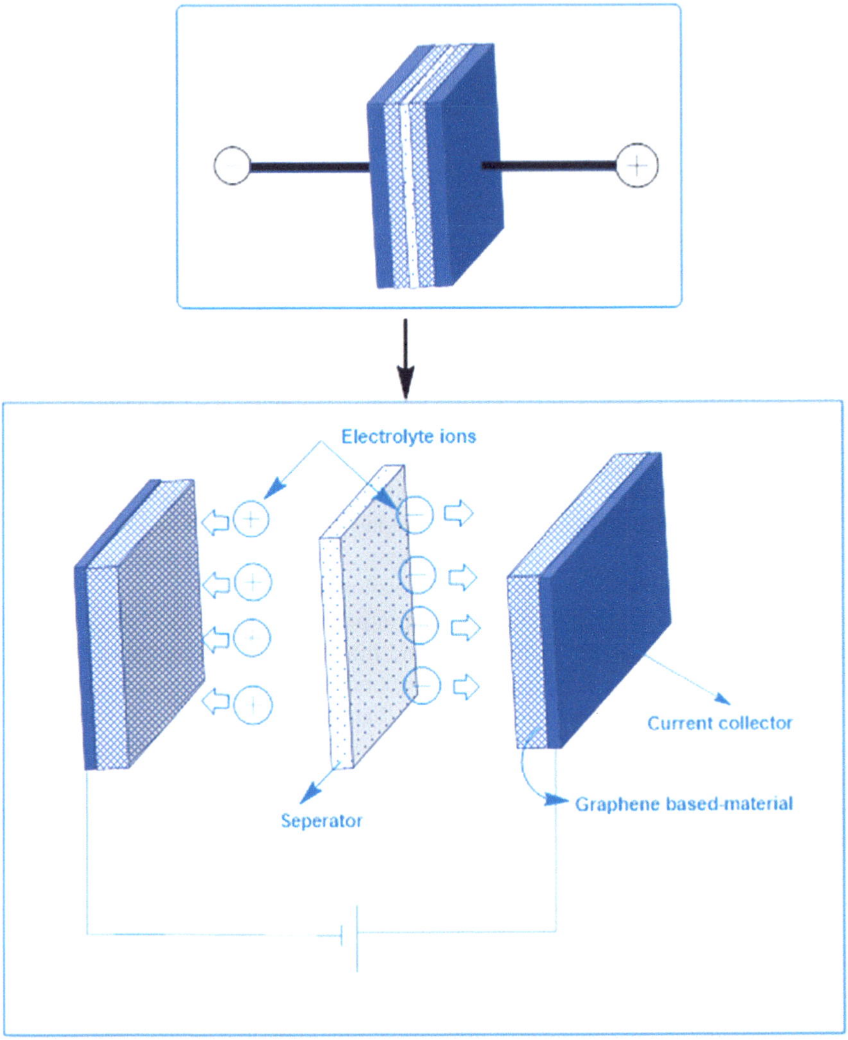

Fig. 8.4 Schematic representation of the mechanism of supercapacitor

Technological advancements involving graphene have changed the narrative since active carbons are now seen as a more substantial substitute for platinum due to their conductivity, large surface area, and adherence to catalyst particles (Devendran and Nagai 2023). Numerous benefits have been shown for using graphene materials as the active parts of fuel cells. Graphene-based materials as electrocatalysts for ORR and fuel oxidation seem promising due to their high surface area and conductive solid characteristics. Enormous tensile strength, minimal fuel permeability, and high ionic conductivity are all characteristics of polymer membranes that have been coupled with graphene; also, bipolar plates' conductivity and corrosion resistance can be improved with graphene (Moyseowicz et al. 2023).

More functional groups in graphene oxide make them suitable for nucleation and the attachment of catalyst nanoparticles to surfaces. Fuel cells employ graphene extensively, primarily as anode catalyst support material and even to replace the cathode catalyst in composite and standalone electrolyte membranes. Additionally, bipolar plates utilize it. Here is an overview of some recent research regarding the function of graphene in each component. The traditional active catalyst in fuel cell electrodes, whether at the anode or cathode in low-temperature fuel cells supplied by hydrogen or other low molecular weight hydrocarbons like methanol and ethanol, is platinum and platinum alloys (Pedrosa et al. 2019).

The intermediates created during the oxidation of various fuels impact Pt, which is likewise scarce and expensive. The catalyst loading at both the anode and cathode of fuel cells has been reduced using multiple techniques, and the Pt catalyst may even have been replaced with non-precious catalysts. ORR controls fuel cells' electrochemical reactions at the cathode. High-priced platinum catalysts are needed because of the slow ORR kinetics. Furthermore, the platinum catalyst's sensitivity to CO poisoning prevents it from being used. As a result, numerous alternative graphene-based materials have been investigated for ORR, both directly as metal-free electrocatalysts and as supports for metal nanoparticles (Pedrosa et al. 2019).

Graphene's enormous surface area allows it to replace Pt catalysts or significantly reduce their loading at the anode. The Pt was placed evenly in the graphene catalyst, increasing activity and electrical conductivity. The Pt catalyst loading in the fuel cell anode was lowered due to its high reactivity, which boosted its stability. Doping Pt with heterogeneous atoms such as nitrogen, sulfur, or boron increased its activity toward alcohol oxidation.

The improved conductivity, wettability, and resistance to CO and boron poisoning of the doped graphene led to a higher number of active sites. Pt activity was boosted by both graphene and boron doping, with the latter being more efficacious (Kuang et al. 2020). The sun provides another readily available and sustainable source, i.e., solar energy. Converting fuel into energy produces no waste, no emissions, and infinite amounts of clean, renewable power. This environmentally friendly electricity strategy mitigates climate change by reducing fossil fuel use. Photovoltaic (PV) solar technology must be refined via considerable research to create a highly efficient solar cell system before it can be used commercially.

According to a recent study, PV cells' efficiency may be improved with each succeeding generation since 2D graphene with nanostructures is a renewable material. Extremely high electron charge carrier mobilities may be achieved in PV cell applications, particularly in organic aspect devices, using a graphene structure with sp^2-hybridized carbon atoms and a thickness of just one bit. Since carbon contains many free pi electrons, it is an excellent electrical conductor. PV efficiency may be improved with graphene and semiconductor metal oxides. Graphene-based metal oxide materials have a reduced bandgap energy, which increases photon absorption and decreases recombination centers (Kuang et al. 2020; Devendran and Nagai 2023).

Graphene is used in industry and research because of its remarkable optical and electrical characteristics. Due to its excellent optical transparency, carrier mobility, zero-bandgap, and mechanical strength, graphene-based PV materials have been extensively studied. The most researched graphene derivatives are graphene oxide (GO) and reduced graphene oxide (rGO). Hydrophilic GO and rGO have hydroxyl, epoxy/ether, carbonyl, and carboxylic acid functional groups from the oxidation process. Graphene-based structures are simple to fabricate because of their agglomerative behavior in protic and polar solvents. Because of its oxygen moiety, GO may be functionalized with organic, inorganic, and small molecule components to form novel compositions. Graphene-based devices are quickly expanding, and it is critical to chronicle their imaginative usage in photovoltaic devices as a top emerging technology (Table 8.1).

Table 8.1 Graphene composites for SCs

Composite	Capacitance	Electrolyte	Energy density (wh/kg)	Power density (w/kg)	References
Graphene	135 Fg^{-1}	KOH	NA	NA	Stoller et al. (2008)
Graphene	2.75 Fg^{-1}	PC with $NaClO_4$	2.75	7800	Rogers (2015)
GO	223.6 Fg^{-1}	$Fe(CN)_6$	NA	NA	Yang and Gunasekaran (2013)
Graphene	80.7 μFcm^{-2}	H_2SO_4-PVA	2.5 mwh cm^{-3}	495 Wcm^{-3}	Wu et al. (2013)
Zn-doped graphene	316 Fg^{-1}	H_2SO_4	36.94	3325	Tatrari et al. (2022)
K-doped graphene	18.2 Fg^{-1}	PVA–H_3PO_4	18.2	1000	Tatrari et al. (2021a, b, c, d)
V_2O_5-graphene	139.7 F g^{-1}	H_3PO_4	19.2	5312	Sahoo et al. (2022)
Graphene	642 F g^{-1}	H_2SO_4	48.9	629.5	Pathak et al. (2022)
GO-PEDOT:PSS	155 F g^{-1}	H_3PO_4	10.79	1529.29	Karakoti et al. (2022)
Graphene	117 Fg^{-1}	H_2SO_4	31.9	NA	Vivekchand et al. (2008)
Graphene	488 Fg^{-1}	H_2SO_4	52.57	350	Pathak et al. (2022)
Graphene-WO_3–MnO_2	657	KOH	48.25	596.06	Tatrari et al. (2023a, b, c)
Graphene	500	H_2SO_4	50.00	1000.00	Tatrariet al. (2023a, b, c)

Each component of solar cells now contains graphene-based materials, which have produced several intriguing findings. Graphene materials have been primarily applied in counter electrodes of DSSCs as a replacement for Pt due to their high conductivity and catalytic properties to facilitate the redox reaction of electrolytes. To compete with charge recombination's effects, graphene has several advantageous qualities that can boost the loading efficiency of the dye molecules, expand the interfacial area, and improve the conductivity of the electrons. The ratio of TiO_2 and Graphene must be balanced to create an effective system. The graphene-TiO_2 contact effectively separates the electrons and holes, stimulating the graphene valence electrons into the TiO_2 conduction band. As a result, only a tiny amount of graphene (about 1%) is needed to enable this separation; however, as graphene concentrations increase in the matrix, transmittance decreases. Adding Graphene to DSSCs increases efficiency by 39% over pure TiO_2 electrodes, enhances light scattering at the photoanode, and effectively disperses dye molecules (Pandey et al. 2021).

For graphene-based counter electrodes, pristine graphene lacks active spots for electro-catalysis processes. GO, rGO, and 3D graphene frameworks are employed as counter electrodes. Oxygen-containing functions catalyze reactions. Due to higher conductivity and charge transfer, DSSC power conversion efficiency (PCE) rises with less GO amount. Hybridizing and doping graphene with metals, metal oxides, and polymers gave it the best DSSC electrode performance. The DSSC-based Pt-graphene counter electrode had a higher PCE of 8.56% than the GO-only device (4.48%) and the sputtered Pt electrode (8.18%) (Pandey et al. 2023).

8.14 Conclusions

Graphene and its derivatives have shown significant potential in electrochemistry and energy storage. They are suitable for electrodes in energy storage devices such as supercapacitors and lithium-ion batteries due to their distinctive properties, including excellent electrical conductivity, a wide surface area, and great mechanical stability. Furthermore, it has been shown that graphene-based materials augment the performance of these devices by speeding the rate of charge transfer and increasing energy and power density. Graphene composite materials are often used as supercapacitor electrodes due to their excellent electrical conductivity, enormous surface area, and great mechanical stability. These properties allow for a considerable charge storage capacity and quick charge transfer, increasing supercapacitor performance. Standard graphene composites comprise graphene oxide, reduced graphene oxide, and nanosheets.

These may be combined with conductive materials such as conductive polymers and metal oxides to produce hybrid electrodes. This enhances the electrochemical performance and stability of the supercapacitor. Current research aims to improve the properties of graphene composites for supercapacitor applications and to commercialize this technology. However, to realize their full potential in these applications, further research must be performed to overcome hurdles such as weak electrochemical activity and concerns over stability. This chapter provided a concise summary of the numerous methods for producing graphene and its composites, as well as the most recent advancements in the area. Graphene hybrid composites have been extensively explored as a viable contender for supercapacitors.

Acknowledgements The Kempe Foundation, in memory of J. C. and Seth M. Kempe, is gratefully acknowledged for the financial support in the form of a stipend for GT (grant number JCK22-0045). The financial support from the European Union and the Department of Science and Technology INSPIRE division (Ref No.: IF180347), New Delhi, India, is gratefully acknowledged.

References

Ahmad, T., Zhang, D.: A critical review of comparative global historical energy consumption and future demand: the story told so far. Energy Rep. **6**, 1973–1991 (2020)

Al Hassan, M.R., et al.: Emergence of graphene as a promising anode material for rechargeable batteries: a review. Mater. Today Chem. **11**, 225–243 (2019)

Ambrosi, A., et al.: Electrochemistry of graphene and related materials. Chem. Rev. **114**(14), 7150–7188 (2014)

Ambrosi, A., Pumera, M.: Precise tuning of surface composition and electron-transfer properties of graphene oxide films through electroreduction. Chem. Euro. J. **19**(15), pp. 4748–4753 (2013)

Bohra, B.S., et al.: The effects of functionalized graphene oxide on the thermal and mechanical properties of liquid crystalline polymers. Soft Matter **18**(20), 3981–3992 (2022)

Bolotin, K.I., et al.: Ultrahigh electron mobility in suspended graphene. Solid State Commun.commun. **146**(9–10), 351–355 (2008)

Brownson, D.A.C., Banks, C.E.: Graphene electrochemistry: an overview of potential applications. Analyst **135**(11), 2768–2778 (2010)

Chaudhuri, A., Chaudhuri, A., Joydhar, A.: Graphene nanocomposites and applications in electrochemical energy storage materials. Mater. Today: Proc. **64**, 1569–1581 (2022)

Chng, E.L.K., Pumera, M.: Solid-state electrochemistry of graphene oxides: Absolute quantification of reducible groups using voltammetry. Chem. Asian J. **6**(11), 2899–2901 (2011)

Dahn, J.R., et al.: Mechanisms for lithium insertion in carbonaceous materials. Science **270**(5236), 590–593 (1995)

Davies, T.J., Hyde, M.E., Compton, R.G.: Nanotrench arrays reveal insight into graphite electrochemistry. Angew. Chem. Int. Ed.. Chem. Int. Ed. **44**(32), 5121–5126 (2005)

Devendran, A., Nagai, A.: Recent advancements in 3D porous graphene-based electrode materials for electrochemical energy storage applications. Mater. Adv. [Preprint] (2023)

Edenhofer, O. et al.: Summary for policymakers (2014)

Geim, A.K., Novoselov, K.S.: The rise of graphene. Nat. Mater. **6**(3), 183–191 (2007)

Grimshaw, J.: Electrochemical Reactions and Mechanisms in Organic Chemistry. Elsevier (2000)

Heersche, H.B., et al.: Bipolar supercurrent in graphene. Nature **446**(7131), 56–59 (2007)

Hossain, S., et al.: Overcharge studies of carbon fiber composite-based lithium-ion cells. J. Power. Sources **161**(1), 640–647 (2006)

Huang, X. et al.: Graphene-based materials: synthesis, characterization, properties, and applications. Small **7**(14), 1876–1902 (2011)

Karakoti, M., Pandey, S., et al.: A waste to energy approach for the effective conversion of solid waste plastics into graphene nanosheets using different catalysts for high performance supercapacitors: a comparative study. Mater. Adv. **3**(4), 2146–2157 (2022a)

Karakoti, M., Tatrari, G., et al.: Tailoring the electrochemical performance of PEDOT: PSS via incorporation of spray dryer processed graphene oxide. Int. J. Energy Res. **46**(13), 18711–18726 (2022b)

Karakoti, M. et al.: Recycling of plastics into advance carbon nanomaterials and their application in energy storage system. In: Recent Developments in Plastic Recycling, pp. 259–281 (2021)

Kuang, P., et al.: 3D graphene-based H_2-production photocatalyst and electrocatalyst. Adv. Energy Mater. **10**(14), 1903802 (2020)

Kumar, R., et al.: Recent progress in the synthesis of graphene and derived materials for next generation electrodes of high performance lithium ion batteries. Prog. Energy Combust. Sci. **75**, 100786 (2019)

Lelieveld, J., et al.: Effects of fossil fuel and total anthropogenic emission removal on public health and climate. Proc. Natl. Acad. Sci. **116**(15), 7192–7197 (2019)

Li, Z., et al.: Activated pyrene decorated graphene with enhanced performance for electrochemical energy storage. Chem. Eng. J. **334**, 845–854 (2018)

Mahmood, N., et al.: Graphene-based nanocomposites for energy storage and conversion in lithium batteries, supercapacitors and fuel cells. J. Mater. Chem. A **2**(1), 15–32 (2014)

Moyseowicz, A., Minta, D., Gryglewicz, G.: Conductive polymer/graphene-based composites for next generation energy storage and sensing applications. ChemElectroChem **10**(9), e202201145 (2023)

Ohzuku, T., Brodd, R.J.: An overview of positive-electrode materials for advanced lithium-ion batteries. J. Power. Sources **174**(2), 449–456 (2007)

Oli, S. et al.: Effects of graphene-based metal composite and urea on seed germination and performance of berberis chitria Buch.-Ham. ex Lindl. Appl. Biochem. Biotechnol. 1–14 (2023)

Paek, S.-M., Yoo, E., Honma, I.: Enhanced cyclic performance and lithium storage capacity of SnO_2/graphene nanoporous electrodes with three-dimensionally delaminated flexible structure. Nano Lett. **9**(1), 72–75 (2009)

Pandey, S., et al.: 3D graphene nanosheets from plastic waste for highly efficient HTM free perovskite solar cells. Nanoscale Adv. **3**(16), 4726–4738 (2021)

Pandey, S. et al.: Recent advances in carbon-based materials for high-performance perovskite solar cells: gaps, challenges and fulfillment. Nanoscale Advances [Preprint] (2023)

Pathak, M., et al.: Few layer graphene nanosheets from kinnow peel waste for high-performance supercapacitors: a comparative study with three different electrolytes. J. Energy Storage **55**, 105729 (2022)

Pedrosa, M., et al.: Metal-free graphene-based catalytic membrane for degradation of organic contaminants by persulfate activation. Chem. Eng. J. **369**, 223–232 (2019)

Pumera, M.: Electrochemistry of graphene: new horizons for sensing and energy storage. Chem. Rec. **9**(4), 211–223 (2009)

Raccichini, R., et al.: The role of graphene for electrochemical energy storage. Nat. Mater. **14**(3), 271–279 (2015)

Razaq, A., et al.: Review on graphene-, graphene oxide-, reduced graphene oxide-based flexible composites: from fabrication to applications. Materials **15**(3), 1012 (2022)

Rogers, L.R.: Developing graphene-based supercapacitors for high temperature applications (2015)

Sahoo, N., et al.: Vanadium pentaoxide-doped waste plastic-derived graphene nanocomposite for supercapacitors: a comparative electrochemical study of low and high metal oxide doping. RSC Adv. **12**(9), 5118–5134 (2022)

Sato, S., et al.: Graphene—novel material for nanoelectronics. Fujitsu Sci. Tech. J **46**(1), 103–110 (2010)

Stoller, M.D., et al.: Graphene-based ultracapacitors. Nano Lett. **8**(10), 3498–3502 (2008)

Sun, Y., Wu, Q., Shi, G.: Graphene based new energy materials. Energy Environ. Sci. **4**(4), 1113–1132 (2011)

Tatrari, G., Tewari, C., Karakoti, M., et al.: Mass production of metal-doped graphene from the agriculture waste of Quercus ilex leaves for supercapacitors: inclusive DFT study. RSC Adv. **11**(18), 10891–10901 (2021a)

Tatrari, G., Karakoti, M., Tewari, C., et al.: Solid waste-derived carbon nanomaterials for supercapacitor applications: a recent overview. Mater. Adv. **2**(5), 1454–1484 (2021c)

Tatrari, G., et al.: Waste plastic derived graphene sheets as nanofillers to enhance mechanical strength of concrete mixture: An inventive approach to deal with universal plastic waste. Clean. Eng. Technol. **5**, 100275 (2021ed)

Tatrari, G., et al.: Bulk production of zinc doped reduced graphene oxide from tire waste for supercapacitor application: Computation and experimental analysis. J. Energy Storage **53**, 105098 (2022)

Tatrari, G., Tewari, C., Pathak, M., Bhatt, D., Karakoti, M., et al.: 3D-graphene hydrogel and tungsten trioxide-MnO_2 composite for ultra-high-capacity asymmetric supercapacitors: a comparative study. J. Energy Storage **68**, 107830 (2023a)

Tatrari, G., Tewari, C., Pathak, M., Bhatt, D., Solanki, M., et al.: Coconut-husk derived graphene for supercapacitor applications: comparative analysis of polymer gel and aqueous electrolytes. Mater. Adv. **4**(15), 3310–3322 (2023c)

Tatrari, G., Ahmed, M., Shah, F.U.: Synthesis, thermoelectric and energy storage performance of transition metal oxides composites. Coord. Chem. Rev. **498**, 215470 (2024)

Tatrari, G., Karakoti, M., Pathak, M., et al.: Quantum dots based materials for new generation supercapacitors application: a recent overview. Mater. Res. Forum LLC 215 (2021)

Tatrari, G., Bhowmick, S., Filippov, A., An, R., Shah, F.U.: Charge storage performance of structurally flexible hybrid ionic liquid electrolyte in supercapacitor. Authorea Preprints [Preprint] (2023)

Tewari, C., et al.: A simple, eco-friendly and green approach to synthesis of blue photoluminescent potassium-doped graphene oxide from agriculture waste for bio-imaging applications. Mater. Sci. Eng. C **104**, 109970 (2019)

Tewari, C., et al.: Green and cost-effective synthesis of 2D and 3D graphene-based nanomaterials from Drepanostachyum falcatum for bio-imaging and water purification applications. Chem. Eng. J. Adv. **10**, 100265 (2022)

Tewari, C., Tatrari, G., et al.: 'Can graphene-based composites and membranes solve current water purification challenges-a comprehensive review. Desalination 116952 (2023)

Tewari, C., Pathak, M., et al.: 'Waste plastics derived reduced graphene oxide-based nanocomposite with Fe_3O_4 for water purification and supercapacitor applications. J. Ind. Eng. Chem. [Preprint] (2023)

Toh, S.Y., et al.: Graphene production via electrochemical reduction of graphene oxide: synthesis and characterisation. Chem. Eng. J. **251**, 422–434 (2014)

Vivekchand, S.R.C., et al.: Graphene-based electrochemical supercapacitors. J. Chem. Sci. **120**, 9–13 (2008)

Wu, Z., et al.: Graphene-based in-plane micro-supercapacitors with high power and energy densities. Nat. Commun.commun. **4**(1), 2487 (2013)

Yang, J., Gunasekaran, S.: Electrochemically reduced graphene oxide sheets for use in high performance supercapacitors. Carbon **51**, 36–44 (2013)

Yang, G., et al.: Structure of graphene and its disorders: a review. Sci. Technol. Adv. Mater. **19**(1), 613–648 (2018)

Zahid, M.U., et al.: Synthesis of carbon nanomaterials from different pyrolysis techniques: a review. Mater. Res. Express **5**(5), 052002 (2018)

Zhang, X., et al.: 2D materials beyond graphene for high-performance energy storage applications. Adv. Energy Mater. **6**(23), 1600671 (2016)

Zhang, Y., et al.: Graphene and its derivatives in lithium–sulfur batteries. Mater. Today Energy **9**, 319–335 (2018)

Zhang, Y., et al.: Multiscale graphene-based materials for applications in sodium ion batteries. Adv. Energy Mater. **9**(8), 1803342 (2019)

Zhao, L., et al.: Facile hydrothermal preparation of recyclable S-doped graphene sponge for Cu2+ adsorption. J. Hazard. Mater. **286**, 449–456 (2015)

Zhu, J., et al.: Graphene and graphene-based materials for energy storage applications. Small **10**(17), 3480–3498 (2014)

Chapter 9
Electrochemically Exfoliated Graphene and Its Derivatives: Applications in Biosensing and Bioimaging

Neeraj Kumar, Arpana Parihar, Mohd. Abubakar Sadique, and Raju Khan

Abstract Graphene and its derivatives exhibit advantageous features when employed as a biosensing platform and for bioimaging purposes. The unique physical attributes of graphene, including high thermal and electrical conductivity, a substantial surface area, distinctive optical properties, favorable biocompatibility, and facile functionalization, render it well-suited for a diverse range of biomedical applications, particularly in biosensing and drug delivery. Various nanostructures derived from graphene have been crafted and implemented in the development of diverse biosensors. This analysis delves into the merits of utilizing electrochemically exfoliated graphene, while also addressing current challenges. Furthermore, the chapter explores the application of electrochemically exfoliated graphene and its derivatives in the scope of bioimaging and biosensing. Owing to its wide applicability, the bulk and cost-effective synthesis of graphene is the current area of interest. Recently, the synthesis of graphene and its derivatives through exfoliation approaches from waste material has gained much attention due to its cost-effectiveness and ease of synthesis. In the chapter, we summarized recent discoveries, deliberated on existing challenges, and delineated future avenues for employing nanostructures based on graphene in biosensing and bioimaging applications. The preceding chapter delved into diverse applications of graphene and its derivatives within the realms of biosensing and bioimaging.

Keywords Electrochemically exfoliation · Graphene · Biosensing · Bioimaging

N. Kumar · A. Parihar · Mohd. A. Sadique · R. Khan (✉)
Industrial Waste Utilization, Nano and Biomaterials, CSIR-Advanced Materials and Processes Research Institute (AMPRI), Hoshangabad Road, Bhopal 462026, MP, India
e-mail: khan.raju@gmail.com; khan.raju@ampri.res.in

N. Kumar · Mohd. A. Sadique · R. Khan
Academy of Scientific and Innovative Research (AcSIR), Ghaziabad 201002, India

R. Khan et al. (eds.), *Electrochemical Exfoliation of Graphene and its Derivatives*, Engineering Materials, https://doi.org/10.1007/978-981-97-2128-3_9

9.1 Introduction

Discovered in 2004, graphene has garnered considerable attention due to its extraordinary mechanical, electrical, thermal, optical, and structural characteristics, positioning it as a promising candidate for diverse applications. The graphene and its derivatives possess oxygen-containing functional groups, including hydroxyl, carbonyl, and carboxyl radicals which provide anchor to various chemical reactions (Morales-Narváez et al. 2012). Owing to their outstanding mechanical, thermal, electrical, and optical properties, graphene and its derivatives have become the focus of growing interest and are currently the subject of active research. There is a rising consensus among researchers about the potential of graphene, leading to the exploration of its applications in various scientific domains, including biosensing and bioimaging (Zhu et al. 2016).

Graphene is a 2D nanomaterial made up of an atomically thin layer of carbon. It shows great potential for various applications (Novoselov et al. 2004; Geim and Novoselov 2007). High theoretical surface area, high Young's modulus (Lee et al. 2008), high intrinsic mobility (Morozov et al. 2008), high thermal and electrical conductivity (Balandin et al. 2008), and optical transmittance (Cai et al. 2009) are unique features owned by graphene. Graphene, initially regarded as an 'academic material', is a free-standing 2D carbon nano-allotrope packed in a patterned honeycomb lattice with extremely high carrier mobility that is suitable for energy applications (Stoller et al. 2008; Avila et al. 2013; Bonaccorso et al. 2015). Due to its high theoretical surface area of 2630 m^2 g^{-1}, optical transparency, substantial mechanical strength, and inherent structural flexibility, it has received significant attention (Geim and Novoselov 2007; Liu et al. 2012a). Producing single or few-layer graphene in a simple, efficient, cost-effective, and scalable manner remains an enormous challenge in materials science (Purkait et al. 2017). This material has gained significant research interest due to its various applications, including catalysis (Shen et al. 2012), electronics (An et al. 2018), transistors (Rodrigues et al. 2023), photonics (Bonaccorso et al. 2010), electrochemical sensors (Facure et al. 2021), biosensors (Tetyana et al. 2021), anticorrosive coatings (Böhm 2014), fuel cells (Choi et al. 2012), energy storage (Dong et al. 2017) and hydrogen storage (Zhou et al. 2016), etc.

Graphene quantum dots (GQDs) are nano-sized fragments of graphene with unique properties. GQDs are planar nanomaterials with lateral dimensions ranging from 2 to 20 nm (Kumar et al. 2021a). Graphene quantum dots (GQDs) exhibit remarkable luminescent properties that are associated with their edge effects or defect states, particularly the surface states. GQDs possess excellent electrochemical properties, making them suitable electrode materials for various applications such as fuel cells, supercapacitors, and photovoltaic cells. Additionally, they are widely used in the field of electrochemical sensors (Khanra et al. 2014). Two-dimensional (2D) graphene oxide (GO) and reduced graphene oxide (rGO), are μm-sized fragments of graphite. Graphene and its derivatives possess functional groups like carboxylic acid (COOH), hydroxyls (OH), amide (NH2), and epoxides (COC) on their surface, particularly in graphene oxide (GO) and reduced graphene oxide (rGO). These functional

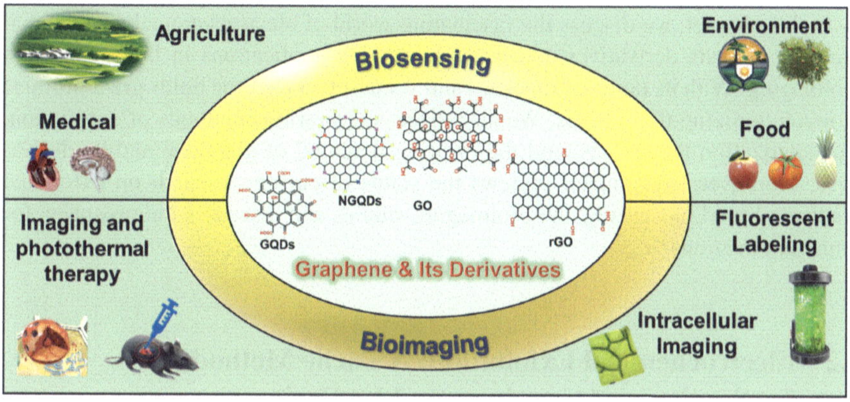

Fig. 9.1 Graphene and its derivatives have various applications in biosensing and bioimaging

groups enable graphene to be coupled with various biomolecules, thereby broadening its biomedical applications (Aliyev et al. 2019; Eckhart et al. 2019; Veclani et al. 2020).

While there is literature that covers the synthesis of graphene (Allen et al. 2010), its chemical and physical properties (Chen et al. 2012; Zhang et al. 2015), and its applications (Chen et al. 2010), the use of graphene and its derivatives in intracellular biosensing and bioimaging is still a relatively new area that has yet to be comprehensively reviewed. Graphene-based materials have unique properties that make them versatile imaging tools suitable for both optical and non-optical imaging studies. This chapter provides an overview of how graphene and its derivatives are being used in intracellular biosensing and bioimaging. The glimpses of the application of graphene have been illustrated in Fig. 9.1.

Biosensing aims to provide accurate and sensitive information about the presence of specific biological analytes, which can include biomolecules such as proteins, nucleic acids, enzymes, hormones, or whole cells in real-time (Bahadir and Sezgintürk 2016). Early disease diagnosis requires highly sensitive biosensors to detect low concentrations of disease-related biomarkers in the biomedical field (Krishnan et al. 2019; Bai et al. 2020; Fahmy et al. 2022).

Various synthesis techniques have been established for the synthesis of graphene and its derivatives, such as Hummer's (Pedrosa et al. 2020), Brodie's methods (Feicht et al. 2019), Staudenmaier's, Hofmann's methods (Stankovich et al. 2007; Poh et al. 2012), chemical vapor deposition (CVD) (Shao et al. 2010; Zhang et al. 2013), Solvent exfoliation (Hernandez et al. 2008; Parvez et al. 2014), and Electrochemical exfoliation (ECE) (Yu et al. 2015). ECE is a cost-effective and environmentally friendly method that involves using an electrolyte and an electrical current to separate graphite layers. This process is mild and does not require a strong oxidizing agent, making it a safer alternative to other methods. The resulting structural delamination of the graphite electrode can be used for various applications (Parvez et al. 2013; Kumar et al. 2021a).

In this chapter, we discuss the fascinating world of electrochemically exfoliated graphene and its derivatives and its wide-ranging applications in biosensing and bioimaging. With its unique properties and versatility, graphene holds great promise in revolutionizing these fields. We also discuss the various methods of exfoliation, characterization techniques, and the exciting potential of graphene and its derivatives. Moreover, this chapter reviews the status of current research on exfoliated graphene-based biosensing and bioimaging studies and discusses the prospects for future applications.

9.2 Electrochemical Exfoliation: A Facile Method for Synthesis of Graphene and Its Derivatives

The ECE method has several advantages, including simplicity, cost-effectiveness, rapidity, and environmental friendliness. Additionally, this method requires only a few chemicals and graphite as the source of graphene. When a voltage is applied to graphite electrodes in a conductive solution, graphene nanosheets can be produced through erosion, which are then functionalized by the solvent (Wicaksono et al. 2020).

Anodic electrochemical exfoliation results in the exfoliation of graphene sheets from the anode through intercalation of negatively charged ions (Yu et al. 2015). Upon initiation of the DC voltage supply, electrons must flow seamlessly between the anode and cathode graphite rods. Incomplete wetting of the electrodes in the electrolyte can disrupt this flow, causing an uneven distribution of electrons and resulting in an irregular voltage gradient along the anode surface. This inconsistency may lead to disproportionate exfoliation of the graphite, yielding sheets with uneven thickness and surface characteristics. In a recent synthesis of graphene quantum dots (GQDs), a careful approach was taken by initially applying a 2 V voltage bias for the first 2 min to facilitate gentle ion intercalation. Subsequently, the voltage was increased to 10 V for 1–3 h (Su et al. 2011; Kumar et al. 2021a). When a voltage is applied, water is reduced at the cathode site, which generates hydroxyl ions (OH–) and causes hydrogen bubbles to form. These OH– ions are strong nucleophiles in electrolytes and they attack specific positions on the graphite rod (Parvez et al. 2014). Initially, oxidation occurs at the edge sites and grain boundaries of graphite. The process of exfoliation of graphene quantum dots using the ECE mechanism is explained in detail in Fig. 9.2.

A table listing the parameters for the electrochemical exfoliation to synthesize graphene and its derivatives is presented in Table 9.1.

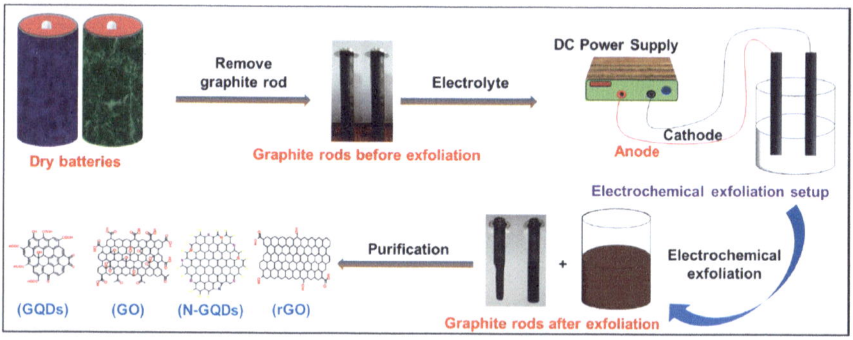

Fig. 9.2 The schematic representation of ECE mechanism and synthesis of graphene and its derivatives

9.3 Properties of Electrochemically Exfoliated Graphene and Its Derivatives

In recent years, there has been outstanding interest surrounding electrochemically exfoliated graphene and its derivatives, primarily driven by their characteristic properties and potential applications across various fields (Fig. 9.3). Some essential characteristics of electrochemically exfoliated graphene and its derivatives have been enlisted below.

High Surface Area: Electrochemically exfoliated graphene and its derivatives typically possess a high surface area, which can be advantageous for various applications such as catalysis, energy storage devices, and sensors. The high surface area allows for increased interactions with other molecules and materials (Kamedulski et al. 2021).

Tunable Thickness: The thickness of graphene layers can be controlled via electrochemical exfoliation, producing single or few-layer sheets with unique properties (Ni et al. 2008).

Electrical Conductivity: Graphene and its derivatives are highly conductive, making them useful for electrical devices such as transparent conductive films and electrodes. Electrochemically exfoliated graphene retains its electrical conductivity (Destiarti et al. 2023).

Mechanical Strength: Graphene is considered one of the most durable materials known, with exceptional mechanical properties. Although electrochemically exfoliated graphene and its derivatives may not be as robust as pristine graphene, they still demonstrate good mechanical strength and flexibility, making them ideal for applications that require both strength and flexibility (Zhao et al. 2019).

Chemical Stability: Graphene and its derivatives, produced through electrochemical exfoliation, are highly stable in diverse chemical environments, making them suitable

Table 9.1 Various parameters of graphene and its derivatives prepared through exfoliation approaches

S.N	Dimensional	Electrolyte	Voltage	Time	Electrode		Product	References
					Cathode (−)	Anode (+)		
1	Zero dimensional	NaOH + Hydrazine hydrate	5 V	8 h	Pt	Graphite	GQDs	Zhang et al. (2012)
2		$K_2S_2O_8$	5 V	–	Graphite	Graphite	RF-GQDs	Tan et al. (2015)
3		Citric acid + NaOH	10 V	1 h	Graphite rod	Graphite rod	GQDs	Ahirwar et al. (2017)
4		NaOH + Semicarbazide	5 V	4 h	Graphite rod	Graphite rod	N-GQDs	Fu et al. (2019)
5		Citric acid + NaOH	10 V	1 h	Graphite rod	Graphite rod	GQDs	Kumar et al. (2021a)
6	Two-dimensional	1-Butyl-3-methylimidazolium tetrafluoroborate	1.5 to 15 V		Pt wire	HOPG	Graphene	Lu et al. (2009)
7		H_2SO_4	−10 to + 10 V	–	Pt wire	HOPG or Graphite	Graphene	Su et al. (2011)
8		9-Anthracene carboxylic acid (9-ACA) + NaOH	20 V	10 h	Graphite rod	Graphite rod	Graphene	Khanra et al. (2012)
9		Graphite + Sodium dodecyl sulfate	1.6 V	12 h	Pt wire	Graphite rod	Graphene sheets	Alanyalioğlu et al. (2012)
10		Na_2SO_4 + Sodium dodecyl sulfate	6 V	1 h	Pt wire	Graphite rod	Graphene film	Liu et al. (2013a)
11		H_2SO_4 or H_3PO_4	+7 V to − 7 V	5–8 Min	Pencil core	Pencil core	GO flakes	Liu et al. (2013b)
12		TCNQ + DMSO	25 V	12 h	Graphite rod	Graphite rod	Graphene	Khanra et al. (2014)

(continued)

Table 9.1 (continued)

S.N	Dimensional	Electrolyte	Voltage	Time	Electrode		Product	References
					Cathode (−)	Anode (+)		
13		$(NH_4)_2SO_4$, Na_2SO_4, and K_2SO_4	10 V	3–5 Min	Pt rod	Graphite flack	Graphene	Parvez et al. (2014)
14		H_2SO_4	20 V	10 Min	Graphite	Graphite powder	Graphene	Chen et al. (2015)
15		Sodium dodecyl benzene sulfonate	4.5 V	5 h	Graphite rod	Graphite rod	Graphene	Tiwari et al. (2017)
16		Ammonium sulfate + reducing agents	10 V	10 s	Pt	Graphite	Graphene	Yang et al. (2015)
17		$(NH_4)_2SO_4$	1 V	2 h	Graphite rod	Graphite rod	GO	Parvez et al. (2016)
18		$(NH4)_2SO_4$	12 V	2–24 h	Pt wire	Graphite rod	Graphene	Marković et al. (2016)
19		(H_2SO_4, Na_2SO_4, and $LiClO_4$)	10 V	–	Pt foil	Graphite foil	Graphene and GO	Ambrosi and Pumera (2016)
20		H_2SO_4	10 V	–	Pt foil	Graphite foil	Graphene	Hu et al. (2017)
21		Sodium dodecyl benzene sulfonate	4.75 V	4 h	Graphite	Graphite	Graphene	Udhaya Sankar et al. (2018)
22		$(NH_4)_2SO_4$,	5 to 10 V	1–2 h	Pt plate	Graphite flack	Graphene	Aawani et al. (2018)
23		H_2SO_4	10 V	1.5–2 h	Graphite rod	Graphite rod	Graphene, Graphene oxide	Bandi et al. (2019)

(continued)

Table 9.1 (continued)

S.N	Dimensional	Electrolyte	Voltage	Time	Electrode		Product	References
					Cathode (−)	Anode (+)		
24		$(NH_4)_2SO_4$	10 V	5 Min	Graphite foil	Graphite foil	Graphene	Parvez et al. (2019)
25		Na_2SO_4	10 V	2 h	Pt foil	Graphite foil	Graphene	Ilnicka et al. (2019)
26		Poly(sodium 4-styrene sulfonate) PSS	3, 5, 8 V	4 h	Graphite rod	Graphite rod	Graphene	Prakoso et al. (2020)
27		$NaOH + NH_4Cl$	5 V	5 h	Graphite rod	Graphite rod	N-Graphene	Jing et al. (2020)
28		$(NH_4)_2SO_4$	5, 10, and 15 V	1–5 h	Graphite	Graphite	Graphene	Wicaksono et al. (2020)

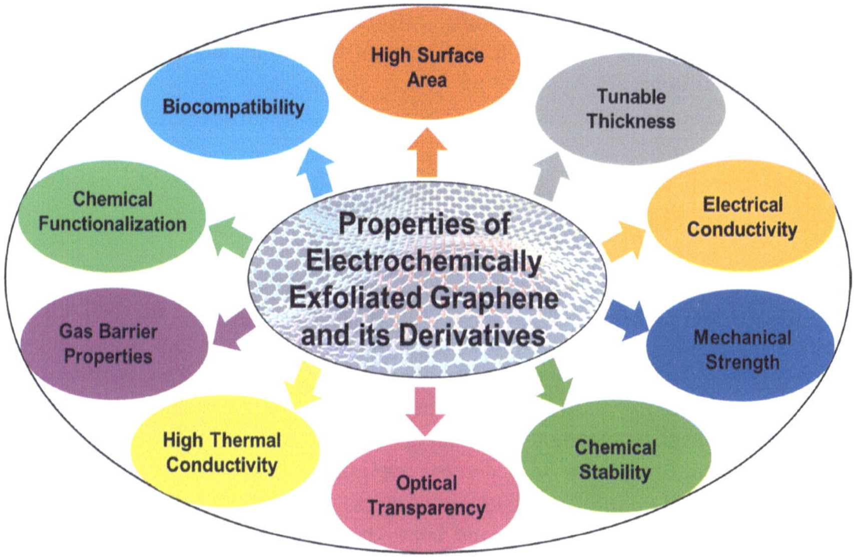

Fig. 9.3 Properties of electrochemically exfoliated graphene and its derivatives

for use in various chemical processes and as support materials for catalysts (Cai et al. 2016).

Optical Transparency: Single-layer graphene is almost transparent in the visible part of the electromagnetic spectrum. It has many applications, including touch screen transparent conductive films and solar cells (Su et al. 2011).

High Thermal Conductivity: Graphene has excellent thermal conductivity that can be maintained even in electrochemically exfoliated graphene and its derivatives. This property is crucial for applications in thermal management and as a filler material in composites (Jagannadham 2012).

Gas Barrier Properties: Graphene and its derivatives have been studied for their ability to prevent the permeation of gases, such as oxygen and moisture, in packaging materials, making them promising candidates for gas barrier applications (Pierleoni et al. 2016).

Chemical Functionalization: Various chemical groups can be added to electrochemically exfoliated graphene and its derivatives, allowing for property modification and targeted functionalities (Ejigu et al. 2017).

Biocompatibility: Graphene and its derivatives are biocompatible, making them promising for biomedical applications. They can be used for drug delivery, biosensing, and bioimaging (Munuera et al. 2015).

The properties of electrochemically exfoliated graphene and its derivatives may vary based on the specific electrochemical technique, starting material, and processing conditions.

9.4 Application of Graphene and Its Derivatives in the Biosensing Field

Over the past few years, there has been a focused investigation into methods for producing large quantities of high-quality graphene to satisfy the increasing demand for electronic devices (Schwierz 2010), energy storage (Stoller et al. 2008), transparent electrodes (Pang et al. 2011), sensing (Hill et al. 2011), catalysis (MacHado and Serp 2011), and drug delivery (Sun et al. 2008).

Among these, biosensors have diverse applications such as disease diagnosis, drug discovery, and monitoring treatment progress. They are also vital for the detection of harmful toxins, pesticides, and heavy metals in the environment, such as water, soil, and air. By using sensitive biosensors, we can ensure the safety of both humans and the environment (Krishnan et al. 2019; Coroş et al. 2020). Graphene and its derivatives have shown remarkable potential in the development of biosensors (Ranjan et al. 2020; Singhal et al. 2022; Parihar et al. 2022; Parihar and Khan 2023; Parihar et al. 2023b, c). This is attributed to their exceptional physical properties such as superior biocompatibility, large surface area, excellent adsorption, electrical conductivity, and high thermal. Furthermore, their unique optical properties make them a promising material for biosensor applications (Bai et al. 2020; Fahmy et al. 2022; Yadav et al. 2023). These properties enhance sensing mechanisms such as electrochemical, fluorescent, surface-enhanced Raman scattering, and surface plasmon resonance (Goodrum et al. 2023).

The widespread application of graphene in sensor development is attributed to its distinctive advantages, encompassing a substantial surface-to-volume ratio, singular optical properties, superior carrier mobility, and remarkable electrical and thermal characteristics relative to other carbon allotropes (Kumar et al. 2021b). These properties persist consistently in both double and multi-layered graphene structures. The utility of these advantages in graphene sensors primarily lies in their adaptability to meet specific application requirements, notwithstanding variations in structure and working conditions (Nag et al. 2018).

At the electrode surface, electrochemical biosensors rely on current changes (Bai et al. 2020). In the field of electrochemical immunoassays, variations in electrochemical signals arise from the interplay between immobilized antibodies and antigens located on the electrode (Assari et al. 2019). This sensing method is cost-effective and provides real-time, sensitive measurements, making it ideal for quantitative detection (Damborska et al. 2017; Hou et al. 2018). Graphene and its derivatives, with large surface areas, and excellent electrical conductivity are well-suited for efficient molecule adsorption and rapid electron transfer. This makes them ideal for electrochemical sensors, ensuring sensitive and selective target detection. Some sensor techniques use chemically unreduced graphene to preserve conduction properties. Widely employed in electrode films, especially working electrodes, graphene enhances biomolecule immobilization, increasing the active surface area and thereby boosting energy conversion (Acharya et al. 2022; Fahmy et al. 2022).

For instance, in a recent study, the electrode modified with GQDs was used as a matrix to immobilize anti-CD44, which allowed CD44 to bind effectively. The label-free quantification of CD44 was achieved across a broad dynamic range (1.0 fg/mL to 1000.0 ng/mL) in both PBS and diluted human serum. Utilizing GQDs, an electrochemical biosensor was developed with high sensitivity (2.12 μA cm^{-2}/(fg/mL)) for CD44 detection. The biosensor exhibited the lowest detection limit (LOD) of 2.11 fg/mL in the linear range of 0.1 pg/mL to 100.0 ng/mL in phosphate buffer saline (PBS). Furthermore, it demonstrated a linear response for CD44 concentrations in the range of 1.0 pg/mL to 100.0 ng/mL with a LOD of 2.71 fg/mL (Kumar et al. 2022) (Fig. 9.4a, b).

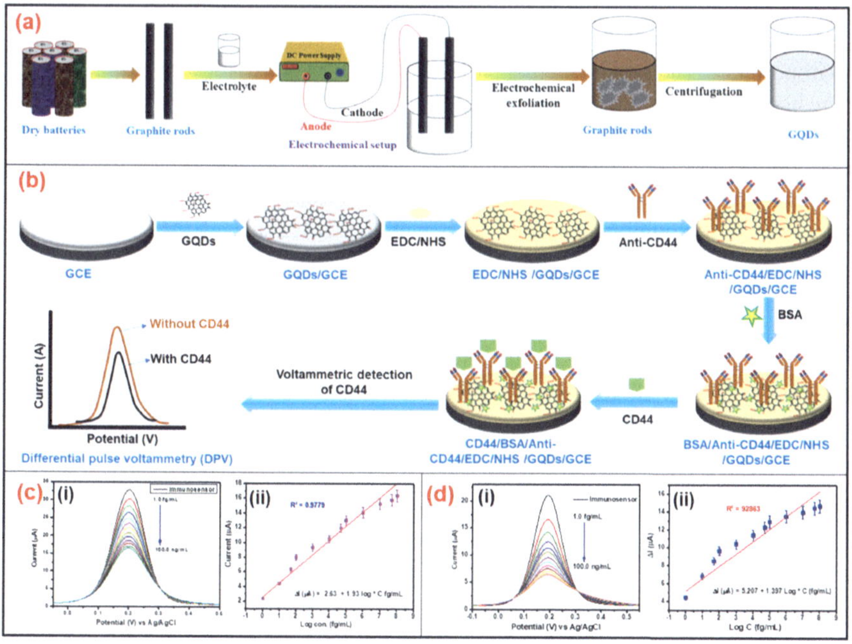

Fig. 9.4 **a** Illustration depicting the synthesis process of GQDs, and **b** Diagram outlining the sensing mechanism of the developed electrochemical biosensor. Reprinted with permission from (Kumar et al. 2022), Copyright 2022 @ MDPI. **c** Detection of the CD44 breast cancer biomarker antigen in PBS and **d** serum, illustrated through **i** DPV detection curves and **ii** corresponding calibration graphs. Reprinted with permission from Kumar et al. (2023), Copyright 2023 @ Elsevier

In another study, the same group developed a highly sensitive electrochemical immunosensor for detecting the breast cancer biomarker CD44. Utilizing electrochemically exfoliated GQDs and synthesized gold nanoparticles (GNPs), a GQDs-GNPs nanocomposite was created, significantly enhancing sensor performance. The sensor demonstrated an impressive limit of detection (LOD) of 1.17 fg/mL in PBS, allowing for label-free detection of CD44 within a concentration range of 1.0 fg/mL to 100.0 ng/mL when diluted in human serum in PBS. In serum samples, the ultra-low detection limit was recorded at 3.17 fg/mL. Overall, the proposed electrochemical immunosensor is characterized by high sensitivity and selectivity in CD44 detection. (Kumar et al. 2023) (Fig. 9.4c, d).

9.5 Application of Graphene and Its Derivatives in the Bioimaging Field

Graphene is an excellent platform for the advancement of bioimaging systems (Liu et al. 2012b; Alava et al. 2013). To create a graphene-based platform for bioimaging, it is important to add special elements to the surface. This will prevent graphene from being harmful to cells and allow for the attachment of other molecules. Surface modification enhances the bioanalytical properties of the platform (Parihar et al. 2023a), improving factors such as biocompatibility, sensitivity, and specificity. This renders it an ideal substrate for the development of a versatile multicomponent platform tailored for bioimaging applications. (Chen et al. 2011; Liu et al. 2012b).

Tan et al. achieved the preparation of water-soluble, 3 nm uniform-sized red-emitting RF-GQDs through electrochemical exfoliation without chemical modification. The results of this study are illustrated in Fig. 9.5(a)i, ii, wherein RF-GQDs effectively label both the cell membrane and cytoplasm of Hela cells with red emission under 488 nm excitation, indicating their translocation from outside the cell membrane to the cytoplasm. Figure 9.5(a)iii demonstrates the sustained viability of Hela cells after 24 h incubation with RF-GQDs at a concentration of 1000 $\mu g\ mL^{-1}$, as assessed by MTT experiments. These findings underscore the significant potential of RF-GQDs for biomedical applications, particularly in direct and efficient cell labeling, suggesting their promise in diverse biomedical applications (Tan et al. 2015).

Similarly, Fu et al. synthesized N-GQDs through an electrochemical method, yielding particles with a 20 nm average diameter. GQDs that only emit blue fluorescence, the N-GQDs emit a wider range of colors, including primary colors, extending into the visible region. These N-GQDs are used to selectively detect Fe^{3+} ions through photoluminescence quenching. They provide a broad linear detection range (0–200 μM) with an LOD (0.87 μM), compliant with EPA standards for drinking water. Post-treatment with 200 $\mu g\ mL^{-1}$ N-GQDs, cell viability remains approximately 96%, indicating minimal toxicity to HeLa cells compared to the control

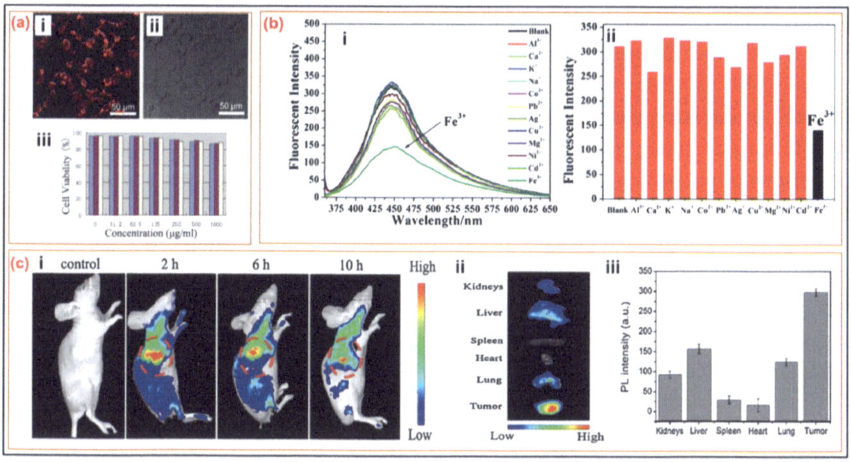

Fig. 9.5 **a**, **i** Hela cell confocal fluorescence microscopy images with RF-GQDs **ii** comparable bright field imaging, **iii** Hela cell viability assay, Reprint with permission from Tan et al. (2015), Copyright 2015 @ RSC. **b**, **i** N-GQD solutions with different metal ions were studied by photoluminescence emission spectra. **ii** Histogram of the fluorescence response of N-GQDs subsequent to the introduction of various metal ions. Reprint with permission from Fu et al. (2019), Copyright 2019 @ RSC. **c**, **i** Imaging in-vivo of nude mice harboring HeLa tumors subsequent to administration of (GQD/DBM)3EuPhen/GQD. **ii** Imaging ex-vivo of isolated mouse organs and **iii** PL intensities of (GQD/ DBM)3EuPhen/GQD from isolated organs, Reprint with permission from Liu et al. (2016), Copyright 2016 @ Springer

(100%) (Fig. 9.5(b)i, ii). Additionally, the synthesized N-GQDs demonstrate favorable physicochemical traits such as red emission, visible region absorption, fluorescence emission in the long-wavelength region, and low cytotoxicity (Fu et al. 2019).

In another study, Liu et al. synthesized high-purity red-emitting composites, water-soluble, denoted as (GQD/DBM)3EuPhen/GQD, through noncovalent functionalization of chelating ligands (DBM and Phen) with GQDs, followed by coordination to Eu(III). Tumor-bearing nude mice were intravenously injected with (GQD/ DBM)3EuPhen/GQD, while control mice received no treatment. Figure 9.5(c)i reveals significant PL signals in the tumor area just 2 h post-injection, highlighting the potential of (GQD/DBM)3EuPhen/GQD for tumor labeling and chemotherapy. Mice were sacrificed at 10 h post-injection, and ex-vivo fluorescence imaging of internal organs (Fig. 9.5(c)ii, iii) showed the strongest signal in the tumor tissue, with weak signals in major organs like kidneys, liver, and lungs, and very low signals in the spleen and heart (Liu et al. 2016).

Dong et al. introduced a novel method for concurrent synthesis of single-layer and multi-layer GQDs by refluxing Vulcan CX-72 carbon black with concentrated nitric acid, followed by separation using acidic centrifugation. MTT assay revealed minimal toxicity of GQDs1 to MCF-7 breast cancer cells, even at 100 μg mL^{-1}. GQDs1, with acceptable photoluminescent yield (PLYQ) and low cytotoxicity, are suitable for biomedical applications. Confocal imaging of MCF-7 cells treated with GQDs1 for 4 h exhibited bright green labeling of the cell membrane, cytoplasm, and nucleus. The resulting multi-layer GQDs showed broad UV–vis absorption, suitable for optoelectronic devices, while the single-layer GQDs demonstrated excellent photoluminescent activity and biocompatibility, presenting an eco-friendly material for various applications like biolabeling, bioimaging, and drug/gene carriers (Dong et al. 2012).

Notably, Yuan et al. successfully produced water-soluble, multicolor fluorescent graphene quantum dots (MCF GQDs) with all-pH responsiveness, using electrochemical exfoliation of graphite followed by refluxing in a mixed nitric and sulfuric acid solution. These GQDs exhibit visible color changes from pH 1 to 14 under UV light. The red photoluminescence arises from quinone structures formed in strongly alkaline conditions, a unique feature. This underscores the practical potential of MCF GQDs in bioimaging and clinical diagnosis applications (Yuan et al. 2015).

Zhang et al. successfully synthesized water-soluble, uniformly sized GQDs with robust yellow emission via electrochemical exfoliation of graphite, followed by room temperature reduction with hydrazine. These GQDs easily penetrate stem cells without compromising viability, differentiation capacity, and proliferation. Inside cells, the GQDs exhibit strong photoluminescence, excellent photostability, and minimal cytotoxicity. MTT assays gauged the toxicity of synthesized GQDs using a Victor3 V Multilabel reader. Absorbance expressed as a percentage of control, was plotted against drug concentration. After 72 h treatment, NSCs and CPCs maintained over 80% viability at 100 $\mu g\ mL^{-1}$, while PPCs showed about 65%. At 25 $\mu g\ mL^{-1}$, all cells were observable, indicating the potential of GQDs as a low cytotoxicity biological imaging platform (Zhang et al. 2012).

9.6 Future Prospects and Challenges

Although graphene that has been electrochemically exfoliated has a lot of potential, more needs to be done to ensure that it is stable and compatible with biological systems. But with continued study and advancement, graphene ought to reach its full potential. In recent years, various approaches have been explored for synthesizing and manufacturing graphene tailored to specific applications. Ongoing research primarily focuses on scaling up the production of graphene-related materials to meet commercial demands. Nevertheless, the challenge lies in achieving high-volume graphene preparation using methods that are both economically viable and environmentally sustainable. While techniques such as chemical vapor deposition (CVD) (Strupinski et al. 2011), micromechanical exfoliation (Sinclair et al. 2019), and epitaxial growth play crucial roles in ensuring material quality, they fall short when it comes to cost-effectiveness and large-scale synthesis (Kum et al. 2019).

One potential avenue for achieving scalable synthesis involves a chemical process that utilizes the exfoliation of graphite oxide. However, this method relies on the use of hazardous and toxic reagents, posing environmental and safety concerns.

Furthermore, the chemically produced reduced graphene oxide exhibits sub-optimal electrical conductivity, making it unsuitable for applications in energy storage devices (Novoselov et al. 2012).

Addressing these challenges, recent research has been dedicated to devising a sustainable method for large-scale production of carbon nanomaterials, including graphene. This innovative approach utilizes natural sources such as waste corn shells, plant leaves, eggshells, biochar, even human hair, and fungus. The goal is to establish a method that not only ensures scalability but also aligns with principles of environmental sustainability, offering a promising avenue for advancing graphene synthesis for diverse applications (Zhu et al. 2011, 2018; Qian et al. 2013, 2015).

Attaining consistent, high-quality graphene exfoliation via electrochemical methods poses challenges, impacting biosensor and bioimaging device performance. Crucial for biocompatibility, stability, and biomolecule interaction, proper graphene functionalization presents challenges in maintaining its properties. Long-term biocompatibility assessment and addressing potential cytotoxicity are vital for biological applications, especially in biosensing and bioimaging (Sharma et al. 2016). Integrating graphene with biological systems while preserving structural integrity poses challenges. Stable interactions with biomolecules and cells are essential for biosensor and bioimaging device reliability. Ongoing challenges involve improving the sensitivity, selectivity, and detection limits of graphene-based biosensors while minimizing interference from other biological components, ensuring accurate and reliable sensing (Suvarnaphaet et al. 2017; Pourmadadi et al. 2022).

9.7 Concluding Remark

The electrochemically exfoliated graphene is a remarkable material that has numerous potential applications in biosensing and bioimaging. Its use can lead to better detection and improved imaging of biological systems and processes. This chapter discussed the electrochemical exfoliation-based approaches for synthesizing graphene and its derivatives along with the advantages and challenges of using an electrochemical method from various precursors and electrolytes. Further, this chapter provided an overview of the remarkable features and potential uses of graphene and its derivatives in various fields such as biosensing and bioimaging. Moreover, the general principles and factors that affect the electrochemical exfoliation method such as the type and concentration of electrolyte, the type and potential of an electrode, the current density, and the water content have been discussed. The synthesis of high-quality graphene is an expensive and complex process and hence the scalability of graphene production is currently limited, which poses a challenge to its commercial viability. Therefore, there is a need for the development of better manufacturing techniques to produce graphene at a larger scale. Further, the performance of graphene as a catalyst can be negatively affected in oxidative environments. To fully utilize graphene, research must be done to address these issues. One helpful method for doing so is the electrochemical exfoliation-based production of graphene and its derivatives.

Acknowledgements The authors express their sincere thanks to the Director, CSIR-AMPRI for his support and encouragement in this work. Mr. Neeraj Kumar is thankful to the University Grants Commission (UGC), India for the generous award of SRF. Arpana Parihar is thankful to the DST-WOS-B fellowship (DST/WOS-B/HN-4/2021) and Mohd Abubakar Sadique is thankful to the SERB-IRPHA for awarding JRF. Raju Khan would like to acknowledge SERB for providing funds in the form of the IPA/2020/000130 project.

References

Aawani, E., Rezagholipour Dizaji, H., et al.: Exfoliation of graphite in various electrolytes and preparation of graphene films via electrochemical approach. Opt. Quant. Electron **50**, 129 (2018). https://doi.org/10.1007/s11082-018

Acharya, P.B., George, A., Shrivastav, P.S.: A status update on the development of polymer and metal-based graphene electrochemical sensors for detection and quantitation of bisphenol A. Crit. Rev. Anal. Chem. (2022). https://doi.org/10.1080/10408347.2022.2094197

Ahirwar, S., Mallick, S., Bahadur, D.: Electrochemical method to prepare graphene quantum dots and graphene oxide quantum dots. ACS Omega **2**, 8343–8353 (2017). https://doi.org/10.1021/acsomega.7b01539

Alanyalioğlu, M., Segura, J.J., Oró-Sol, J., Casañ-Pastor, N.: The synthesis of graphene sheets with controlled thickness and order using surfactant-assisted electrochemical processes. In: Carbon, pp. 142–152. Elsevier Ltd. (2012)

Alava, T., Mann, J.A., Théodore, C., et al.: Control of the graphene-protein interface is required to preserve adsorbed protein function. Anal. Chem. **85**, 2754–2759 (2013). https://doi.org/10.1021/AC303268Z/SUPPL_FILE/AC303268Z_SI_002.PDF

Aliyev, E., Filiz, V., Khan, M.M., et al.: Structural Characterization of graphene oxide: surface functional groups and fractionated oxidative debris. Nanomaterials **9**, 1180 (2019). https://doi.org/10.3390/NANO9081180

Allen, M.J., Tung, V.C., Kaner, R.B.: Honeycomb carbon: a review of graphene. Chem. Rev. **110**(1), 132–145 (2010). https://doi.org/10.1021/CR900070D/ASSET/CR900070D.FP.PNG_V03

Ambrosi, A., Pumera, M.: Electrochemically exfoliated graphene and graphene oxide for energy storage and electrochemistry applications. Chem. Eur. J. **22**, 153–159 (2016). https://doi.org/10.1002/chem.201503110

An, H., Habib, T., Shah, S., et al.: Stretchable, bendable, and foldable conductive coatings are crucial for wearable electronics and biometric, pp. 1–9 (2018)

Assari, P., Rafati, A.A., Feizollahi, A., Asadpour Joghani, R.: An electrochemical immunosensor for the prostate specific antigen based on the use of reduced graphene oxide decorated with gold nanoparticles. Microchim. Acta. Acta **186**, 1–9 (2019). https://doi.org/10.1007/S00604-019-3565-8/TABLES/2

Avila, J., Razado, I., Lorcy, S., et al.: Exploring electronic structure of one-atom thick polycrystalline graphene films: a nano angle resolved photoemission study. Sci. Rep. **3**(1), 1–8 (2013). https://doi.org/10.1038/srep02439

Bahadir, E.B., Sezgintürk, M.K.: Applications of graphene in electrochemical sensing and biosensing. TrAC, Trends Anal. Chem. **76**, 1–14 (2016). https://doi.org/10.1016/J.TRAC.2015.07.008

Bai, Y., Xu, T., Zhang, X.: Graphene-based biosensors for detection of biomarkers. Micromachines (Basel) **11** (2020). https://doi.org/10.3390/MI11010060

Balandin, A.A., Ghosh, S., Bao, W., et al.: Superior thermal conductivity of single-layer graphene. Nano Lett. **8**, 902–907 (2008). https://doi.org/10.1021/NL0731872/ASSET/IMAGES/MEDIUM/NL-2007-031872_0005.GIF

Bandi, S., Ravuri, S., Peshwe, D.R., Srivastav, A.K.: Graphene from discharged dry cell battery electrodes. J. Hazard. Mater. **366**, 358–369 (2019). https://doi.org/10.1016/j.jhazmat.2018.12.005

Böhm, S.: Graphene against corrosion. Nat. Nanotechnol. **9**(10), 741–742 (2014). https://doi.org/10.1038/nnano.2014.220

Bonaccorso, F., Sun, Z., Hasan, T., Ferrari, A.C.: Graphene photonics and optoelectronics. Nat. Photonics **4**(9), 611–622 (2010). https://doi.org/10.1038/nphoton.2010.186

Bonaccorso, F., Colombo, L., Yu, G., et al.: Graphene, related two-dimensional crystals, and hybrid systems for energy conversion and storage. Science (1979) **347** (2015). https://doi.org/10.1126/SCIENCE.1246501/ASSET/F43DC9DE-35A1-4E26-B700-818C38EE1489/ASSETS/GRAPHIC/347_1246501_FA.JPEG

Cai, W., Feng, X., Hu, W., et al.: Functionalized graphene from electrochemical exfoliation for thermoplastic polyurethane: thermal stability, mechanical properties, and flame retardancy. Ind. Eng. Chem. Res. **55**, 10681–10689 (2016). https://doi.org/10.1021/ACS.IECR.6B02579/ASSET/IMAGES/MEDIUM/IE-2016-02579N_0012.GIF

Cai, W., Zhu, Y., Li, X., et al.: Large area few-layer graphene/graphite films as transparent thin conducting electrodes. Appl. Phys. Lett. **95** (2009). https://doi.org/10.1063/1.3220807/338025

Chen, D., Feng, H., Li, J.: Graphene oxide: preparation, functionalization, and electrochemical applications. Chem. Rev. **112**(11), 6027–6053 (2012). https://doi.org/10.1021/CR300115G/ASSET/CR300115G.FP.PNG_V03

Chen, D., Tang, L., Li, J.: Graphene-based materials in electrochemistry. Chem. Soc. Rev. **39**(8), 3157–3180 (2010). https://doi.org/10.1039/B923596E

Chen, Q., Wei, W., Lin, J.M.: Homogeneous detection of concanavalin a using pyrene-conjugated maltose assembled graphene based on fluorescence resonance energy transfer. Biosens. Bioelectron.. Bioelectron. **26**, 4497–4502 (2011). https://doi.org/10.1016/J.BIOS.2011.05.009

Chen, C.H., Yang, S.W., Chuang, M.C., et al.: Towards the continuous production of high crystallinity graphene via electrochemical exfoliation with molecular in situ encapsulation. Nanoscale **7**, 15362–15373 (2015). https://doi.org/10.1039/c5nr03669k

Choi, H.J., Jung, S.M., Seo, J.M., et al.: Graphene for energy conversion and storage in fuel cells and supercapacitors. Nano Energy **1**, 534–551 (2012). https://doi.org/10.1016/J.NANOEN.2012.05.001

Coroş, M., Pruneanu, S., Stefan-van Staden, R.-I.: Review—recent progress in the graphene-based electrochemical sensors and biosensors. J. Electrochem. Soc.electrochem. Soc. **167**, 037528 (2020). https://doi.org/10.1149/2.0282003JES/XML

Damborska, D., Bertok, T., Dosekova, E., et al.: Nanomaterial-based biosensors for detection of prostate specific antigen. Microchimica Acta **184**(9), 3049–3067 (2017). https://doi.org/10.1007/S00604-017-2410-1

Destiarti, L., Riyanto, R., Roto, R., Mudasir, M.: Electrolyte effect in electrochemical exfoliation of graphite. Mater. Chem. Phys. **302**, 127713 (2023). https://doi.org/10.1016/J.MATCHEMPHYS.2023.127713

Dong, Y., Chen, C., Zheng, X., et al.: One-step and high yield simultaneous preparation of single- and multi-layer graphene quantum dots from CX-72 carbon black. J. Mater. Chem. **22**, 8764–8766 (2012). https://doi.org/10.1039/c2jm30658a

Dong, Y., Wu, Z.S., Ren, W., et al.: Graphene: a promising 2D material for electrochemical energy storage. Sci. Bull. (beijing) **62**, 724–740 (2017). https://doi.org/10.1016/J.SCIB.2017.04.010

Eckhart, K.E., Holt, B.D., Laurencin, M.G., Sydlik, S.A.: Covalent conjugation of bioactive peptides to graphene oxide for biomedical applications. Biomater. Sci. **7**, 3876–3885 (2019). https://doi.org/10.1039/C9BM00867E

Ejigu, A., Kinloch, I.A., Dryfe, R.A.W.: Single stage simultaneous electrochemical exfoliation and functionalization of graphene. ACS Appl. Mater. Interfaces **9**, 710–721 (2017). https://doi.org/10.1021/ACSAMI.6B12868/SUPPL_FILE/AM6B12868_SI_001.PDF

Facure, M.H.M., Schneider, R., Lima, J.B.S., et al.: Graphene quantum dots-based nanocomposites applied in electrochemical sensors: a recent survey. Electrochem **2**, 490–519 (2021). https://doi.org/10.3390/ELECTROCHEM2030032

Fahmy, H.M., Abu Serea, E.S., Salah-Eldin, R.E., et al.: Recent progress in graphene- and related carbon-nanomaterial-based electrochemical biosensors for early disease detection. ACS Biomater. Sci. Eng.biomater. Sci. Eng. **8**, 964–1000 (2022). https://doi.org/10.1021/ACSBIOMATERIALS.1C00710/ASSET/IMAGES/MEDIUM/AB1C00710_0026.GIF

Feicht, P., Biskupek, J., Gorelik, T.E., et al.: Brodie's or Hummers' method: oxidation conditions determine the structure of graphene oxide. Chem. Euro. J. **25**, 8955–8959 (2019). https://doi.org/10.1002/CHEM.201901499

Fu, Y., Gao, G., Zhi, J.: Electrochemical synthesis of multicolor fluorescent N-doped graphene quantum dots as a ferric ion sensor and their application in bioimaging. J. Mater. Chem. B **7**, 1494–1502 (2019). https://doi.org/10.1039/c8tb03103g

Geim, A.K., Novoselov, K.S.: The rise of graphene. Nat. Mater. **6**(3), 183–191 (2007). https://doi.org/10.1038/nmat1849

Goodrum, R., Weldekidan, H., Li, H., et al.: Graphene-based nanostructures from green processes and their applications in biomedical sensors. Adv. Ind. Eng. Polymer Res. (2023)

Hernandez, Y., Nicolosi, V., Lotya, M., et al.: High-yield production of graphene by liquid-phase exfoliation of graphite. Nat. Nanotechnol. **3**(9), 563–568 (2008). https://doi.org/10.1038/nnano.2008.215

Hill, E.W., Vijayaragahvan, A., Novoselov, K.: Graphene sensors. IEEE Sens. J. **11**, 3161–3170 (2011). https://doi.org/10.1109/JSEN.2011.2167608

Hou, B., Liu, H., Qi, S., et al.: Preparation of pristine graphene in ethanol assisted by organic salts for nonenzymatic detection of hydrogen peroxide. J. Colloid Interface Sci. **510**, 103–110 (2018). https://doi.org/10.1016/J.JCIS.2017.09.052

Hu, L., Peng, X., Li, Y., et al.: Direct anodic exfoliation of graphite onto high-density aligned graphene for large capacity supercapacitors. Nano Energy **34**, 515–523 (2017). https://doi.org/10.1016/j.nanoen.2017.03.007

Ilnicka, A., Skorupska, M., Kamedulski, P., Lukaszewicz, J.P.: Electro-exfoliation of graphite to graphene in an aqueous solution of inorganic salt and the stabilization of its sponge structure with poly(Furfuryl alcohol). Nanomaterials **9** (2019). https://doi.org/10.3390/nano9070971

Jagannadham, K.: Thermal conductivity of copper-graphene composite films synthesized by electrochemical deposition with exfoliated graphene platelets. Metall. Mater. Trans. B **43**, 316–324 (2012). https://doi.org/10.1007/S11663-011-9597-Z/TABLES/6

Jing, M., Wu, T., Zhou, Y., et al.: Nitrogen-doped graphene via in-situ alternating voltage electrochemical exfoliation for supercapacitor application. Front. Chem. **8** (2020). https://doi.org/10.3389/fchem.2020.00428

Kamedulski, P., Skorupska, M., Binkowski, P., et al.: High surface area micro-mesoporous graphene for electrochemical applications. Sci. Rep. **11**(1), 1–12 (2021). https://doi.org/10.1038/s41598-021-01154-0

Khanra, P., Kuila, T., Bae, S.H., et al.: Electrochemically exfoliated graphene using 9-anthracene carboxylic acid for supercapacitor application. J. Mater. Chem. **22**, 24403–24410 (2012). https://doi.org/10.1039/c2jm34838a

Khanra, P., Lee, C.N., Kuila, T., et al.: 7,7,8,8-Tetracyanoquinodimethane-assisted one-step electrochemical exfoliation of graphite and its performance as an electrode material. Nanoscale **6**, 4864–4873 (2014). https://doi.org/10.1039/c3nr05307e

Krishnan, S.K., Singh, E., Singh, P., et al.: A review on graphene-based nanocomposites for electrochemical and fluorescent biosensors. RSC Adv. **9**, 8778–8881 (2019). https://doi.org/10.1039/C8RA09577A

Kum, H., Lee, D., Kong, W., et al.: Epitaxial growth and layer-transfer techniques for heterogeneous integration of materials for electronic and photonic devices. Nat. Electron. **2**(10), 439–450 (2019). https://doi.org/10.1038/s41928-019-0314-2

Kumar, N., Sadique, M.A., Khan, R., et al.: Immunosensor for breast cancer CD44 biomarker detection based on exfoliated graphene quantum dots integrated gold nanoparticles. Hybrid Adv. **3**, 100065 (2023). https://doi.org/10.1016/J.HYBADV.2023.100065

Kumar, N., Abubakar Sadique, M., Khan, R.: Electrochemical exfoliation of graphene quantum dots from waste dry cell battery for biosensor applications. Mater. Lett. **305** (2021a). https://doi.org/10.1016/j.matlet.2021.130829

Kumar, N., Sorna Gowri, V., Ranjan, P., et al.: Efficiency of nanomaterials for electrochemical diagnostics based point-of-care detection of non-invasive oral cancer biomarkers. Adv. Mater. Lett. **12**, 1–20 (2021b). https://doi.org/10.5185/amlett.2021.081651

Kumar, N., Yadav, S., Sadique, M.A., Khan, R.: Electrochemically exfoliated graphene quantum dots based biosensor for CD44 breast cancer biomarker. Biosensors (Basel) **12** (2022). https://doi.org/10.3390/bios12110966

Lee, C., Wei, X., Kysar, J.W., Hone, J.: Measurement of the elastic properties and intrinsic strength of monolayer graphene. Science **321**, 385–388 (2008). https://doi.org/10.1126/SCIENCE.1157996/SUPPL_FILE/LEE-SOM.PDF

Liu, J., Xue, Y., Zhang, M., Dai, L.: Graphene-based materials for energy applications. MRS Bull. **37**, 1265–1272 (2012a). https://doi.org/10.1557/MRS.2012.179

Liu, Y., Dong, X., Chen, P.: Biological and chemical sensors based on graphene materials. Chem. Soc. Rev. **41**, 2283–2307 (2012b). https://doi.org/10.1039/C1CS15270J

Liu, J., Notarianni, M., Will, G., et al.: Electrochemically exfoliated graphene for electrode films: effect of graphene flake thickness on the sheet resistance and capacitive properties. Langmuir **29**, 13307–13314 (2013a). https://doi.org/10.1021/la403159n

Liu, J., Yang, H., Zhen, S.G., et al.: A green approach to the synthesis of high-quality graphene oxide flakes via electrochemical exfoliation of pencil core. RSC Adv. **3**, 11745–11750 (2013b). https://doi.org/10.1039/c3ra41366g

Liu, Y., Zhou, S., Fan, L., Fan, H.: Synthesis of red fluorescent graphene quantum dot-europium complex composites as a viable bioimaging platform. Microchim. Acta. Acta **183**, 2605–2613 (2016). https://doi.org/10.1007/s00604-016-1909-1

Lu, J., Yang, J.X., Wang, J., et al.: One-pot synthesis of fluorescent carbon nanoribbons, nanoparticles, and graphene by the exfoliation of graphite in ionic liquids. ACS Nano **3**, 2367–2375 (2009). https://doi.org/10.1021/nn900546b

MacHado, B.F., Serp, P.: Graphene-based materials for catalysis. Catal. Sci. Technol.. Sci. Technol. **2**, 54–75 (2011). https://doi.org/10.1039/C1CY00361E

Marković, Z.M., Budimir, M.D., Kepić, D.P., et al.: Semi-transparent, conductive thin films of electrochemical exfoliated graphene. RSC Adv. **6**, 39275–39283 (2016). https://doi.org/10.1039/c6ra04250c

Morales-Narváez, E., Merkoçi, A., Merkoçi, A., Morales-Narváez, E.: Graphene oxide as an optical biosensing platform. Adv. Mater. **24**, 3298–3308 (2012). https://doi.org/10.1002/ADMA.201200373

Morozov, S.V., Novoselov, K.S., Katsnelson, M.I., et al.: Giant intrinsic carrier mobilities in graphene and its bilayer. Phys. Rev. Lett. **100**, 016602 (2008). https://doi.org/10.1103/PHYSREVLETT.100.016602/FIGURES/4/MEDIUM

Munuera, J.M., Paredes, J.I., Villar-Rodil, S., et al.: High quality, low oxygen content and biocompatible graphene nanosheets obtained by anodic exfoliation of different graphite types. Carbon N Y **94**, 729–739 (2015). https://doi.org/10.1016/J.CARBON.2015.07.053

Nag, A., Mitra, A., Mukhopadhyay, S.C.: Graphene and its sensor-based applications: a review. Sens. Actuators A Phys. **270**, 177–194 (2018)

Ni, Z.H., Wang, H.M., Ma, Y., et al.: Tunable stress and controlled thickness modification in graphene by annealing. ACS Nano **2**, 1033–1039 (2008). https://doi.org/10.1021/NN800031M/ASSET/IMAGES/MEDIUM/NN-2008-00031M_0009.GIF

Novoselov, K.S., Geim, A.K., Morozov, S.V., et al.: Electric field effect in atomically thin carbon films. Science **306**(5696), 666–669 (2004). https://doi.org/10.1126/SCIENCE.1102896/SUPPL_FILE/NOVOSELOV.SOM.PDF

Novoselov, K.S., Fal'Ko, V.I., Colombo, L., et al.: A roadmap for graphene. Nature **490**(7419), 192–200 (2012). https://doi.org/10.1038/nature11458

Pang, S., Hernandez, Y., Feng, X., et al.: Graphene as transparent electrode material for organic electronics. Adv. Mater. **23**, 2779–2795 (2011). https://doi.org/10.1002/ADMA.201100304

Parihar, A., Khan, R.: Yttrium functionalized reduced graphene oxide nanocomposite-based aptasensor for ultrasensitive detection of a breast cancer biomarker. ACS Appl. Nano Mater. (2023). https://doi.org/10.1021/ACSANM.3C03234/ASSET/IMAGES/MEDIUM/AN3C03234_0009.GIF

Parihar, A., Ahmed, S.S., Sharma, P., et al.: Plant-based bioactive molecules for targeting of endoribonuclease using steered molecular dynamic simulation approach: a highly conserved therapeutic target against variants of SARS-CoV-2. Mol. Simul. **49**, 1267–1279 (2023a). https://doi.org/10.1080/08927022.2022.2113811

Parihar, A., Sharma, P., Choudhary, N.K., et al.: Green synthesis of CdS and CdS/rGO nanocomposites: evaluation of electrochemical, antimicrobial, and photocatalytic properties. ACS Appl. Bio Mater. **6**, 3706–3716 (2023b). https://doi.org/10.1021/ACSABM.3C00390/ASSET/IMAGES/MEDIUM/MT3C00390_0012.GIF

Parihar, A., Yadav, S., Sadique, M.A., et al.: Internet-of-medical-things integrated point-of-care biosensing devices for infectious diseases: toward better preparedness for futuristic pandemics. Bioeng. Transl. Med. **8**, e10481 (2023c). https://doi.org/10.1002/BTM2.10481

Parihar, A., Singhal, A., Kumar, N., et al.: Next-Generation intelligent MXene-based electrochemical aptasensors for point-of-care cancer diagnostics. Nano-Micro Lett. **14**(1), 1–34 (2022). https://doi.org/10.1007/S40820-022-00845-1

Parvez, K., Li, R., Puniredd, S.R., et al.: Electrochemically exfoliated graphene as solution-processable, highly conductive electrodes for organic electronics. ACS Nano **7**, 3598–3606 (2013). https://doi.org/10.1021/NN400576V/SUPPL_FILE/NN400576V_SI_001.PDF

Parvez, K., Wu, Z.S., Li, R., et al.: Exfoliation of graphite into graphene in aqueous solutions of inorganic salts. J. Am. Chem. Soc. **136**, 6083–6091 (2014). https://doi.org/10.1021/JA5017156/SUPPL_FILE/JA5017156_SI_001.PDF

Parvez, K., Rincón, R.A., Weber, N.E., et al.: One-step electrochemical synthesis of nitrogen and sulfur co-doped, high-quality graphene oxide. Chem. Commun.commun. **52**, 5714–5717 (2016). https://doi.org/10.1039/c6cc01250g

Parvez, K., Worsley, R., Alieva, A., et al.: Water-based and inkjet printable inks made by electrochemically exfoliated graphene. Carbon N Y **149**, 213–221 (2019). https://doi.org/10.1016/j.carbon.2019.04.047

Pedrosa, M., Da Silva, E.S., Pastrana-Martínez, L.M., et al.: Hummers' and Brodie's graphene oxides as photocatalysts for phenol degradation. J. Colloid Interface Sci. **567**, 243–255 (2020). https://doi.org/10.1016/J.JCIS.2020.01.093

Pierleoni, D., Xia, Z.Y., Christian, M., et al.: Graphene-based coatings on polymer films for gas barrier applications. Carbon N Y **96**, 503–512 (2016). https://doi.org/10.1016/J.CARBON.2015.09.090

Poh, H.L., Šaněk, F., Ambrosi, A., et al.: Graphenes prepared by Staudenmaier, Hofmann and Hummers methods with consequent thermal exfoliation exhibit very different electrochemical properties. Nanoscale **4**, 3515–3522 (2012). https://doi.org/10.1039/C2NR30490B

Pourmadadi, M., Dinani, H.S., Tabar, F.S., et al.: Properties and applications of graphene and its derivatives in biosensors for cancer detection: a comprehensive review. Biosensors (Basel) **12** (2022). https://doi.org/10.3390/BIOS12050269

Prakoso, B., Ma, Y., Stephanie, R., et al.: Facile synthesis of battery waste-derived graphene for transparent and conductive film application by an electrochemical exfoliation method. RSC Adv. **10**, 10322–10328 (2020). https://doi.org/10.1039/d0ra01100b

Purkait, T., Singh, G., Singh, M., et al.: Large area few-layer graphene with scalable preparation from waste biomass for high-performance supercapacitor. Sci. Rep. **7** (2017). https://doi.org/10.1038/s41598-017-15463-w

Qian, W., Sun, F., Xu, Y., et al.: Human hair-derived carbon flakes for electrochemical supercapacitors. Energy Environ. Sci. **7**, 379–386 (2013). https://doi.org/10.1039/C3EE43111H

Qian, K., Kumar, A., Zhang, H., et al.: Recent advances in utilization of biochar. Renew. Sustain. Energy Rev. **42**, 1055–1064 (2015). https://doi.org/10.1016/J.RSER.2014.10.074

Ranjan, P., Parihar, A., Jain, S., et al.: Biosensor-based diagnostic approaches for various cellular biomarkers of breast cancer: a comprehensive review. Anal. Biochem.biochem. **610**, 113996 (2020). https://doi.org/10.1016/J.AB.2020.113996

Rodrigues, T., Curti, F., Leroux, Y.R., et al.: Discovery of a Peptide Nucleic Acid (PNA) aptamer for cardiac troponin I: Substituting DNA with neutral PNA maintains picomolar affinity and improves performances for electronic sensing with graphene field-effect transistors (gFET). Nano Today **50**, 101840 (2023). https://doi.org/10.1016/J.NANTOD.2023.101840

Schwierz, F.: Graphene transistors. Nat. Nanotechnol. **5**(7), 487–496 (2010). https://doi.org/10.1038/nnano.2010.89

Shao, Y., Wang, J., Wu, H., et al.: Graphene based electrochemical sensors and biosensors: a review. Electroanalysis **22**, 1027–1036 (2010). https://doi.org/10.1002/ELAN.200900571

Sharma, D., Kanchi, S., Sabela, M.I., Bisetty, K.: Insight into the biosensing of graphene oxide: present and future prospects. Arab. J. Chem. **9**, 238–261 (2016). https://doi.org/10.1016/J.ARABJC.2015.07.015

Shen, J., Zhu, Y., Yang, X., Li, C.: Graphene quantum dots: emergent nanolights for bioimaging, sensors, catalysis and photovoltaic devices. Chem. Commun.commun. **48**, 3686–3699 (2012). https://doi.org/10.1039/C2CC00110A

Sinclair, R.C., Suter, J.L., Coveney, P.V.: Micromechanical exfoliation of graphene on the atomistic scale. Phys. Chem. Chem. Phys. **21**, 5716–5722 (2019). https://doi.org/10.1039/C8CP07796G

Singhal, A., Parihar, A., Kumar, N., Khan, R.: High throughput molecularly imprinted polymers based electrochemical nanosensors for point-of-care diagnostics of COVID-19. Mater. Lett. **306**, 130898 (2022). https://doi.org/10.1016/J.MATLET.2021.130898

Stankovich, S., Dikin, D.A., Piner, R.D., et al.: Synthesis of graphene-based nanosheets via chemical reduction of exfoliated graphite oxide. Carbon N Y **45**, 1558–1565 (2007). https://doi.org/10.1016/J.CARBON.2007.02.034

Stoller, M.D., Park, S., Yanwu, Z., et al.: Graphene-based ultracapacitors. Nano Lett. **8**, 3498–3502 (2008). https://doi.org/10.1021/NL802558Y/ASSET/IMAGES/MEDIUM/NL-2008-02558Y_0003.GIF

Strupinski, W., Grodecki, K., Wysmolek, A., et al.: Graphene epitaxy by chemical vapor deposition on SiC. Nano Lett. **11**, 1786–1791 (2011). https://doi.org/10.1021/NL200390E/ASSET/IMAGES/MEDIUM/NL-2011-00390E_0006.GIF

Su, C.Y., Lu, A.Y., Xu, Y., et al.: High-quality thin graphene films from fast electrochemical exfoliation. ACS Nano **5**, 2332–2339 (2011). https://doi.org/10.1021/NN200025P/SUPPL_FILE/NN200025P_SI_003.ZIP

Sun, X., Liu, Z., Welsher, K., et al.: Nano-graphene oxide for cellular imaging and drug delivery. Nano Res. **1**(3), 203–212 (2008). https://doi.org/10.1007/S12274-008-8021-8

Suvarnaphaet, P., Pechprasarn, S.: Graphene-based materials for biosensors: a review. Sensors **17**, 2161 (2017). https://doi.org/10.3390/S17102161

Tan, X., Li, Y., Li, X., et al.: Electrochemical synthesis of small-sized red fluorescent graphene quantum dots as a bioimaging platform. Chem. Commun.commun. **51**, 2544–2546 (2015). https://doi.org/10.1039/c4cc09332a

Tetyana, P., Shumbula, P.M., Njengele-Tetyana, Z., et al.: Biosensors: design, development and applications. Nanopores (2021). https://doi.org/10.5772/INTECHOPEN.97576

Tiwari, S.K., Huczko, A., Oraon, R., et al.: Facile electrochemical synthesis of few layered graphene from discharged battery electrode and its application for energy storage. Arab. J. Chem. **10**, 556–565 (2017). https://doi.org/10.1016/j.arabjc.2015.08.016

Udhaya Sankar, G., Ganesa Moorthy, C., RajKumar, G.: Synthesizing graphene from waste mosquito repellent graphite rod by using electrochemical exfoliation for battery/supercapacitor applications. Energy Sources, Part a: Recovery, Utilization Environ. Effects **40**, 1209–1214 (2018). https://doi.org/10.1080/15567036.2018.1476609

Veclani, D., Tolazzi, M., Melchior, A.: Molecular interpretation of pharmaceuticals' adsorption on carbon nanomaterials: theory meets experiments. Processes **8**, 642 (2020). https://doi.org/10.3390/PR8060642

Wicaksono, W.P., Hutama, R.D., Ridwanto, M., et al.: A facile electrochemical synthesis of graphene from battery waste for sensing application. In: AIP Conference Proceedings. American Institute of Physics Inc. (2020)

Yadav, S., Parihar, A., Sadique, M.A., et al.: Emerging point-of-care optical biosensing technologies for diagnostics of microbial infections. ACS Appl. Opt. Mater. **1**, 1245–1262 (2023). https://doi.org/10.1021/ACSAOM.3C00129

Yang, S., Brüller, S., Wu, Z.S., et al.: Organic radical-assisted electrochemical exfoliation for the scalable production of high-quality graphene. J. Am. Chem. Soc. **137**, 13927–13932 (2015). https://doi.org/10.1021/jacs.5b09000

Yu, P., Lowe, S.E., Simon, G.P., Zhong, Y.L.: Electrochemical exfoliation of graphite and production of functional graphene. Curr. Opin. Colloid Interface Sci.opin. Colloid Interface Sci. **20**, 329–338 (2015). https://doi.org/10.1016/J.COCIS.2015.10.007

Yuan, F., Ding, L., Li, Y., et al.: Multicolor fluorescent graphene quantum dots colorimetrically responsive to all-pH and a wide temperature range. Nanoscale **7**, 11727–11733 (2015). https://doi.org/10.1039/c5nr02007g

Zhang, M., Bai, L., Shang, W., et al.: Facile synthesis of water-soluble, highly fluorescent graphene quantum dots as a robust biological label for stem cells. J. Mater. Chem. **22**, 7461–7467 (2012). https://doi.org/10.1039/c2jm16835a

Zhang, W., Wu, L., Li, Z., Liu, Y.: Doped graphene: synthesis, properties and bioanalysis. RSC Adv. **5**(61), 49521–49533 (2015). https://doi.org/10.1039/C5RA05051K

Zhang, Y., Zhang, L., Zhou, C.: Review of chemical vapor deposition of graphene and related applications. Acc. Chem. Res. **46**, 2329–2339 (2013). https://doi.org/10.1021/AR300203N/ASSET/IMAGES/MEDIUM/AR-2012-00203N_0002.GIF

Zhao, X., Papageorgiou, D.G., Zhu, L., et al.: The Strength of Mechanically-Exfoliated Monolayer Graphene (2019)

Zhou, C., Szpunar, J.A., Cui, X.: Synthesis of Ni/graphene nanocomposite for hydrogen storage. ACS Appl. Mater. Interfaces **8**, 15232–15241 (2016). https://doi.org/10.1021/ACSAMI.6B02607/SUPPL_FILE/AM6B02607_SI_001.PDF

Zhu, H., Wang, X., Yang, F., Yang, X.: Promising carbons for supercapacitors derived from fungi. Adv. Mater. **23**, 2745–2748 (2011). https://doi.org/10.1002/ADMA.201100901

Zhu, X., Liu, Y., Li, P., et al.: Applications of graphene and its derivatives in intracellular biosensing and bioimaging. Analyst **141**, 4541–4553 (2016). https://doi.org/10.1039/c6an01090c

Zhu, B., Liu, B., Qu, C., et al.: Tailoring biomass-derived carbon for high-performance supercapacitors from controllably cultivated algae microspheres. J. Mater. Chem. A Mater. **6**, 1523–1530 (2018). https://doi.org/10.1039/C7TA09608A

Chapter 10
Electrochemical Exfoliation of Graphene and Its Derivatives and Its Extended Applications in Therapeutics

Sharon J. Paul, Prakash Chandra, Neeraj Kumar, and Raju Khan

Abstract Graphene possesses numerous exceptional and promising physical, chemical, and mechanical characteristics, making it a compelling candidate for a wide array of therapeutic applications. These applications span from basic drug or gene delivery systems to a diverse platform of multiple therapeutic modalities, comprising tissue engineering and cancer therapies. Graphene is a valuable scaffold in tissue engineering and has potential in photomedicine for wound healing and cancer therapy due to its photosensitizing properties. Graphene and its derivatives are currently being tested as carriers of therapeutic agents for cells or tissues, both in vivo and in vitro. These nanocarriers can have therapeutic effects which are influenced by various microenvironments. These include elevated glutathione and acidic pH as endogenous stimuli, and ultrasonic or light signals as exogenous stimuli. Graphene-based materials have paved the way for drug delivery and tumour treatment therapies due to their unique features and responsiveness. This chapter highlights the properties and unique structure of graphene that make it ideal for therapeutics. It also embodies a better and clearer understanding of therapeutic covering drug delivery, tissue engineering, gene delivery, several cancer therapies photodynamic therapy (PDT), radiotherapy (RT), photothermal therapy (PTT), etc. In this emerging field, we have discussed the significant challenges that need to be addressed, and we have also talked about the future prospects that it holds.

S. J. Paul
Photonic Materials Metrology Sub Division, Advanced Materials and Device Metrology Division, CSIR-National Physical Laboratory, New Delhi 110012, India

P. Chandra (✉)
Department of Chemistry, Institute of Basic Science, Bundelkhand University, Jhansi 284128, Uttar Pradesh, India
e-mail: drprakashcy@gmail.com; prakash@bujhansi.ac.in

N. Kumar · R. Khan
Industrial Waste Utilization, Nano and Biomaterials, CSIR-Advanced Materials and Processes Research Institute (AMPRI), Hoshangabad Road, Bhopal 462026, MP, India

Academy of Scientific and Innovative Research (AcSIR), Ghaziabad 201002, India

R. Khan et al. (eds.), *Electrochemical Exfoliation of Graphene and its Derivatives*, Engineering Materials, https://doi.org/10.1007/978-981-97-2128-3_10

Keywords Therapeutic · Theranostic · Gene delivery · Drug delivery · Tissue engineering · Cancer therapy

10.1 Introduction

Graphene has received massive curiosity since 2004 when it was discovered by Andre Geim and Konstantin Novoselov by micromechanical exfoliation of graphite using Scotch Tape method. Graphene is a wonder material causing significant impact on several modern industries like electronics, batteries, catalysis, solar cells, medical, bioengineering as well as energy storage and conversion. A monolayer of graphene consists of a hexagonal lattice of carbon atoms, each bonded in a sp^2 arrangement. Graphene exhibits captivating optical, thermal, electronic, magnetic, chemical, and mechanical properties owing to the unique structure of substantial π-electron conjugation (Singh et al. 2011). Graphene serves as an innovative foundational element for all graphitic multifunctional materials. It has a very flexible structure therefore it can be fabricated into spheres, ellipsoids, or even tubular 0-D fullerenes, 1-D cylinders as carbon nanotubes, and 3-D graphite structures.

10.1.1 The Graphene Family of Nanomaterials

To understand the derivatives of graphene, it is important to understand the structure of 3-D bulk graphite from which graphene is extracted. Graphite is composed of numerous stacked layers of graphene honeycomb trusses held together by weak Van der Waals forces. It contains over one million graphene layers with an interlayer spacing of 0.354 nm. This material is naturally abundant and readily accessible. Single-layer graphene has a monoatomic graphite layer while bilayer and tri-layer graphene consist of two to three graphite layers, respectively. Graphene oxide (GO) and reduced graphene oxide (rGO) are amongst the most captivating derivatives of graphene. Graphene oxide is a chemically modified crystalline graphene 2-D structure with abundant oxygen group functionalities on its edges and basal planes, typically having lower C/O (carbon/oxygen) atomic ratios. GO can be effortlessly produced by exfoliating and oxidizing graphite in a solution while stirring constantly. The Hummers method is a well-established method for the synthesis of graphene oxide from graphite powder (Marcano et al. 2010). rGO is the graphitic film extracted from GO by chemical, thermal, or electrical treatments (Pei and Cheng 2012). rGO has significantly lower oxygen content compared to GO.

Graphene quantum dots (GQDs), refers to a quantum variant of graphene comprising only a few layers, typically at diminutive dimensions (ideally below 10 nm) (Kumar et al. 2022). This derivative of graphene possesses enhanced electroluminescence with high optical absorption in the UV region (Kumar et al. 2023). Graphene has a derivative called GQDs. It is a quantum form of this material that has small sizes, preferably less than 10 nm, and contains some layers of the material. This particular type of graphene exhibits high optical absorption in the UV region, which enhances electroluminescence. Graphane is a graphene derivative that is fully hydrogenated. The addition of hydrogen changes the sp^2 to the sp^3 geometry, causing the carbon atoms to become tetrahedral instead of flat. This modification opens the zero bandgap of graphene to obtain dielectric graphene by transformation of the conducting graphene. Graphane holds enormous applications in areas of nanoelectronics and hydrogen-fuel technologies. *Graphone* is a partially hydrogenated derivative of graphene that has potential applications in spintronics, magnetism, and organic ferroelectrics. Fluorographene is the fluorinated derivative of graphene comprising a monolayer of graphite fluoride. It has a wide electronic band gap and is considered to be the thinnest insulator. Graphite fluoride is synthesized by fluorination of graphene, then followed by mechanical or chemical exfoliation, and is considered the counterpart of polytetrafluoroethylene (Teflon). Its applications include electronics, energy technologies, and wide-ranging fields spanning sensing and bioimaging.

Graphyne and graphdiyne are two-dimensional derivatives of graphene, consisting of one-atom thick sheets of sp and sp^2 bonded carbon atoms arranged in a crystal lattice. These sheets have acetylenic linkages (sp hybridization) that connect the hexagons of graphene (benzene rings with sp^2 hybridization). Graphdiyne is made up of two di-acetylenic chains that connect the carbon hexagons, while graphyne only has one linear acetylenic chain between carbon hexagons. Both the derivatives of graphene are semiconductors and have found application in inorganic, organic sensors, desalinator, other bio-related processes, and nanoelectronics.

Doped graphene is another type of graphene derivative that has opened new horizons for diverse applications. Various metals (Ni, Co, Fe, Mn, Pd, Au, Pt, Ti, Rh, Ag, Ru, Al, etc.), non-metals (P, Si, B, Se, N, O, I, S, etc.) and polymers have been introduced as dopants into graphene, and they have interesting properties such as ferromagnetism, superconductivity, improved mechanical attributes, and enhanced chemical/electrochemical activity.

10.2 Structural Properties of Graphene and Its Derivatives Favouring Therapeutic Applications

Graphene is a unique 2-D material where each carbon atom is connected to three other carbon atoms in a flat hexagonal structure, creating a giant aromatic polymolecule. This makes it the first known material of its kind. The unique properties like enhanced surface-area/volume ratio, polyaromatic structure, and ease of functionalization offer wide capacity and flexibility to graphene and its derivatives for therapeutic applications.

Aromatic molecules bind to pristine graphene (graphene without any functional groups) through non-covalent (weak) relations between rings of carbon (p–p assembling) or hydrophobic interactions. Hydrophobic solubilization is achieved by exfoliating graphite layers in a non-polar solvent or with surfactants. Nonetheless, pristine graphene has a drawback, as surfactants and residual non-polar solvent molecules may become adsorbed on the surface of the flakes, leading to steric hindrance and contributing significantly to toxicity.

A graphene form with additional chemical functional groups offers a wider range of possible interactions. The most common and preferred alternative is GO. There are two acceptable structures for GO: one structure depicts GO flake as a flat sp^2-bonded honeycomb lattice with occasional defects, including sp^3 bonds containing epoxides and alcohols on the planar surface. Simultaneously, carboxylic acids are bonded at the edges in the other structure type. The other structure of GO suggests carboxylic acids bonded at the edges but the planar surface is covered with highly oxidized debris of aromatic carbonaceous fragments that help to solubilize the flakes in water. Functional groups on GO facilitate enhanced solubility and offer the chance to attach other functional molecules via covalent bonding or electrostatic interactions.

The solubility and applicability of graphene derivatives, such as graphene quantum dots, in both diagnostic and therapeutic applications, are heightened through chemical modifications involving biocompatible polymers. Therefore, by controlling the structure of graphene, the behaviour of the derivative changes causing substantial improvements in the desired properties.

Also, graphene possess extraordinary electronic properties arising from the unique crystal structure which is sp^2-hybridized. Graphene possesses a honeycomb lattice structure, comprising sp^2-hybridized carbon atoms, each having 4 valence bonds consisting of 1 s-orbital and 3 p-orbitals (Novoselov et al. 2004). The hybridization of one s-orbital and two p-orbitals (px and py) with neighbouring carbons results in the creation of covalent σ-bonds within the basal plane. Simultaneously, the p-orbitals in the out-of-plane direction (pz) overlap to generate delocalized π-bonds, playing a crucial role in the remarkable electronic properties of graphene. Additionally, the absorbance spectrum of graphene and its derivatives spans from ultraviolet to the infrared region, imparting them with stimuli-responsive characteristics.

10.3 Graphene-Based Therapeutics

Therapeutic generally means curing, healing, or restoring a medical condition or disease. Therapeutics is derived from the Greek word therapeutikos further derived from therapeuein, which means "to treat". Figure 10.1 shows a schematic illustration of the typical therapeutic approaches that include drug delivery, gene delivery, and cancer therapy like photodynamic therapy (PDT), photothermal therapy (PTT), sonodynamic therapy (SDT), RT, ultrasound hyperthermia therapy (SHT) and magnetic hyperthermia therapy (MHT) or a combination of the therapies.

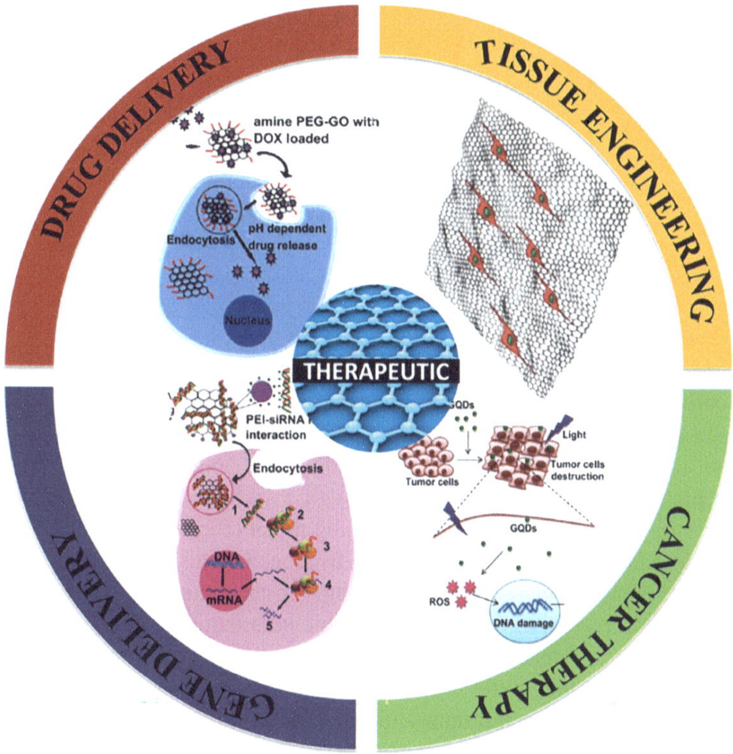

Fig. 10.1 A schematic representation of the typical therapeutic approaches. Reprinted with permission from Syama and Mohanan (2019) copyright @ Springer

10.4 Graphene and Its Derivatives as Therapeutics Drug Delivery Vehicle

A perfect drug delivery system guarantees precise transportation of therapeutic agents to their intended action site. It also increases the efficacy of the drug by potentially reducing its degradation and protecting it from dilution or off-target effects (OTEs). The physicochemical properties enable the formation of a stable and inert vehicle system, allowing it to navigate complex physiological environments. The specific features of this product enable safe absorption and delivery, which can be customized according to the specific needs of individual cells. Graphene and its derivatives are capable of providing a stable and inert drug delivery vehicle. Graphene nanoparticles are biocompatible and have lower toxicity (Ali-Boucetta et al. 2013). Different functional groups of graphene impart highly flexible physicochemical properties. Functional flexibility is the fundamental attribute that equips and provides versatility to GO in therapeutics delivery. The presence of oxygen-containing functional groups, such as carboxyls, hydroxyls, and epoxides, inherently imparts hydrophilicity to it, thereby minimizing the requirement for additional solubilization. Additionally, these functional groups create opportunities for binding additional molecules through covalent bonding or electrostatic interactions.

When we discuss about drug release it can be either "in vivo" meaning the research is done on a living organism, or "in vitro" meaning the research is done in a laboratory dish or test tube. The drug delivery also involves sequential events beginning from the cargo attachment, cell targeting, and delivery at its site of action.

10.4.1 Cargo Attachment

The attachment of cargo refers to the method by which a payload is loaded onto or into therapeutic delivery vehicles. This process affects various aspects, including cargo protection, targeted release, and delivery vehicle size and shape. Graphene offers a flexible platform for cargo attachment through π–π bonding or hydrophobic and electrostatic interactions without compromising the potency or efficiency of drugs.

As an example, doxorubicin (DOX) forms hydrogen bonds with oxygenated functional groups, enabling pH-controlled loading. On the other hand, SN38 binds to graphene through π–π bonding and hydrophobic interactions (Liu et al. 2008). Irrespective of the linkage, the drug effectiveness improves after attaching it to graphene as compared to exposure to the drug alone. DOX is an anti-cancer drug that was incorporated into a pristine and unmodified GO via simple mixing, and sonication process at environmental temperature (Yang et al. 2008). The use of PEGylated graphene oxide (PEG-GO) for loading DOX through π–π interactions demonstrates a favourable real-time drug release upon outside triggering by glutathione (GSH) (Dembereldorj et al. 2012). GO-loaded DOX demonstrated improved cellular toxicity and promising inhibition of tumour growth, with cell death rates ranging between 66 and 91% (Yang et al. 2009, 2011). When loaded onto GO via π–π interactions and amide bonding, drugs like Methotrexate and Paclitaxel have shown promising results in inhibiting tumour growth, with a 66–90% success rate against lung and breast cancer (Wojtoniszak et al. 2013; Xu et al. 2014). Conjugating Ibuprofen with GO functionalized via amide linkages improved biocompatibility by 20% compared to GO sheets for MCF-7 and CEM cell lines (Rana et al. 2011).

10.4.2 Cell Targeting

After securely attaching to graphene, it is crucial to transport the drug to its intended tissue or target cell. This is where the drug can be absorbed and released most effectively. The versatility of the graphene platform has proven essential in supporting various mechanisms and approaches for cell targeting. Passive targeting mechanisms have garnered significant attention, as they eliminate the need for extensive functionalization. The enhanced retention (EPR) effect and increased permeability represent a widely employed mechanism wherein graphene derivatives, when delivered through the bloodstream, tend to accumulate preferentially in tumours. This phenomenon is due to the overabundance of blood vessels, which are often leaky, and the decreased flow of lymphatic fluid which are characteristics of cancerous tumours. Researchers often use labeled graphene derivatives for in vivo imaging, such as magnetic nanoparticles for magnetic resonance imaging (MRI) (Yang et al. 2012) or fluorescent moieties for near-infrared fluorescence (NIRF) imaging. Furthermore, composites of graphene derivatives and magnetic nanoparticles have been formulated to enable the specific targeting of graphene particles using localized magnetic fields (Feng et al. 2014; Ma et al. 2012). An alternative strategy involves leveraging the disparity in pH within the tumour microenvironment to modify the uptake propensity (Liang et al. 2019). Graphene flakes were conjugated with poly(β-malic acid) (PMLA) with a pH responsive polymer and DOX was loaded to generate a smart drug delivery system. As a result, the flakes were negatively charged in neutral or basic environments, upon exposure to an acidic environment, they undergo a positive shift, interacting with the negatively charged cell membrane and subsequently initiating endocytosis (Bina et al. 2021). By attaching ligands that can be identified by cell-surface receptors, we can achieve a more precise and interesting interaction. This method involves guiding graphene/drug complexes to particular tissues or cells, enhancing cell absorption, and promoting particular intracellular trafficking modes. The folate receptor serves as a targeting ligand, proving to be a paradigmatic approach for intracellular drug delivery and receptor-mediated targeting. A group of researchers has presented a promising practical approach for targeting breast cancer through the utilization of sulfonated and folate-conjugated carboxylate graphene (Zhang et al. 2010). Graphene's sensitivity to radiofrequency (RF) radiation has also been used as an alternate approach for cell targeting in a much similar way as IR absorption for selectively killing cancer cells. RF has been preferred over IR due to its increased tissue penetration depth.

10.4.3 Cargo Delivery

Cargo delivery implies that the effective and targeted drug effect can be at its site of action. The factors responsible for achieving this involve the preparation of graphene and the method of linkage. For minimally or non-functionalized graphene derivatives the cargo is generally bound by π–π interactions or hydrophobic. These involve in determination of variation in physiological characteristics in the cell during its interaction with ligands. For instance, the variations in pH were used to alter bonding affinity. In the case of functionalized graphene derivatives, the delivery of cargo needs to be customized according to the binding procedures used for cargo attachment, as well as any particular functionalization approaches employed to enhance cell targeting (Moore et al. 2014; Yin et al. 2013). Various graphene and its derivatives for drug delivery are shown in Table 10.1.

Table 10.1 The table shows some examples of graphene and its derivatives for drug delivery

Graphene derivatives	Drug	Application	References
GO	DOX	Results: released 15–20% increase	Dembereldorj et al. (2012)
		In vivo: Cg–Foxn1nu/CrljOri nude mice	
		In vitro: A549 cells	
		Results: higher drug release at pH 5.3	Depan et al. (2011)
		In vitro: drug release	
		Results: 71% tumour growth inhibition	Yang et al. (2012)
		In vivo: BALB/c nude mice	
		In vitro: HeLa cells and OCT-1 mouse osteoblasts	
		Results: pH-triggered controlled magnetic behaviour	Yang et al. (2009)
		Results: enhanced cellular toxicity	Yang et al. (2011)
		In vitro: SK3 cells	
	Paclitaxel	Results: 90%	Xu et al. (2014)
		In vitro: A549 and MCF-7 cells	
	Methotrexate	Results: 66.1%	Wojtoniszak et al. (2013)
		In vitro: MCF7 cells	
	Ibuprofen	Results: about 95%	Rana et al. (2011)
		In vitro: MCF7 and CEM cells	
	Ce6	Results: 98%	Tian et al. (2011)
		In vitro: KB cells	
		Results: 90%	Huang et al. (2011)
		In vitro: MGC803 cells	
Nano-GO	DOX	Results: 80% cell growth inhibition	Sun et al. (2008)
		In vitro: CEM.NK T-cells and Raji B-cells	
	SN38	Results: 80% cell growth inhibition	Liu et al. (2008)
		In vitro: HCT-116 cells	

(continued)

Table 10.1 (continued)

Graphene derivatives	Drug	Application	References
Reduced GO	DOX	Results: 80% cell growth inhibition	Kim et al. (2013)
		In vitro: PC-3 and HeLa cells	
Graphene nanosheet	DOX	Results: The cytotoxicity enhanced gradually	Liu et al. (2011)
		In vitro: MCF-7cells	
		Results: 55%	Wang et al. (2013)
		In vitro: U251 and 1800 cells	
		Results: 90%	Pan et al. (2011)
		In vitro: A-5RT3 cells	
	Methotrexate	Results: 70.2%	An et al. (2013)
		In vitro: A549 cells	
	5FU	Results: 72% growth inhibition	Fan et al. (2013)
		In vitro: HepG2 cells	
Graphene quantum dot	DOX	Results: 95%	Chen et al. (2014)
		In vitro: A549 cells	
		Results: 60%	Abdullah Al et al. (2013)
		In vivo: BALB/c mice	
		In vitro: A549 cells	
		Results: 60%	Wang et al. (2014)
		In vitro: A549, HeLa, and HEK293A cells	

Reproduced with permission from Nurunnabi et al. (2015). Copyright 2015 @ RSC Adv

10.5 Graphene and Its Derivatives for Gene Delivery

After the success of graphene as a multi-drug delivery vehicle, graphene has been employed as a carrier vehicle for other biopharmaceuticals including genetic materials. Graphene derivatives serve as promising carrier materials due to their substantial gene packing density, attributed to their increased surface area. These derivatives are modified with polymers to render the graphene surface cationic, facilitating interaction with anionic genes. Polymers commonly employed to modify the surface to a cationic state and facilitate interaction with anionic genes include polyethyleneimine (PEI) (Ren et al. 2012), PEG (Chen et al. 2011) and poly(sodium 4-styrenesulfonates (PSS) through graphene derivatives (Wang et al. 2014) for gene delivery via graphene derivatives.

Single-stranded DNA/RNA have been adsorbed on graphene's polyaromatic basal plane through nucleobase interactions to reduce their degradation (Manna and Pati 2013). A molecular beacon was adsorbed to GO and protected from enzymatic degradation by DNAse I, thereafter released successfully within a HeLa cell model. To deliver double-stranded nucleic acids functionalization of GO is essential. Conjugation strategies and polymeric modifications additionally contribute to improved biocompatibility and prolonged circulation times in vivo. These strategies enhance the complexation and condensation of nucleic acids, while also promoting interactions with negatively charged cell membranes. In a specific instance, low molecular weight PDMAEMA (Poly(2-dimethylamino-ethylmethacrylate)) was affixed to GO flakes via disulfide bonds, with pDNA macromolecules inserted between the sheets. Another efficient interaction was the layered assembled GO carried siRNA and Adriamycin simultaneously that overcame multi-drug miR-21 targeted resistance. This assembly also enables increased uptake in MCF-7 cells (Zhi et al. 2013). Khademhosseini group demonstrated a promising study on potential cardiac implants for vasculogenesis. They used GO-based injectable hydrogel to deliver angiogenic genes (Paul et al. 2014).

10.6 Derivatives of Graphene for Tissue Engineering

Graphene has been considered a versatile carbon-based material for tissue engineering. Tissue engineering represents an interdisciplinary field that seamlessly integrates knowledge from medicine, biology, and engineering. Its goal is to develop biomimetic tissue scaffolds for applications such as organ transplantation, as well as for use in diagnostic and therapeutic research. Biomaterials play a crucial role in tissue engineering by modulating cell–cell interactions, inducing specific cell functions, and guiding direct cell differentiation (Freed et al. 1994). Individual materials often fall short in replicating both the physical and biological characteristics of native tissues. Consequently, hybrid materials have been investigated to construct artificial tissues and their derivatives that meet diverse requirements in tissue engineering. This includes applications in cardiac tissue engineering, bone tissue engineering, neural tissue engineering, skeletal muscle tissue engineering, cartilage tissue engineering, skin/adipose tissue engineering, and regeneration.

Cardiac tissue injuries arise from coronary artery diseases or myocardial infarction. Researchers have developed 2-dimensional or 3-dimensional scaffolds for cardiac tissue engineering by incorporating electrically conductive nanomaterials such as carbon-based and gold nanoparticles into biomaterials. These scaffolds have shown improvements in cellular function (Dvir et al. 2011). Stem cells have great potential for repairing and regenerating damaged heart muscle. However, they are strongly affected by various physical and biochemical signals in their surrounding environment. In this regard, graphene and its derivatives can be used to influence the behaviour of stem cells inside the body. By providing structural guidance and improving the electrical conductivity of the surrounding matrix, graphene can play an important role in controlling the fate of stem cells. A study has shown that culturing mesenchymal stem cells (MSCs) on graphene monolayers can stimulate cardiomyogenic differentiation in MSCs (Bressan et al. 2014). Additionally, the co-implantation of MSCs with graphene oxide (GO) flakes resulted in a substantial improvement in their survival rate. This enhancement is attributed to the protective effect of GO on the cells against in vivo reactive oxygen species (Park et al. 2015). The nervous system, a complex network of neural lineage cells that transmit signals to coordinate communication between the brain and different parts of the body, can find advantages in graphene's conductive properties and biocompatibility. This characteristic makes graphene a fitting candidate for recapitulating the intricate signaling processes associated with the nervous system.

Graphene-modified materials are biocompatible, allowing cells to adhere and proliferate, thereby enhancing stem cell differentiation into osteogenic lineage. The biocompatibility of the graphene-modified substrates shows improved viability, cell adhesion, and proliferation rate (Naujoks et al. 2011). Graphene-based coatings are noncytotoxic and allow for the attachment and proliferation of fibroblasts, osteoblasts, and MSC, enhancing stem cell differentiation (Ryoo et al. 2010).

Materials based on graphene can function as "growth factor factories," supplying both cells and proteins crucial for cartilage tissue engineering. MSCs are commonly utilized in cellular therapies to prompt chondrogenic differentiation at sites with cartilage defects. When integrated with graphene flakes, MSCs create bio-composites that serve as a pre-concentrate platform for cartilage tissue engineering (Bressan et al. 2014). Graphene, incorporated as a scaffold in stem cell therapy for cartilage treatment, plays a role in initiating the cellular repair process. This is attributed to its stimulating effects on cellular growth and differentiation, along with its outstanding mechanical properties.

Due to the limited regenerative capacities of skeletal muscle tissues, it is imperative to develop alternatives to mitigate permanent loss of tissue function. Materials based on graphene have shown the ability to boost the proliferation, adhesion, and myogenesis of skeletal muscle cells. Notably, C2C12 myoblasts exhibited increased attachment and improved spreading behaviour when cultured on ultrathin thermally reduced (TR) graphene-based films (Ahadian et al. 2014). Stem cell therapies are frequently utilized for the regeneration of skeletal muscle tissue. When combined with nanomaterials, stem cells act as ideal substrates for functional muscle tissue engineering. They function as efficient scaffolds for promoting cellular viability and differentiation. Graphene possesses unique properties that make it an excellent material for the regeneration of skeletal muscle tissue. It has excellent flexibility, conductivity (0.6 S/m), and strong mechanical properties with an ultralow density of 0.16 mg/cm^3. A graphene aerogel prepared with ethylenediamine is ultralight and highly compressible, with excellent elastic stiffness. In cyclic compression tests, the aerogel fully recovered after 90% compression without fracturing. Moreover, it showed lifetime stability of up to 70% under 1000 compression cycles.

10.7 Graphene and Its Derivatives for Cancer Therapy

Apart from responding to microenvironment and/or intracellular activated platforms such as enzymes and pH level, graphene-based materials also effectively respond to physical stimuli such as light, ultrasound, and magnetic fields. This property makes them a potential candidate for cancer treatment through various therapies like photodynamic therapy, photothermal therapy (PTT), photothermal therapy (PTT), magnetic hyperthermia therapy (MHT), ultrasound hyperthermia therapy (SHT), sonodynamic therapy (SDT), RT, and ion-interference therapy (IIT), or a combination of the therapies. Additionally, functional nanoparticles can be attached to the surface of graphene to provide additional properties for both bioimaging and tumour therapy.

The remarkable capacity for NIR absorption and efficient photothermal conversion renders graphene-based materials well-suited for PTT applications. In PTT, the exposure of a photothermal agent to electromagnetic radiation, such as NIR or visible light results in a localized temperature increase. The conversion of light energy into heat induces tumour cell death. PTT kills the cancer cells and enhances the intertumoural blood flow to enhance tumour oxygenation. Recently a thermo-responsive hydrogel comprising of injectable chitosan-agarose rGO (thermogel-rGO) when irradiated with NIR, the thermogel-rGO decreased the breast cancer cells' viability to 60% (Lima-Sousa et al. 2020). The combination of AuNR assembled with L-cysteine-rGO (AuNR@Lcyst-rGO) showed superior biocompatibility, excellent stability, and high efficiency in NIR-triggered photothermal therapy due to synergistic effects between gold rods and rGO (Zhang et al. 2022).

PDT (photodynamic therapy) is a non-invasive treatment option for tumours that involves the use of real-time detection and tracking of malignant tissue through fluorescence imaging. During PDT, a non-toxic photosensitizing substance is activated by light, which triggers a photochemical reaction that generates and localizes highly toxic reactive oxygen species (ROS) at the targeted cancerous tissue site. Recently, nanoconjugate of green luminescent GQDs was covalently attached to the amino BODIPY derivative (BDPA) via 1-ethyl-3-(3-dimethylaminopropyl)carbodiimide–*N*-hydroxysuccinimide coupling. This nanoconjugate showed remarkable singlet oxygen generation efficiency with 90% for highly efficient PDT applications (Mangalath et al. 2021).

Due to the increasing demand, a "**theragnostic**" platform is now used. This platform involves targeted diagnostic imaging followed by targeted delivery of therapies. Also, a combination of therapies is adopted for better therapeutic outcomes. Chemotherapy is the first-line treatment in clinical cancer, but its effectiveness is hindered by drug resistance. Therefore, various modalities in combination with chemotherapy have been employed for better results. A combination of photothermal and chemo-therapies revealed effective inhibition of tumour growth, where a hybrid vesicle with DOX loaded (rGO)-loaded ultra-small plasmonic gold nanorod vesicle (rGO-AuNRVe-DOX) was used as shown in Fig. 10.2a (Song et al. 2015).

A single nanoplatform was developed for synergistic targeted PTT and chemotherapy using DOX. The nanoplatform was made by functionalizing graphene

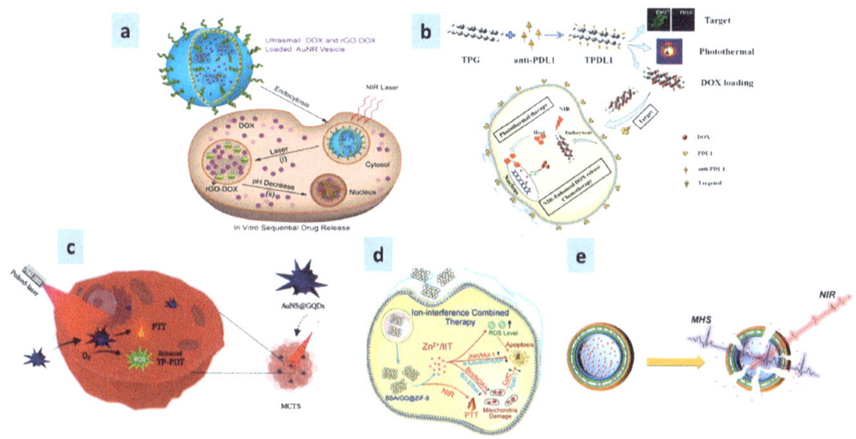

Fig. 10.2 Diagram illustrating the use of graphene and its derivatives as highly effective carriers for combined cancer therapy, incorporating **a** the combination of chemotherapy and PDT. Reproduced with permission from Song et al. (2015), Copyright 2015@ ACS Nano. **b** Combination of chemotherapy and PTT. Reproduced with permission from Hao et al. (2020), copyright 2020@*ACS Applied Bio Materials*. **c** Combination of PDT and PTT. Reproduced with permission from Soleimany et al. (2023), copyright 2020@*ACS Applied Materials and Interfaces*. **d** Combination of IIT and PTT. Reproduced with permission from Lv et al. (2022), copyright 2022@*ACS Nano*. **e** Combination of MHT and chemotherapy. Reproduced with permission from Deng et al. (2016), copyright 2016@*ACS Applied Materials and Interfaces*

oxide with tea polyphenol and conjugating it with anti-PDL1. This nanoplatform, called TPDL1, exhibited NIR-enhanced, pH-responsive, and stable drug release properties. It also showed efficient photothermal conversion under NIR irradiation. The nanoplatform demonstrated excellent targeting ability, making it a promising candidate for cancer therapy. A detailed illustration of this nanoplatform can be seen in Fig. 10.2b (Hao et al. 2020).

An innovative nano-hybrid, formed by combining plasmonic metal and QD for simultaneous TP-PDT/PTT, demonstrated markedly enhanced phototherapeutic effects compared to individual strategies when tested against both 2-D monolayer cells and 3-D multicellular tumour spheroids. Thiolate chitosan-coated gold nano-stars (AuNS–TCS) nano-hybrid comprised used as the N, S-doped graphene quantum dot (Rf-N, S-GQD)conjugated with riboflavin and photothermal agent as the two-photon photosensitizer (TP-PS) as shown in Fig. 10.2c (Soleimany et al. 2023).

A nanocomplex named BSArGO@ZIF-8 NSs, based on Zinc, is used for ion-interference and photothermal treatments of various cancer cells. Figure 10.2d shows the results (Lv et al. 2022). The nanocomplex is made up of metal–organic framework nanoparticles (ZIF-8 NPs) that are loaded onto a GO surface. This is then reduced by ascorbic acid and modified with bovine serum albumin. Ion-interference therapy is a new treatment that is aimed at inhibiting the proliferation of tumour cells by artificially interfering with the ion homeostasis in tumour cells. Several, inorganic nanoparticles such as Ca_2O, ZnO, $CaCO_3$, NaCl, and BPS (black phosphorus) have been used for ion-interference treatment as they are biocompatible, long internal circulation, possess strong specific recognition ability, and provide efficient inhibition of tumour development (Zhang et al. 2019; Zhao et al. 2020; Jiang et al. 2019).

Magnetic hyperthermia (MHT) is another therapy used for tumour treatment. This technique utilizes iron oxide nanoparticles (magnetic nanoparticles) to generate heat by a magnetic field with external alternating current. Heat is generated due to hysteresis loss or relaxation behaviour. This therapeutic approach facilitates the targeted penetration of cancer cells within the biological system, minimizing damage to adjacent normal tissue. Nonetheless, a primary drawback of this therapy is the tendency of magnetic nanoparticles to aggregate within biological vessels. Graphene and its derivatives offer a stable material for anchoring magnetic nanoparticles, thereby improving their physiological stability. The integration of graphene-based materials with magnetic nanoparticles creates a unified system that encompasses both imaging and therapeutic functionalities. A hybrid microcapsule, consisting of polysaccharides (hyaluronic acid, sodium alginate, and chitosan), iron oxide, and GO, was used to demonstrate the synergistic effect of hyperthermia and chemotherapy in vitro and in vivo, as shown in Fig. 10.2e. Fe_3O_4@GO component triggered dual high-energy and high-penetration hyperthermia therapy along with drug release using an alternative magnetic field and near-infrared laser (Deng et al. 2016).

RT uses high-energy radiation such as x-rays, gamma rays, neutrons, and protons to destroy shrink tumours, and cancer cells. However, the oxygen-deficient tumour microenvironment decreases the cancer cells' sensitivity to x-ray/γ-ray irradiation, reducing the effectiveness of therapy in the treatment of hypoxic solid tumours (Thompson and Maity 2014). To overcome this issue, the implementation of PTT before RT helps to improve tumour oxygenation via increasing the intratumoural blood flow (Yi et al. 2016).

Ultrasound has been commonly used in diagnostic imaging purposes as it is a non-toxic mechanical wave; this can also be used for therapeutic effects on tumour treatment. Ultrasound waves can penetrate deep through biological tissues without depth limitation as compared to light waves used in PDT. This makes ultrasound an effective way to activate sonosensitizers to produce ROS for eradicating cancer cells. Graphene and its derivatives are extensively used to enhance sonodynamic therapy (SDT) due to their large surface-to-volume ratio, high photothermal-conversion capability, and high electroconductivity (Chen et al. 2016). A 2-D graphene integrated with TiO_2 nano-sonosensitizers enhanced SDT efficacy and resulted in a higher rate of inhibition of tumour growth (Dai et al. 2017). Graphene-based materials have shown potential in delivering ultrasound hyperthermia for cancer therapy, in addition to SDT. A synergistic chemotherapy and ultrasound hyperthermia therapy (SHT) showed profound improvement in antitumour efficacy for gliomas. A smart magnetic nanocomposite based on GO was able to accurately transport and deliver the drug epirubicin to the brain tumour site with a super high drug concentration. Also, the high concentration of GO served as a heat-conducting base to increase the local temperature upon application of focused ultrasound (Feng et al. 2018).

10.8 Future Prospects and Challenges

Significant progress has been made in using graphene and its derivatives for therapeutic purposes, as they have been found to be biocompatible and have a low toxicity profile. However, only a few graphene-based materials have been tested in animal models, and even fewer have been used in clinical practice (Reina et al. 2017). According to widespread knowledge, graphene tends to accumulate in the spleen and liver (Tonelli et al. 2015). Therefore, more efforts need to be made to improve the targeting ability of graphene and its derivatives via surface modification or altering the administration pathway. One potential solution to overcome the biodegradation difficulties of nanoparticles is to optimize their composition and reduce their particle sizes for more efficient biodegradability (Lin et al. 2018) For facilitating convenient elimination through the renal pathway, it is advisable to apply a coating of biocompatible moieties to graphene and its nanocomposites, with sizes smaller than 100 nm. This approach assists in the efficient clearance of the material from the body following systemic administration, minimizing the likelihood of any discernible toxicity in the system (Rahman et al. 2015). Therefore, much more research is yet to be done to understand the biodegradability, biocompatibility, pharmacokinetics, and acute/chronic toxicity levels. The use of imaging and therapy together can produce a combined effect that enhances the targeting ability, while minimizing side effects. Graphene has the potential to be functionalized and conjugated, which can improve its biocompatibility and make it a promising material for various applications (Gu et al. 2019). By gaining a deeper understanding of how graphene interacts with cells (especially in terms of reactive oxygen species generation), we can develop new platforms for both therapy and diagnosis in the future.

One of the major challenges in the field of graphene-based therapeutics is the lack of clinical testing, as obtaining approval for clinical trials is a time-consuming and difficult process. Graphene-based materials are not anticipated to be commercially available for biological applications until after 2030. Consequently, additional in vivo studies are imperative to fully understand and unlock the practical potential of graphene in real-world applications.

References

Abdullah Al, N., Lee, J.E., Jeong, J.H., Park, S.Y.: Photoresponsive fluorescent reduced graphene oxide by spiropyran conjugated hyaluronic acid for in vivo imaging and target delivery. Biomacromolecules **14**, 4082–4090 (2013)

An, J., Gou, Y., Yang, C., Hu, F., Wang, C.: Synthesis of a biocompatible gelatin functionalized graphene nanosheets and its application for drug delivery. Mater. Sci. Eng.: C **33**, 2827–2837 (2013)

Ahadian, S., Ramón-Azcón, J., Chang, H., Liang, X., Kaji, H., Shiku, H., Nakajima, K., Ramalingam, M., Wu, H., Matsue, T., Khademhosseini, A.: Electrically regulated differentiation of skeletal muscle cells on ultrathin graphene-based films. RSC Adv. **4**(19), 9534–9541 (2014)

Ali-Boucetta, H., Bitounis, D., Raveendran-Nair, R., Servant, A., Van den Bossche, J., Kostarelos, K.: Purified graphene oxide dispersions lack in vitro cytotoxicity and in vivo pathogenicity. Adv. Healthc. Mater. **2**(3), 433–441 (2013)

Bina, L., Romano, V., Hoogland, T.M., Bosman, L.W., De Zeeuw, C.I.: Purkinje cells translate subjective salience into readiness to act and choice performance. Cell Rep. **37**(11) (2021)

Bressan, E., Ferroni, L., Gardin, C., Sbricoli, L., Gobbato, L., Ludovichetti, F.S., Tocco, I., Carraro, A., Piattelli, A., Zavan, B.: Graphene based scaffolds effects on stem cells commitment. J. Transl. Med. **12**, 1–15 (2014)

Chen, B., Liu, M., Zhang, L., Huang, J., Yao, J., Zhang, Z.: Polyethylenimine-functionalized graphene oxide as an efficient gene delivery vector. J. Mater. Chem. **21**(21), 7736–7741 (2011)

Chen, T., Yu, H., Yang, N., Wang, M., Ding, C., Fu, J.: Graphene quantum dot-capped mesoporous silica nanoparticles through an acid-cleavable acetal bond for intracellular drug delivery and imaging. J. Mater. Chem. B **2**, 4979–4982 (2014)

Chen, Y.W., Liu, T.Y., Chang, P.H., Hsu, P.H., Liu, H.L., Lin, H.C., Chen, S.Y.: A theragnostic nrGO@ MSN-ION nanocarrier developed to enhance the combination effect of sonodynamic therapy and ultrasound hyperthermia for treating tumor. Nanoscale **8**, 12648–12657 (2016)

Dai, C., Zhang, S., Liu, Z., Wu, R., Chen, Y.: Two-dimensional graphene augments nanosonosensitized sonocatalytic tumor eradication. ACS Nano **11**, 9467–9480 (2017)

Dembereldorj, U., Kim, M., Kim, S., Ganbold, E.O., Lee, S.Y., Joo, S.W.: A spatiotemporal anticancer drug release platform of PEGylated graphene oxide triggered by glutathione in vitro and in vivo. J. Mater. Chem. **22**, 23845–23851 (2012)

Deng, L., Li, Q., Al-Rehili, S.A., Omar, H., Almalik, A., Alshamsan, A., Khashab, N.M.: Hybrid iron oxide–graphene oxide–polysaccharides microcapsule: a micro-matryoshka for on-demand drug release and antitumor therapy in vivo. ACS Appl. Mater. Interfaces **8**, 6859–6868 (2016)

Depan, D., Shah, J., Misra, R.D.K.: Controlled release of drug from folate-decorated and graphene mediated drug delivery system: Synthesis, loading efficiency, and drug release response. Mater. Sci. Eng. C **31**, 1305–1312 (2011)

Dvir, T., Timko, B.P., Brigham, M.D., Naik, S.R., Karajanagi, S.S., Levy, O., Jin, H., Parker, K.K., Langer, R., Kohane, D.S.: Nanowired three-dimensional cardiac patches. Nat. Nanotechnol. **6**(11), 720–725 (2011)

Fan, X., Jiao, G., Gao, L., Jin, P., Li, X.: The preparation and drug delivery of a graphene–carbon nanotube–Fe_3O_4 nanoparticle hybrid. J. Mater. Chem. B **1**, 2658–2664 (2013)

Feng, X., Chen, L., Guo, W., Zhang, Y., Lai, X., Shao, L., Li, Y.: Graphene oxide induces p62/SQSTM-dependent apoptosis through the impairment of autophagic flux and lysosomal dysfunction in PC12 cells. Acta Biomater.biomater **81**, 278–292 (2018)

Feng, L., Zhang, Z., Ren, J., Qu, X.: Functionalized graphene as sensitive electrochemical label in target-dependent linkage of split aptasensor for dual detection. Biosens. Bioelectron. **62**, 52–58 (2014)

Freed, L.E., Vunjak-Novakovic, G., Biron, R.J., Eagles, D.B., Lesnoy, D.C., Barlow, S.K., Langer, R.: Biodegradable polymer scaffolds for tissue engineering. Bio/technology **12**(7), 689–693 (1994)

Gu, Z., Zhu, S., Yan, L., Zhao, F., Zhao, Y.: Graphene-based smart platforms for combined cancer therapy. Adv. Mater. **31**, 1800662 (2019)

Hao, L., Song, H., Zhan, Z., Lv, Y.: Multifunctional reduced graphene oxide-based nanoplatform for synergistic targeted chemo-photothermal therapy. ACS Appl. Bio Mater. **3**(8), 5213–5222 (2020)

Huang, P., Xu, C., Lin, J., Wang, C., Wang, X., Zhang, C., Cui, D.: Folic acid-conjugated graphene oxide loaded with photosensitizers for targeting photodynamic therapy. Theranostics **1**, 240 (2011)

Jiang, W., Yin, L., Chen, H., Paschall, A.V., Zhang, L., Fu, W., Xie, J.: NaCl nanoparticles as a cancer therapeutic. Adv. Mater. **31**, 1904058 (2019)

Kim, H., Lee, D., Kim, J., Kim, T.I., Kim, W.J.: Photothermally triggered cytosolic drug delivery via endosome disruption using a functionalized reduced graphene oxide. ACS Nano **7**, 6735–6746 (2013)

Kumar, N., Yadav, S., Sadique, M.A., Khan, R.: Electrochemically exfoliated graphene quantum dots based biosensor for CD44 breast cancer biomarker. Biosensors **12**(11), 966 (2022)

Kumar, N., Sadique, M.A., Khan, R., Gowri, V.S., Kumar, S., Ashiq, M., Natarajan, S.: Immunosensor for breast cancer CD44 biomarker detection based on exfoliated graphene quantum dots integrated gold nanoparticles. Hybrid Adv. **3**, 100065 (2023)

Liang, X., Chen, L., Nguyen, D., Zhou, Z., Gu, X., Yang, M., Wang, J., Jiang, S.: Generating synthesized computed tomography (CT) from cone-beam computed tomography (CBCT) using CycleGAN for adaptive radiation therapy. Phys. Med. Biol. **64**(12), 125002 (2019)

Lima-Sousa, R., de Melo-Diogo, D., Alves, C.G., Cabral, C.S., Miguel, S.P., Mendonça, A.G., Correia, I.J.: Injectable in situ forming thermo-responsive graphene based hydrogels for cancer chemo-photothermal therapy and NIR light-enhanced antibacterial applications. Mater. Sci. Eng.: C **117**, 111294 (2020)

Lin, H., Chen, Y., Shi, J.: Nanoparticle-triggered in situ catalytic chemical reactions for tumour-specific therapy. Chem. Soc. Rev. **47**, 1938–1958 (2018)

Liu, Z., Robinson, J.T., Sun, X., Dai, H.: PEGylated nanographene oxide for delivery of water-insoluble cancer drugs. J. Am. Chem. Soc. **130**, 10876–10877 (2008)

Liu, K., Zhang, J.J., Cheng, F.F., Zheng, T.T., Wang, C., Zhu, J.J.: Green and facile synthesis of highly biocompatible graphene nanosheets and its application for cellular imaging and drug delivery. J. Mater. Chem. **21**, 12034–12040 (2011)

Lv, C., Kang, W., Liu, S., Yang, P., Nishina, Y., Ge, S., Bianco, A., Ma, B.: Growth of ZIF-8 nanoparticles in situ on graphene oxide nanosheets: a multifunctional nanoplatform for combined ion-interference and photothermal therapy. ACS Nano **16**(7), 11428–11443 (2022)

Ma, C.S., Deenick, E.K., Batten, M., Tangye, S.G.: The origins, function, and regulation of T follicular helper cells. J. Exp. Med. **209**(7), 1241–1253 (2012)

Manna, A.K., Pati, S.K.: Theoretical understanding of single-stranded DNA assisted dispersion of graphene. J. Mater. Chem. B **1**(1), 91–100 (2013)

Marcano, D.C., Kosynkin, D.V., Berlin, J.M., Sinitskii, A., Sun, Z., Slesarev, A., Tour, J.M.: Improved synthesis of graphene oxide. ACS Nano **4**, 4806–4814 (2010)

Mangalath, S., Saneesh Babu, P.S., Nair, R.R., Manu, P.M., Krishna, S., Nair, S.A., Joseph, J.: Graphene quantum dots decorated with boron dipyrromethene dye derivatives for photodynamic therapy. ACS Appl. Nano Mater. **4**(4), 4162–4171 (2021)

Moore, T.L., Podilakrishna, R., Rao, A., Alexis, F.: Systemic administration of polymer-coated nano-graphene to deliver drugs to glioblastoma. Part. Part. Syst. Charact. **31**, 886–894 (2014)

Naujoks, C., Langenbach, F., Berr, K., Depprich, R., Kübler, N., Meyer, U., Handschel, J., Kögler, G.: Biocompatibility of osteogenic predifferentiated human cord blood stem cells with biomaterials and the influence of the biomaterial on the process of differentiation. J. Biomater. Appl. **25**(5), 497–512 (2011)

Novoselov, K.S., Geim, A.K., Morozov, S.V., Jiang, D.E., Zhang, Y., Dubonos, S.V., Grigorieva, I.V. and Firsov, A.A.: Electric field effect in atomically thin carbon films. Science **306**(5696), 666–669 (2004)

Nurunnabi, M., Parvez, K., Nafiujjaman, M., Revuri, V., Khan, H. A., Feng, X., Lee, Y.K.: Bioapplication of graphene oxide derivatives: drug/gene delivery, imaging, polymeric modification, toxicology, therapeutics and challenges. RSC Adv. **5**, 42141–42161 (2015)

Pan, Y., Bao, H., Sahoo, N.G., Wu, T., Li, L.: Water soluble poly (N isopropylacrylamide)–graphene sheets synthesized via click chemistry for drug delivery. Adv. Funct. Mater.funct. Mater. **21**, 2754–2763 (2011)

Park, J., Kim, B., Han, J., Oh, J., Park, S., Ryu, S., Jung, S., Shin, J.Y., Lee, B.S., Hong, B.H., Choi, D.: Graphene oxide flakes as a cellular adhesive: prevention of reactive oxygen species mediated death of implanted cells for cardiac repair. ACS Nano **9**(5), 4987–4999 (2015)

Paul, A., Hasan, A., Kindi, H.A.: Injectable graphene oxide/hydrogel-based angiogenic gene delivery system for. ACS Nano **8**(8) (2014)

Pei, S., Cheng, H.M.: The reduction of graphene oxide. Carbon **50**, 3210–3228 (2012)

Rahman, M., Zaki Ahmad, M., Ahmad, J., Firdous, J., Jalees Ahmad, F., Mushtaq, G., Akhter, S.: Role of graphene nano-composites in cancer therapy: theranostic applications, metabolic fate and toxicity issues. Curr. Drug Metab.metab. **16**, 397–409 (2015)

Rana, V.K., Choi, M.C., Kong, J.Y., Kim, G.Y., Kim, M.J., Kim, S.H., Ha, C.S.: Synthesis and drug delivery behavior of chitosan functionalized graphene oxide hybrid nanosheets. Macromol. Mater. Eng. Mater. Eng. **296**, 131–140 (2011)

Reina, G., González-Domínguez, J.M., Criado, A., Vázquez, E., Bianco, A., Prato, M.: Promises, facts and challenges for graphene in biomedical applications. Chem. Soc. Rev. **46**, 4400–4416 (2017)

Ren, T., Li, L., Cai, X., Dong, H., Liu, S., Li, Y.: Engineered polyethylenimine/graphene oxide nanocomposite for nuclear localized gene delivery. Polym. Chem. **3**(9), 2561–2569 (2012)

Ryoo, S.R., Kim, Y.K., Kim, M.H., Min, D.H.: Behaviors of NIH-3T3 fibroblasts on graphene/ carbon nanotubes: proliferation, focal adhesion, and gene transfection studies. ACS Nano **4**(11), 6587–6598 (2010)

Singh, V., Joung, D., Zhai, L., Das, S., Khondaker, S.I., Seal, S.: Graphene based materials: past, present and future. Prog. Mater. Sci. **56**, 1178–1271 (2011)

Song, J., Yang, X., Jacobson, O., Lin, L., Huang, P., Niu, G., Ma, Q., Chen, X.: Sequential drug release and enhanced photothermal and photoacoustic effect of hybrid reduced graphene oxide-loaded ultrasmall gold nanorod vesicles for cancer therapy. ACS Nano **22; 9**(9), 9199–209 (2015)

Soleimany, A., Khoee, S., Dias, S., Sarmento, B.: Exploring low-power single-pulsed laser-triggered two-photon photodynamic/photothermal combination therapy using a gold nanostar/graphene quantum dot nanohybrid. ACS Appl. Mater. Interfaces **15**(17), 20811–20821 (2023)

Sun, X., Liu, Z., Welsher, K., Robinson, J.T., Goodwin, A., Zaric, S., Dai, H.: Nano-graphene oxide for cellular imaging and drug delivery. Nano Res. **1**, 203–212 (2008)

Syama, S., Mohanan, P.V.: Comprehensive application of graphene: emphasis on biomedical concerns. Nano-micro Letters **11**, 1–31 (2019)

Thompson, R.F., Maity, A.: Radiotherapy and the tumor microenvironment: mutual influence and clinical implications. In: Tumor Microenvironment and Cellular Stress: Signaling, Metabolism, Imaging, and Therapeutic Targets, pp. 147–165 (2014)

Tian, B., Wang, C., Zhang, S., Feng, L., Liu, Z.: Photothermally enhanced photodynamic therapy delivered by nano-graphene oxide. ACS Nano **5**, 7000–7009 (2011)

Tonelli, F.M., Goulart, V.A., Gomes, K.N., Ladeira, M.S., Santos, A.K., Lorençon, E., Resende, R.R.: Graphene-based nanomaterials: biological and medical applications and toxicity. Nanomedicine **10**, 2423–2450 (2015)

Wang, X., Sun, X., Lao, J., He, H., Cheng, T., Wang, M., Huang, F.: Multifunctional graphene quantum dots for simultaneous targeted cellular imaging and drug delivery. Colloids Surf. B: Biointerfaces **122**: 638–644 (2014)

Wang, Y., Wang, K., Zhao, J., Liu, X., Bu, J., Yan, X., Huang, R.: Multifunctional mesoporous silica-coated graphene nanosheet used for chemo-photothermal synergistic targeted therapy of glioma. J. Am. Chem. Soc. **135**, 4799–4804 (2013)

Wojtoniszak, M., Urbas, K., Perużyńska, M., Kurzawski, M., Droździk, M., Mijowska, E.: Covalent conjugation of graphene oxide with methotrexate and its antitumor activity. Chem. Phys. Lett. **568**, 151–156 (2013)

Xu, Z., Wang, S., Li, Y., Wang, M., Shi, P., Huang, X.: Covalent functionalization of graphene oxide with biocompatible poly (ethylene glycol) for delivery of paclitaxel. ACS Appl. Mater. Interfaces **6**, 17268–17276 (2014)

Yang, X., Zhang, X., Liu, Z., Ma, Y., Huang, Y., Chen, Y.: High-efficiency loading and controlled release of doxorubicin hydrochloride on graphene oxide. J. Phys. Chem. C **112**, 17554–17558 (2008)

Yang, X., Zhang, X., Ma, Y., Huang, Y., Wang, Y., Chen, Y.: Superparamagnetic graphene oxide–Fe_3O_4 nanoparticles hybrid for controlled targeted drug carriers. J. Mater. Chem. **19**, 2710–2714 (2009)

Yang, X., Wang, Y., Huang, X., Ma, Y., Huang, Y., Yang, R., Chen, Y.: Multi-functionalized graphene oxide based anticancer drug-carrier with dual-targeting function and pH-sensitivity. J. Mater. Chem. **21**, 3448–3454 (2011)

Yang, Y., Zhang, Y.M., Chen, Y., Zhao, D., Chen, J.T., Liu, Y.: Construction of a graphene oxide based noncovalent multiple nanosupramolecular assembly as a scaffold for drug delivery. Chem. Eur. J. **18**, 4208–4215 (2012)

Yi, X., Chen, L., Zhong, X., Gao, R., Qian, Y., Wu, F., Yang, K.: Core–shell Au@ MnO_2 nanoparticles for enhanced radiotherapy via improving the tumor oxygenation. Nano Res. **9**, 3267–3278 (2016)

Yin, D., Li, Y., Lin, H., Guo, B., Du, Y., Li, X., Zhang, L.: Functional graphene oxide as a plasmid-based Stat3 siRNA carrier inhibits mouse malignant melanoma growth in vivo. Nanotechnology **24**, 105102 (2013)

Zhang, L., Ren, X., Alt, E., Bai, X., Huang, S., Xu, Z., Wu, X.: Chemoprevention of colorectal cancer by targeting APC-deficient cells for apoptosis. Nature **464**, 1058–1061 (2010)

Zhang, M., Song, R., Liu, Y., Yi, Z., Meng, X., Zhang, J., Bu, W.: Calcium-overload-mediated tumor therapy by calcium peroxide nanoparticles. Chem **5**, 2171–2182 (2019)

Zhang, M., Zhang, X., Zhao, K., Dong, Y., Yang, W., Liu, J., Li, D.: Assembly of gold nanorods with L-cysteine reduced graphene oxide for highly efficient NIR-triggered photothermal therapy. Spectrochim. Acta Part A: Mol. Biomol. Spectrosc. **266**, 120458 (2022)

Zhao, Y.Z., Lin, M.T., Lan, Q.H., Zhai, Y.Y., Xu, H.L., Xiao, J., Yao, Q.: Silk fibroin-modified disulfiram/zinc oxide nanocomposites for pH triggered release of Zn2+ and synergistic antitumor efficacy. Mol Pharmaceutics **17**, 3857–3869 (2020)

Zhi, F., Dong, H., Jia, X., Guo, W., Lu, H., Yang, Y., Ju, H., Zhang, X., Hu, Y.: Functionalized graphene oxide mediated adriamycin delivery and miR-21 gene silencing to overcome tumor multidrug resistance in vitro. PLoS One **8**(3), e60034 (2013)

Chapter 11
Methodology Advancements for Bulk Production and Commercialization of Graphene and Its Derivatives

Kalpesh V. Bhavsar, Uday P. Jagtap, Kevin Jose, and Hardik S. Churi

Abstract Graphene, a 2-D material made up of carbon atoms arranged in a lattice resembling a hexagon, has received significant attention due to its extraordinary qualities and prospective uses in a variety of sectors. However, considering the high expenses and complexity of the synthesis processes, the bulk production and commercialization of graphene and its derivatives continue to be a challenge. In recent times, significant advancement is being performed in establishing novel approaches for the bulk production and commercialization of graphene and its derivatives. Roll-to-roll manufacturing, optimized chemical vapour deposition (CVD) techniques, laser-assisted synthesis, exfoliation by chemical means utilizing ionic fluids, hybrid approaches, and green synthesis are a few of these developments. These techniques hold the promise of generating high-grade graphene and its derivatives in significant numbers, with superior qualities, and at a reduced cost. In this article, we cover the most recent developments in this area as well as the state-of-the-art techniques for producing graphene and its derivatives in large quantities for commercial use. We also discuss each approach's strengths and weaknesses, as well as the way they might impact applications in various ways. For graphene and its derivatives to reach their full potential in a variety of industries, including electronics, energy storage, and healthcare, efficient and affordable procedures must be developed for their bulk production and commercialization.

Keywords Graphene · Commercialization · Chemical vapour deposition · Hybrid approaches · Flash graphene · Green synthesis

K. V. Bhavsar (✉) · U. P. Jagtap · K. Jose
National Centre for Nanosciences and Nanotechnology, University of Mumbai, Kalina Campus, Kalina, Mumbai 400 098, Santacruz (E), India
e-mail: kalpeshbhavsar1@gmail.com; kalpesh.bhavsar@nano.mu.ac.in

H. S. Churi
Sonopant Dandekar Shikshan Mandali (SDSM) College, College Road, Tembhode, Palghar 401404, Maharashtra, India

R. Khan et al. (eds.), *Electrochemical Exfoliation of Graphene and its Derivatives*, Engineering Materials, https://doi.org/10.1007/978-981-97-2128-3_11

11.1 Introduction

The most promising material of the twenty-first century, made of carbon, is graphene. Carbon is found naturally in numerous products and used for the production of synthetic compounds. It is also found in precious products like diamonds and inexpensive substances like coal. Graphite is a common allotrope of carbon. Graphene is a promising carbon-based material with tremendous applications in various industries. Graphene is a highly stable two-dimensional ultrathin carbon atoms layer with sp_2 bonds as shown in Fig. 11.1. Carbon atoms of the graphene layer are tightly packed in a hexagonal pattern with each other. Generally, graphene is a building block of graphite and can be converted into any desired shape. Graphene was first synthesized in 2004 by researchers from Manchester and received a Nobel Prize in 2010 (Novoselov et al. 2004). Various fields are now focusing on graphene and its derivatives.

The unique combination of high mechanical strength, transparency, flexibility, high electrical and thermal conductivity is making it an optimal material for diverse range of applications in fields like electronics, biomedical, and energy sectors (Tiwari et al. 2016). The mechanical strength of graphene is good and because of that graphene can be one of the strongest materials. Graphene is the strongest as well as flexible materials having a large surface area that makes it an ideal material for energy storage and catalysis. Graphene is especially used for the manufacture of super capacitors and electrodes (Ameen et al. 2020). Transparent graphene material is also employed in solar panels. Graphene has the potential of high electrical conductivity and hence widely used in electronic industries for manufacturing transistors where it can be used to produce high-performance, high-frequency devices

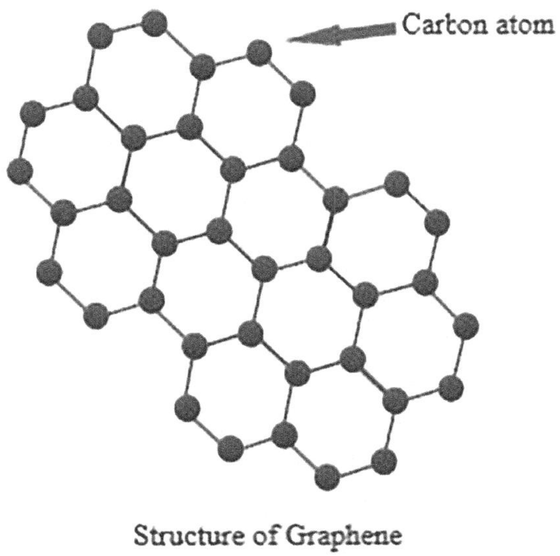

Fig. 11.1 Structure of graphene

(Wei et al. 2016). Graphene also possesses excellent thermal conductivity which is used for cooling electronic devices (Dreyer et al. 2010). These properties are quickly recognized by the world as a breakthrough in material science, and hence researchers are focusing on the bulk production of graphene and its derivatives (Ameen et al. 2020). Despite its potential graphene is in its early stage of commercialization and widespread adoption. Various challenges are present in front of researchers because the methods are tedious and expensive. The methods that are available to produce graphene and its derivatives are viable but they do create a lot of defects in the material produced. Extensive research is required for the bulk production of defect-free inexpensive graphene. Various start-up industries are focusing on the fabrication of low-cost graphene.

Researchers developed a variety of derivatives of graphene that possess high potential for various activities in specific sectors refer Table 11.1. Derivatives of graphene are modified in order to increase its chemical, physical, and electrochemical properties. Commercially available graphene derivatives are (GQD) Graphene quantum dots, (GO) Graphene Oxide, (rGO) Reduced Graphene Oxide, and functionalized graphene. For the fabrication of graphene oxide, primarily graphene is treated with strong oxidizing agents like potassium permanganate and sulphuric acid. Graphene Oxide is a 2D material containing various oxygen functional groups intact within it. GO is hydrophilic with a large surface area and possesses high UV absorption properties (Mindivan 2016). rGO is produced by reducing GO with the removal of a few oxygen-containing functional groups. The reduction of GO to rGO increases mechanical strength and increases its electrical conductivity (Mindivan 2016). GQD is an advanced material that is used in biomedical research. To produce GQDs, graphene is cut into smaller pieces (Lee et al. 2013). GQDs are versatile with unique properties such as fluorescence and photoluminescence.

The selection of the correct precursor materials and fabrication technique helps with the production of defect-free high-quality graphene material. Various methods are used for the fabrication of graphene of which some are used at the laboratory level for qualitative analysis while some are exclusively used for bulk production. Mechanical exfoliation is a process in which scotch tape is used for peeling off graphene layers from graphite (Rudrapati 2020). For bulk production of graphene generally the Chemical Vapour Deposition (CVD) process is used (Mattevi et al. 2011). This process uses organic materials and carbon sources such as methane, ethylene, and carbon-containing gases. Sometimes CVD required specific instruments which increases the overall cost of the process. Another method used to fabricate graphene is epitaxy. In epitaxy, a metal substrate is used on which carbon is decomposed and the graphene layer is formed (Gamo et al. 1997). Electrochemical exfoliation is a modified version of the exfoliation technique that involves the use of electric current to exfoliate graphene monolayers from graphite. The Hummer method is a specific technique used to make GO (Mindivan 2016). Several other methods are widely used to synthesize graphene but they still require more research to increase the fabrication of defect-free graphene production. Numerous methods are employed to characterize the quality of graphene and its derivatives such as

Table 11.1 Derivatives of graphene

Sr. no	Graphene derivatives	Type	Synthesis method	Application	Methodology advancements	References
1	Graphene oxide	Oxidized graphene	Hummers method	Batteries, Sensors, Composites, Biomedical	Electrochemical exfoliation, Laser reduction	Ray (2015)
2	Reduced graphene oxide	Reduced graphene	Chemical reduction	Energy storage, Electronics, Inks, Catalysis	Microwave reduction, Hydrothermal reduction	Ray (2015)
3	Graphene nanoplatelets	Few layer graphene	Liquid phase exfoliation	Conductive polymers, Coatings, Aerospace	Supercritical fluid exfoliation, Sonic exfoliation	Ray (2015)
4	Graphene quantum dots	Quantum dots	Electrochemical exfoliation	Bioimaging, Photovoltaics, Sensors, Displays	Electrochemical synthesis, Laser ablation	Peng et al. (2014)
5	Graphene foam	3D graphene structure	Chemical Vapor Deposition (CVD)	Energy storage, Thermal management, Aerospace	Freeze casting, Template-assisted assembly	Ray (2015)
6	Graphene nanoribbons	One Dimensional graphene	Unzipping of carbon nanotubes	Electronics, Optoelectronics, Nanodevices	Bottom-up synthesis, Chemical vapor deposition	Peng et al. (2014)
7	Graphene aerogel	Porous graphene	Supercritical drying, Gelation	Energy storage, Aerospace, Sensors	Freeze drying, Template-assisted assembly	Ray (2015)
8	Graphene polymer composites	Composite material	Solution mixing, In-situ polymerization	Structural materials, Electronics	Layer-by-layer assembly, Interfacial polymerization	Ray (2015)
9	Graphene membranes	Thin graphene films	Chemical Vapor Deposition (CVD), Epitaxial growth	Water filtration, Gas separation	Graphene oxide templating, Self-assembly	Peng et al. (2014)
10	Graphene nanosheets	Thin graphene layers	Liquid phase exfoliation, CVD	Energy storage, Coatings, Sensor	Chemical Vapour deposition (CVD)	Peng et al. (2014)
11	Graphene quantum well	Confined graphene layers	Molecular beam epitaxy	Optoelectronics, Spintronics	Pattern growth, Interfacial engineering	Ray (2015)

(continued)

Table 11.1 (continued)

Sr. no	Graphene derivatives	Type	Synthesis method	Application	Methodology advancements	References
12	Graphene ink	Graphene-based ink	Liquid phase exfoliation, Stabilization	Printed electronics, Flexible devices	Liquid phase exfoliation with surface modification	Ray (2015)
13	Graphene nanocomposites	Graphene-based composites	Solution mixing, In-situ polymerization	Automotive aerospace, Structural materials	Electrospinning, In situ reinforcement	Ray (2015)
14	Graphene supercapacitors	Energy storage devices	Electrochemical deposition	High-power storage, Electric vehicles	Ionogel electrolytes, Pseudocapacitive-materials	Peng et al. (2014)
15	Graphene polymer films	Flexible graphene films	Solution casting, Spin coatings	Flexible electronics, Wearable devices	Layer-by-layer assembly, Interfacial polymerization	Peng et al. (2014)
16	Graphene silver composites	Hybrid material	Electrochemical deposition, Solution mixing	Antimicrobial coatings, Biomedical applications	Galvanic replacement, Chemical reduction	Peng et al. (2014)
17	Graphene thermal interface materials	Thermal management	Chemical Vapour Deposition (CVD)	Electronics, Power electronics, LED lighting		Peng et al. (2014)

Scanning Electron Microscopy, UV and IR spectroscopy, X-ray diffraction, thermal analysis, chemical analysis, and surface area analysis.

Researchers and industrialists focus on the bulk production of graphene and its derivatives as well as focus on decreasing the scale-up costs. Across the globe, start-up companies are interested in graphene production through commercial fabrication methods. Graphene shows promising results from space to deep marine technologies. Physical and chemical stability along with a large surface area of graphene increases its demands in biomedical fields (Bhavsar and Jagtap 2021). Various advanced techniques of material engineering help with the fabrication of graphene to produce high-quality, stable, inexpensive, and defect-free material. The real challenge is to determine which methods out of the many qualify to be used in commercial production. This chapter focuses on the various commercial methods employed for the bulk production of graphene and its various derivatives, which emphasizes the principles, benefits, and drawbacks of each approach. These methodologies include chemical vapour deposition (CVD), liquid-phase exfoliation, epitaxial growth, and more. By deeply understanding the various bulk production methodologies, researchers, engineers, and industry experts may acquire insights into the possibilities and challenges associated with the large-scale production of graphene.

11.2 Basic Methodologies for Bulk Production of Graphene and Its Derivatives

Several researchers are currently working on graphene synthesis; as a result, many techniques for exfoliating graphite into graphene are now known. They are classified into two types: Bottom-up and Top-down approach (Fig. 11.2). A network of thin layers of graphene forms up graphite, and the forces between them must be overcome in order to split them into different layers of graphene (Edwards and Coleman 2013). The term "top-down approach" is used to describe this method. This method presents several challenges, including defects in the surface that appear during the separation of sheets and its subsequent re-agglomeration. These techniques often have low yields and require lengthy procedures (Edwards and Coleman 2013). Although carbon molecules are used as building blocks in the bottom-up approach and generally it is obtained from various sources, the top-down method focuses on splitting the graphene precursor (graphite) into atomic sheets from the stack. Although the bottom-up method is ineffective in producing graphene sheets with high surface areas, it does allow for the mass manufacture of graphene nanoribbons and dots (also known as Nano flakes) (Warner et al. 2012). Initially after its discovery, the most widely utilized method for producing high-quality, flawless graphene was the micromechanical splitting of graphite, which is often referred to as the "scotch tape," (Novoselov et al. 2004).

Chemical vapour deposition (CVD) came into use as an effective substitute soon after. However, it quickly became evident that these processes had a lower yield and

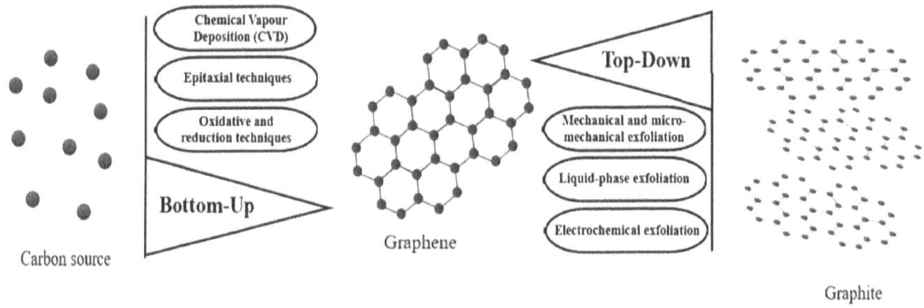

Fig. 11.2 Top-down and bottom-up techniques for bulk production of graphene

as for micromechanical cleavage, it was tedious which prevented the effective and complete use of these materials (Verdejo et al. 2011). An alternative approach to micromechanical cleavage is the exfoliation of graphite or its derivatives, primarily graphite oxide, which enables an increased yield while making the process feasible (Park and Ruoff 2009). These are examples of top-down techniques in which precursors of the material are reduced by chemical, mechanical, or thermal processes to a thin sheet of graphite.

11.3 Top-Down Approach

11.3.1 Mechanical or Micromechanical Exfoliation Method

The micromechanical exfoliation or the "Scotch tape" method was the first process used to isolate graphene. It involves isolating individual graphene sheets by repeatedly removing tiny graphite layers with the use of adhesive tape. In the development of graphene, this process has been a revelation. Despite being straightforward, it is not feasible enough for the bulk synthesis of graphene (Rudrapati 2020). The van der Waals force is responsible for the attachment of many graphene layers in graphite. An adhesive scotch tape is used to strip out the individual layers of graphene from graphite. Graphene flakes are present in layers attached to adhesive tape and multiple layers are peeled off from graphite. These various layers are subjected to repeated exfoliation to obtain a single graphene layer (Geim and MacDonald 2007). When the scotch tape is removed the flakes of graphene are present on the surface of the substrate. Some of these flakes are nm in size that can be identified by using colour images from an optical microscope (Abhervé et al. 2015). The wedge-based method uses a sharp single-crystal diamond wedge that penetrates graphite for exfoliation (Mannix et al. 2022). These mechanical methods generate graphene layers in small amounts therefore for bulk production such methods could not be efficient. The graphite is exposed to potassium solution which converts it into graphite oxides. The space between the layers increases in graphite and is further treated with dilute sodium hydroxide that forms graphene oxide flakes. The size of these flakes is determined by electron microscopy which can be used as graphene layers (Yamada et al. 2013).

Similar experimental techniques were used (Novoselov et al. 2005) to generate atomic planes of dichalcogenides, mica, boron nitride, and oxides that are highly complicated, although none of them have so far received a lot of attention. Keith et al. described the method of shear exfoliation, producing graphene by immersing and mixing graphite at local shear rates of more than 104 s^1 in stabilizing liquids (N-methyl-2-pyrrolidone, sodium cholate, etc.). This method seeks to replace the use of sonication as a technique to create individual layers, making it an energy-efficient method (Paton et al. 2014). As a result, defect-free and unoxidized graphene may be produced. According to Keith et al., the rate of shearing has to be much more than 104 s^1, which is also feasible in a standard blender used in kitchens. In terms of characteristics, the graphene produced using this approach is identical to that produced by sonicating graphite in solvents and surfactants. This process is simple, and it appears to be effective for producing graphene on an industrial scale. The three-roll mill machine with a polymer adhesive is another micromechanical technology modelled after the "Scotch tape" approach. The adhesive is made of polyvinyl chloride (PVC) soaked in a solution of dioctyl phthalate (DOP), which acts similarly to the initial Scotch tape in the micromechanical method. Exfoliation and dispersion take place in the adhesive. The continual exfoliation is caused by the moving rollers' ability to move graphite flakes in an inverted S curve from the feed to the apron roll, then back again towards the feed. The total removal of leftover PVC and DOP to generate graphene is not simple and adds extra complexity, despite the three-roll mill machine being a fairly typical industrial procedure in the rubber sector and being readily available (Bourlinos et al. 2009).

11.3.2 Liquid Phase Exfoliation Method

Liquid phase exfoliation is a method that is greatly employed for the bulk synthesis of defect-free graphene. Surface energy is an important factor in the production of graphene. Researchers observed that the surface energy of graphene is similar to carbon nanotubes and hence liquid exfoliation can be implemented for bulk production of graphene from graphite. For liquid exfoliation of graphene, graphite powder is sonicated with different solvents like N-methyl-pyrrolidone and dimethylformamide (Coleman 2013). This mixture is centrifugated to eliminate the undispersed and the dispersed material is filtered. The surface tension and surface energy of the filtrate are measured, similar to graphene (Khan et al. 2010). After the process of liquid exfoliation large quantities of monolayer and multilayer 2D objects of graphene were observed with the help of the electron microscopy technique (Lotya et al. 2009). Some of the researchers are now focusing on aqueous solutions with surfactants instead of reagents for the production of graphene. The surfactants like sodium cholate and sodium dodecylbenzene-sulfonate are used for aqueous graphene flakes production at room temperature (Lotya et al. 2009). Most of the process of graphene production creates some defects in graphene flakes. The immiscible liquids exfoliation is a process that peculiarly helps produce inexpensive pristine graphene flakes at the laboratory level. The defect-less layer of graphene is termed pristine graphene. An immiscible liquid combination such as water-heptane is used for the production of graphene flakes (Woltornist et al. 2013). In this technique, graphite flakes are mixed with a water-heptane solution and further sonicated slightly. The graphene is immiscible in both the solvents and therefore monolayer of graphene forms in between interfaces of two liquids. In this method, the graphene layer shows climbing phenomena where a glass slide is introduced in the mixture of water-heptane solution (Ruths and Israelachvili 2011). The graphene layer forms at the interface of water-heptane liquids rising the hydrophilic glass which helps for extraction of graphene film. This one-step technique produces a macroscopic monolayer of graphene (Woltornist et al. 2013). Molten salts can be used for the production of electrolytic carbon nanomaterial like graphene (Churi and Dave 2023). The graphite flakes undergo degradation when treated with molten salts. Most probably lithium chloride reacts with graphite to form different forms of graphene at high temperatures from which graphene layers can be achieved by a micro-mechanical exfoliation process (Kamali and Fray 2013).

11.3.3 Electrochemical Exfoliation Method

Electrochemical exfoliation (ECE) is a promising method for producing graphene and other two-dimensional materials. This process involves the use of an electrolyte solution to exfoliate layers of graphite, resulting in high-quality graphene flakes with multiple applications. The advantages of ECE include low cost, scalability, and environmental friendliness. A rising number of fabrication methods for graphene use electrochemical exfoliation. The modified Hummers process used to create graphene oxide is one example of how the electrochemical strategy is connected to wet chemical exfoliation strategies. Graphene can be synthesized by electrochemical method with the help of solvent intercalation. The electrochemical process is promising for the large-scale manufacturing of graphene and has numerous potential benefits over conventional chemical techniques. Utilizing electrochemical activation might lead to a quicker process for product purification, and electrochemical manufacturing prevents the usage of harsh chemicals. By exfoliating graphite in ionic liquids, Lu et al. (2009) presented a direct approach to creating various nano forms of carbon, even graphene. They conducted an experiment in which 1-butyl-3-methylimidazolium tetraflouroborate [BMIm][BF_4] was added to water, which was utilized as an electrolyte for the exfoliation of graphite. Static potentials ranging from 1.5 to 15 V were used. To reach a pH of neutral, the exfoliated particles were rinsed with ethanol and water. An intriguing technique for the exfoliation of graphene from graphite by making use of an electrochemical process was described in which they applied platinum filaments as the cathode, flakes of graphite as the anode, and 0.1 M H_2SO_4 as the electrolyte solution (Parvez et al. 2013). Graphite flakes dissolved in the solution when the system was provided with a positive charge of +10 V. After two minutes, the power was switched off, and the exfoliated material was collected using vacuum filtering. To ensure that any residual acidic material was removed, the sample was repeatedly rinsed with water. The resultant powder was mixed with DMF to produce graphene sheets. The type of voltage used—either anodic or cathodic-determines how electrochemical exfoliation works. In anodic exfoliation, anions and any co-intercalating species from the reaction mixture are intercalated into graphite. The working anode made of graphite gains a positive charge as electrons are drawn out of it. This charge promotes the intercalation of heavy negative ions, such as sulphate anions, which expands the space between the graphene sheets and aids in the exfoliation of the sheets ultimately. A negative potential at the graphite an active electrode draws ions that are positively charged in the solution (for example, Li^+) as well as any co-intercalating compounds during cathodic exfoliation. In addition, these intercalating species break apart the graphene sheets, resulting in expansion and exfoliation. In their investigation, they used Pure Graphite as the electrode material, and PSS (Polysodium 4-Styrenesulfonate) was dissolved in deionized water which is used as an electrolyte (Bourlinos et al. 2009). The graphite rods were placed in an electrochemical cell with the electrolyte solution and 5 V of current was applied. After few minutes of electrolysis, a black-coloured substance accumulated at the anode surface. In order to separate the substance from the cell, the substance had to be exfoliated for 4 h. After a 1000 rpm centrifugation, the substance was poured out slowly. The resultant dispersion was found in an extremely stable state. The dispersion was cleaned with a mixture of deionized water and alcohol, then later vacuum dried to produce powdered dry graphene. The dry powder and sediment were then weighed to determine the yield.

11.4 Bottom-Up Approach

11.4.1 Chemical Vapour Deposition (CVD)

Chemical Vapour Deposition (CVD) is a technique performed to produce graphene in bulk quantities. In CVD a vacuum chamber is used for the synthesis of high-quality products. The transition metals like Nickel, Lead, Rubidium, Iridium, or Copper are utilized in synthesizing graphene. In this process, graphite flakes are chemically treated and deposited on a heating substrate that forms a monolayer of crystalline particles on the surface of the substrate (Mattevi et al. 2011). Nickel and Copper are employed in the large-scale manufacture of graphene. In this CVD process, methane gas is decomposed at a high temperature on a copper substrate. Graphene monolayer forms on copper foil with the help of hexane and organic solvents that are used as a precursor (Srivastava et al. 2010). High quality with a bulk amount of graphene is obtained from the CVD process.

There are various crystalline substrates used to make graphene. Graphene atoms attach to the substrate material which helps in the creation of the graphene layer. The interaction between graphene atoms and substrates varies because the size of the graphene layer differs with different substrates. Most scientists use silicon carbide for the synthesis of 2D graphene film (Sutter 2009). The silicon carbide is reduced to graphene by heating at high temperatures under low pressure. The dimensions of the graphene film depend on the wafer. The van der Waals attraction between graphene and wafer is weak and graphene layers are easily separated by the drawing method. The reusable silicon wafer is made by coating germanium oxide on which the graphene layer is deposited by using CVD (Lee et al. 2014). The uniform layer of graphene obtained from this reusable silicon wafer can be directly used for various activities. The graphene layer obtained by this technique is of high quality with minute defects (Lee et al. 2014). Numerous metals are widely used as a substrate to synthesize graphene. A single layer of metal has a smooth uniform surface and contains chemical uniformity. These metal structures are coated with oxides which form a hexagonal close-packed surface that is similar to the geometry of a graphene single-layer structure. The distance between the atoms in the metal crystal monolayer is different concerning the type of metal that results in the formation of different geometrical patterns. Large numbers of metal crystal monolayers are required for the bulk production of graphene. The synthesis of metal substrates or wafers was found by Czochralski and Bridgman-Stockbarger (Brückner and Schwerdtfeger 1994) which helps for bulk production of metal wafers. Most probably ruthenium and gold are preferred for manufacturing wafers but these metals are rare and expensive. The graphene layer is synthesized on the ruthenium surface by using the CVD method. In this process, the ruthenium surface is heated at a very high temperature and then exposed to organic compounds like methane or ethylene. The carbon present in alkanes attaches covalently with the ruthenium surface and forms a monolayer of graphene. Multiple layers can be formed on this surface but the interaction between other layers is weak as compared to the first layer that is covalently attached to the ruthenium surface therefore it is easy to obtain a monolayer of graphene (Sutter et al. 2008; Gutzwiller 1963). Iridium is also used for graphene production. The iridium wafer is heated at high temperatures. The heated wafer is exposed to ethylene with the help of CVD and the carbon atoms adsorbs on the wafer surface that forms a graphene layer (Hämäläinen et al. 2013). Platinum is an expensive metal that can be used for graphene production. The platinum substrate is heated at more than 1000 °C on which ethylene is decomposed by using the ultra-high vacuum method (Kim et al. 2017a, b). The decomposed ethylene forms graphene islands on the platinum surface due to the annealing temperature (Kim et al. 2017a, b). These micro islands can be visualized by an electron microscope to observe different patterns. Multiple techniques have been reported to produce graphene on nickel film by utilizing CVD. The film is heated to a temperature of 1000 °C and exposed to argon gas and after that methane gas is allowed to flow on the surface. The carbon atoms from methane adsorbed on the surface form a graphene film (GAMO et al. 1996). The identical structure of graphene is also synthesized on a cobalt wafer with a similar technique used for nickel. The propylene is deposited on the cobalt surface at 450 °C (Varykhalov and Rader 2009).

Graphene production on copper is the easiest way. For this process, copper foil is treated at room temperature with low pressure in the presence of methane. The single layer is only formed on copper foil because methane contains less number of carbons (Li et al. 2009). After the formation of the single layer the growth of graphene stops therefore this technique is used for the synthesis of large films. Roll-to-roll is a method used for the bulk production of graphene. In this process, copper is used as a substrate on which graphene is synthesized (Bae et al. 2010). The cold wall CVD system is a promising technique far better than the conventional CVD system but the research is still going on.

11.4.2 Epitaxial Technique

Epitaxy refers to the process of depositing a crystalline layer on a substrate with a matching crystal structure. Due to the thin graphene films (>50 m) that can be produced using this technique, epitaxially grown graphene is suitable for usage in transistors and circuits. In this process, silicon carbide (SiC) is heated to 1100 °C to produce graphene. In this method, the size of SiC wafers continues to determine the size of graphene flakes. According to studies, the SiC surface affects the mobility, thickness, and carrier density of the graphene produced by this system (Ohta et al. 2007). Graphene produced using this method often exhibits lower anti-localization, in contrast to graphene that has been exfoliated. However, SiC-epitaxial graphene exhibits extraordinarily large, temperature independent mobility, comparable to graphene generated by drawing or peeling off technique, but not as high as exfoliated graphene (Bae et al. 2010). Graphene made using this technique exhibits massless Dirac fermions even without transfer. The electrical properties of individual sheets within a stack are not necessarily influenced by the van der Waals forces that preserve multilayer cohesion in multi-layered epitaxial graphene. The symmetry of interlayer interactions is relevant to this phenomenon. Other instances of this phenomena do not happen, and it has an impact on the electrical characteristics of materials such bulk graphite. The 100 GHz cutoff frequencies may be found on a 2-inch SiC wafer (Jobst et al. 2009). Due to the high cost of the SiC substrate and the limited yield, this method produces high-quality graphene at a high manufacturing cost. Therefore, this approach is not appropriate for industrial manufacturing.

Various other methods are available and used by researchers for the bulk production of graphene. Flame synthesis is a process used to produce nanoparticles. Researchers are focusing on the production of graphene material using flame synthesis. Pulsed laser desorption assay is peculiarly used for the production of various materials. In this process, carbon material or organic compounds are deposited at a 45° angle on the substrate at a low temperature which helps to produce a defect-free graphene layer. Some researchers used the nanotube slicing method to synthesize graphene.

11.5 Challenges in the Bulk Synthesis of Graphene and Graphene Derivatives

The current Graphene production techniques significantly affect production by quality, quantity, and price with process standardization and its industrialization. Several techniques are able to be adopted to create a variety of graphene. Currently, chemical vapour deposition on metallic substrates is used to create graphene films. Chemical vapour deposition can create large-area films that may be applied to a number of substrates, making it a manufacturing process with industrial relevance (Kobayashi et al. 2013). The synthesis of graphene films can be carried out in a batch or continuous fashion. Manufacture of light-emitting diodes (LEDs) as well as transistors and solar cells for example makes the use of batch operations, which are highly significant from an industrial standpoint. Graphene is now produced in batches on 6 inch wafers, while 12 inch wafers may be made accessible in the coming years. On the other hand, continuous or semicontinuous manufacturing techniques are now being developed and should be accessible in the upcoming years to address the increased demand for graphene films. Although graphene is highly promising due to its exceptional properties, achieving low-cost large-scale production with consistent quality remains a significant hurdle. Some of the key challenges in the bulk production of graphene and graphene derivatives include.

11.5.1 Scalable Synthesis Methods

The first difficulty in producing graphene and its derivatives in large quantities is creating scalable synthesis techniques. Although there are several ways to make graphene, including mechanical exfoliation, CVD, and epitaxial growth, each method has scaling issues.

It takes a lot of time and effort to mechanically exfoliate graphite, which includes peeling away small layers of graphene. Using this approach, it is challenging to create substantial amounts of graphene. Contrarily, CVD enables the creation of graphene on metal substrates, but the size of the graphene domains is often constrained, making it difficult to transfer the graphene to further substrates without introducing flaws (Zurutuza and Marinelli 2014). Researchers are looking at other synthesis techniques that might manufacture graphene in greater numbers in order to get over these restrictions. One such method is liquid-phase exfoliation, which involves dispersing graphite in a solvent and applying shear or sonic forces to the mixture to separate the graphene layers. This process can produce liquid graphene, which may be used to create graphene inks or coatings. However, obtaining a high yield and preserving the graphene's purity during the exfoliation procedure continue to be an active research topic.

Chemical synthesis, which produces graphene by reducing graphene oxide or other carbon precursors, is another viable technique. Using this method, graphene in solution may be created, which can then be processed for a variety of uses. To produce high-quality graphene, the difficulty is in optimizing the reaction conditions, managing the level of reduction, and eliminating impurities. Another component of scalability, in addition to synthesis methods, is the creation of cost-effective strategies (Kobayashi et al. 2013). Many of the present synthesis methods call on pricey precursors, including high-purity graphite or certain catalysts, which can greatly raise the cost of manufacturing. Researchers are investigating cheaper, more accessible carbon sources and catalysts that may still produce high-quality graphene.

11.5.2 Cost of Production

This is one of the crucial variable that might delay the widespread use of graphene. Since graphene was originally offered to the research community, costs have already decreased significantly.

Even on a lab scale, the cost per square centimetre and cost per gram has decreased over the last two years to a third and a fourth of their initial costs, respectively. A greater material cost is expected to increase the acceptance of technologies that harness several graphene characteristics. The cost of the material necessary to achieve commercialization is also dependent on it. The adoption of graphene is costly, high-volume goods and applications as well as using it for material replacements, cost reductions, or medium-level improvements in a single parameter will be delayed by a slower decline in graphene's price (Zurutuza and Marinelli 2014). However, within certain bounds, a significant enhancement in single characteristics or a novel mixture of attributes definitely ought to redefine the price of some items to render the cost of grapheme is irrelevant.

The raw ingredients required for the production of graphene are a significant economic factor. High-purity graphite is often needed as a starting material to produce high-quality graphene (Lowe and Zhong 2016). Obtaining such graphite can be expensive, particularly when taking into account the vast amounts required for bulk manufacturing. Research is ongoing to find cheaper, more accessible carbon sources that can produce high-quality graphene (Zurutuza and Marinelli 2014). Additionally, certain synthesis processes need the use of pricey catalysts or growth substrates, which raises the overall cost of manufacturing. Choosing affordable catalysts and substrates without sacrificing graphene quality is essential for cutting manufacturing costs.

Another expense factor is energy use. Some synthesis technologies, such as CVD, need high temperatures and energy-intensive operations, which raises the cost of operation. Costs associated with manufacturing can be decreased by creating energy-efficient synthesis techniques or improving the current processes to use as little energy as possible. Cost-effectiveness is also influenced by the synthesis methods' scalability. It may not be economically possible to use some techniques for large-scale manufacturing because they are difficult to scale or ask for specialized equipment. Finding scalable techniques that are simple to apply in an industrial context might aid in reducing production costs. As previously indicated, researchers are currently looking into alternate synthesis processes that employ more easily accessible and reasonably priced raw materials to address the cost-effectiveness dilemma. Automation and process optimization can also assist to boost productivity, save labour costs, and simplify manufacturing.

In order to solve the cost problem, industry and academia must work together. Through information exchange and technology transfer, partnerships may promote the development of efficient synthesis methods, enable economies of scale, and improve manufacturing procedures.

11.5.3 Handling, Storage, and Transfer

For practical uses, graphene must be placed onto a chosen substrate or carrier material once it has been manufactured and is often grown on a substrate or created as a film (Novoselov et al. 2012). The size of the graphene sheet increases the difficulty of the transfer procedure. For the transfer to be effective and damage-free, several difficulties must be resolved.

Mechanical integrity: Two-dimensional substance called graphene has excellent mechanical qualities, but it is extremely fragile and easily rips or wrinkles when handled. Maintaining mechanical integrity becomes essential while moving large-area graphene sheets in order to prevent structural damage (Zurutuza and Marinelli 2014). In order to protect the integrity of the graphene throughout the transfer process, approaches include utilizing support films, reducing interaction with foreign particles, and using controlled transfer procedures can be used.

Contamination management: To preserve the quality and characteristics of graphene during the transfer process, contamination management is essential. The transferred graphene may have imperfections and flaws owing to particles, dust, and residue from the transfer materials or substrates. Risks of contamination can be reduced by implementing cleanroom facilities, taking appropriate safety precautions, and improving transfer processes (Lowe and Zhong 2016).

Scalability and uniformity: It is difficult to transmit graphene in a consistent, scalable manner over vast regions. The thickness, structural integrity, and electrical characteristics of the transferred graphene may vary as a result of inhomogeneous transfer (Lowe and Zhong 2016). The manufacturing of high-quality graphene in bulk requires the development of methods that guarantee uniform covering and dependable transmission across huge surfaces.

In order to produce graphene on an industrial scale and to integrate it into diverse sectors, efficient large-area transfer and handling techniques are essential for realizing its special features in practical applications. This obstacle will eventually be overcome, allowing graphene and its derivatives to reach their full potential. This will be accomplished by further improvements in transfer techniques and protocol optimization.

11.5.4 Standardization and Quality Control

Last but not least given its wide variety, efforts must be undertaken in order to standardize the official definition of various varieties of graphene and associated materials that will speed up adoption and commercialization of graphene by assuring prospective consumers of the nature and amazing quality of these materials. Graphene may be synthesized using a variety of techniques, including chemical reduction, mechanical exfoliation, and CVD. The characteristics and calibre of the manufactured graphene may differ depending on the method used. Additionally, several synthesis variables, including temperature, pressure, and precursor materials, might affect the properties of the finished product (Novoselov et al. 2012). To ensure consistent quality, standardization of synthesis methodologies and process parameter optimization are required.

Impurities and flaws have a significant impact on graphene's characteristics. Its electrical, mechanical, and thermal characteristics can all be greatly impacted by even the smallest impurities or structural flaws. It is essential to regulate impurity levels and reduce errors throughout the production process. To find and measure impurities and flaws, quality control procedures including sophisticated characterization techniques (such Raman spectroscopy and scanning electron microscopy) and extensive quality evaluation methodologies must be used (Zurutuza and Marinelli 2014). Variations in the structure of graphene, such as the number of layers, the location of grain boundaries, and the size of the domains, can have an impact on its characteristics. It is difficult to achieve homogeneity in the structural properties of graphene generated in large quantities. To evaluate the structural consistency and find any deviations, tools including statistical analysis, high-tech imaging, and spectroscopic procedures might be used.

11.5.5 Impact on the Environment

Some techniques used to produce graphene employ dangerous chemicals, high temperatures, and energy-intensive procedures. It is crucial to create ecologically sound and long-lasting synthesis processes that reduce their impact on the environment and manage any possible risks to people's health and safety from industrial-scale graphene manufacturing (Berktas et al. 2020).

11.6 Advancements in Methodologies for Bulk Production of Graphene

11.6.1 Improved CVD Techniques

Extensive research has been done on a number of techniques, including but not limited to mechanical exfoliation, liquid exfoliation, epitaxy, and chemical vapour deposition to create graphene with a variety of forms, sizes, domains, and sample areas. Considering various demands such as pristine quality, size, and affordability, graphene films generated by CVD exhibit the highest performance among the various techniques (Novoselov et al. 2012). It is imperative to advance research on graphene films at laboratory and industrial scales. We are able to control the main characteristics of graphene, such as the number of layers and the domain size, to create application-specific requirements for properties, such as sheet resistance, transparency, and as well as stability, by engineering the fabrication conditions (process, equipment, and critical parameters). Consideration of all these factors for application directs the flow of the production process.

11.6.2 PECVD

PECVD has become a useful technique for creating carbon-based compounds, including diamonds, carbon nanotubes, and graphene. By accelerating the excitation, ionization, and complete dissociation of carbon sources, the extremely energetic electrons in the plasma are able to generate graphene at low temperatures right on the required substrates. Three components make up most PECVD equipment: a supply of gas, a generator for plasma generation, and a vacuum heating chamber (Qi et al. 2018). According to the power source, there are three different types of plasma generators that may be used in PECVD systems: Direct-Current (DC), Radio-Frequency (13.56 MHz), and Microwave (MW) plasma (2.45 GHz). Low pressure and gases are necessary for PECVD (Li et al. 2016). Graphene growth requires carbon-containing gaseous precursors, and for high-quality graphene synthesis, it is necessary to add gases like H_2 and O_2 to etch amorphous carbon. PECVD has a low growth temperature and quick growth, which is advantageous for very high productivity and perfect for the electronics sector.

As the substrate is submerged in the plasma during PECVD development, the electrical field causes the graphene sheets to align vertically to the substrate. In order to make graphene films that are planar with regard to the substrate, (RP CVD) upstream Remote Plasma CVD is employed to get rid of this orientation dependence. Using RF-PECVD at 650–700 °C, graphene films was produced on Nickel substrates using CH_4 and H_2; the inclusion of H_2 encouraged the formation of a much more high-quality graphene (Nandamuri et al. 2010). Peng et al. had looked into RF-PECVD's ability to generate graphene on Ni thin films without using hydrogen.

It is well understood that the temperature of the substrate, the thickness of the Ni film, and the deposition time affect the uniformity of the layer of graphene and its overall crystalline structure. They discovered that a temperature of 475 °C is the critical temperature required for the formation of graphene (Peng et al. 2013); below which, it is impossible for carbon species synthesized by the Radiofrequency plasma to dissolve into the Ni thin layer. Additionally, it was found that the host Ni layer's essential thickness was 30 nm; a thinner coating causes the surface to generate amorphous carbon or discontinuous graphene.

A PECVD procedure that combines the development of graphene along with the Cu surface preparation surface in one step was proven to improve the graphene's quality. Three distinct mechanisms may be identified in PECVD growth. By adding trace cyano radicals to the microwave hydrogen plasma (MHP), native oxide was first removed and the surface of Cu was polished (Boyd et al. 2015). Afterwards, graphene started to develop on the Cu substrates on both sides. After long exposure with the plasma, the top and bottom surfaces of the Cu foil were finally covered by both the monolayer graphene and the disordered graphite. With the help of this method, graphene was able to develop within minutes at temperatures much lower than 420 °C with the roughness of sub-nanometers and superb crystalline structure. Electrical mobility values for the as-grown graphene reached 60,000 $cm^2 V^1 s^1$, which is substantially greater compared to those earlier reported using conventional PECVD techniques.

11.6.3 Thermal-Chemical Vapour Deposition (T-CVD)

The carbon source must be activated at high processing temperatures in order to create graphene. Hot-wall CVD is the most used Thermal-CVD technique in research, where the substrate is heated with radiation emitting from the chamber walls while the chamber is heated to a higher temperature by an external energy source. The lengthy heating and cooling processes often severely constrain the manufacturing of graphene. It requires a lot of energy to maintain the high temperature throughout the chamber. As a result, numerous cold-wall CVD techniques, in which the substrate is directly heated while the chamber walls are at a lower temperature, have been researched for the formation of graphene. Shorter growing periods, quicker heating and cooling, and reduced gas use are all made possible by cold-wall CVD. Additionally, cold-wall CVD restricts chemical reactions that are known to contaminate graphene in its gaseous phase at high temperatures. Joule heating, induction, thermal radiation, and stage heating are the four cold-wall heating modalities that can be used (Deng et al. 2019). This is made possible by the Cu substrates' high electrical and thermal conductivity.

Copper foils high conductivity makes it possible to heat directly using the Joule method by running an electric current through it. Joule heating was used in the construction of a CVD system by (Kobayashi et al. 2013). Two current-feeding electrodes, two winders, and a stainless-steel vacuum chamber were the components of the system. By allowing electricity to flow through them, copper foil suspended between two rollers is heated to a temperature of around 1000 °C during operation. A temperature of 950 °C is achieved with a continuous direct current of 82 A mm^{-2}. Using this technique, 100 m long graphene sheet was created and applied to a substrate. Due to heat loss at the margins, when a Cu foil is heated by Joules, the temperature distribution may not be uniform. As a result, the middle has greater graphene film coverage than the edge.

Bointon et al. also announced resistively heated stage CVD (RHCVD)-produced high-quality monolayer graphene. A resistively heated stage with an integrated thermocouple that could maintain steady temperatures of up to 1100 °C was located inside the reaction chamber. The state placed the Cu foil on top of it and heated it there. Although the operating chamber temperature was only around 100 °C, the Cu foil had a maximum temperature of 1000 °C (Bointon et al. 2015). The synthetic graphene was homogeneous and shown mobility at 1.4 K of 3300 cm^2 V/s^1. In contrast to the conventional development technique, this approach forms a thick carbon coating on the surface of Cu early on, which gradually gets thinner with longer growth times and eventually transforms into graphene.

A rapid thermal CVD (RTCVD) technique was created by Ryu et al. to achieve very quick heating and cooling (Ryu et al. 2014). The 24 halogen lamps made up the heating unit that produced light at wavelengths ranging from visible to infrared (IR), as well as a susceptor of graphite that effectively converted near-Infrared light to thermal radiation for a more improved distribution of heat on the Cu surface. In order to prevent distortion brought on by thermal expansion and gravity, the foils of Cu were loaded in a vertical manner. According to reports, this synthesis procedure does not use hydrogen, which is advantageous for mass manufacturing when considering security concerns. The synthesized graphene had excellent electronic characteristics equivalent to those of graphene produced using conventional thermal CVD.

11.6.4 Flash Graphene

Flash Graphene, a new and revolutionary method of making high-quality graphene very quickly and efficiently with minimal and inexpensive equipment that can be easily procured by any research lab. The ease of making flash graphene and the fact that it can be synthesized from even waste materials makes it a highly preferable option to help curb pollution, especially reducing and reusing the materials in landfills that are not properly utilized as they are a goldmine of materials that can help us mass produce graphene and apply its amazing properties in many of our existing products (Wyss et al. 2021).

Flash Graphene, as the name suggests, is the production of high-quality graphene in an instant which is accomplished by a process called Flash joule heating which involves the use of a high current to heat the carbon material and convert it into graphene. In Flash joule heating, any carbon material e.g., pyrolyzed rubber tire which is mixed with carbon black to make it more conductive is placed between two electrodes, and then a high electrical pulse is quickly discharged throughout the material placed in a quartz tube between the electrodes, raising its temperature to 3000 K in 0.5 s (Advincula et al. 2021). The carbon bonds break and all the other unwanted material is removed in the form of gas. The rapid cooling after that leads to the formation of graphene or Turbostratic flash graphene. Turbostratic graphene differs from the conventional AB stacking of graphene synthesized by other methods like CVD (Stanford et al. 2020). In AB stacking, the center of the hexagon of the 1st layer of graphene is above the carbon atom in the 2nd layer below it. While in turbostratic stacking, each graphene layer is rotated about the axis normal to the individual graphene sheets which means it has more interlayer spacing making it much easier to exfoliate it into single sheets compared to AB stacking structure. The increased space helps turbostratic graphene maintain its 2D properties even if the layers are increased because the electron mobility is decreased between the individual layers of graphene.

Now it is important to understand that improvements in a revolutionary process can be achieved to make it more efficient and commercially applicable. This has been shown in a few studies that have come up since the introduction of Flash graphene. Typical Flash joule heating systems to make graphene requires the reduction of carbonaceous materials such as Plastic, rubber, and other carbon material into ash which can then be converted into graphene with Flash joule heating. This results in increased expenditure on pyrolyzing equipment required to pyrolyze the carbon materials into ash. It is also a very energy exhaustive process as much of the energy given as input is lost raising expenses even more. This is a major challenge when it comes to commercializing this technology.

A new study came up with a new method of Flash Joule Heating known as Alternating Current Flash Joule Heating (AC-FJH) which is far more effective and less expensive than the previously mentioned method (Algozeeb et al. 2020). This method is very useful when it comes to plastic waste. It can be sustained for a few seconds to 8 s, which forms an intermediate form of graphene but with low quality. This intermediate form of graphene is then further converted into high-quality graphene by a DC pulse. The beauty of combining AC and DC pulses to produce Flash Graphene is that by applying an AC pulse first to the plastic sample, all the volatiles from the sample is released along with the sample being pyrolyzed the plastic is effectively reduced into a much smaller and simpler form which can then be easily converted into proper quality graphene which is fit for industrial use (Wyss et al. 2021). This alternative method of Flash Joule heating involves just a slight variation in the original process making it a far more effective, economical, and energy-efficient process suitable for large-scale manufacturing. AC-FJH coupled with DC-FJH can help solve the problem of plastic waste accumulation in landfills and the oceans and help create and sustain an entirely new industry that can change the face of technology around the world. Carbon is an element that sublimes at the temperature of 3900 K while other metals and elements easily sublime at temperatures below 2900 K proving two things, one is that high temperatures provided by high voltages are required to remove contaminants from a material that is about to be converted to graphene and it also shows that a temperature of 3000 K must be achieved to obtain best quality graphene (Advincula et al. 2021).

Economically and environmentally, flash graphene is quite a cheap process that can be easily implemented as a large-scale manufacturing process as the major components required for the production setup such as electrodes, carbon material holding tubes such as a Quartz tube, a DC and AC power supply, etc. are already readily available in the market and the only obstacle is to customize it enough so that it can produce Flash Graphene in large batches fit for customer demands (Berktas et al. 2020). It also makes use of any kind of carbon material especially waste materials such as plastic, rubber, food waste, etc. that are typically found in extremely large amounts in landfills. This makes it an excellent strategy to reduce pollution and gives access to very cheap and affordable raw materials that will easily be provided by municipal corporations that are looking for ways to safely reduce the number of waste materials in landfills (Algozeeb et al. 2020).

One of science's biggest problems is how to manage waste plastic and turn it into goods with added value. A study describes a process for upcycling solid plastic trash into a value-added product i.e. graphene which is affordable, environmentally beneficial, and capable of mass manufacturing. Incorporating nano-clays into waste plastic offers a different, cutting-edge, and affordable way to create high-quality graphene. Recently, one team of researchers showed how to produce high-quality graphene nanoflakes using polypropylene and montmorillonite clay that has undergone organic modification (Gong et al. 2014). Researchers use a two-step pyrolysis procedure, first shredding waste plastic into tiny bits, then washing, drying, and combining with bentonite, to show the production of graphene nanosheets from waste plastic. After that, pyrolysis is performed on the bentonite and waste plastic combination, which is then heated to high temperatures (about 400 °C in the presence of nanoclay, followed by 750 °C under a nitrogen environment) to produce graphene Nanosheets (Pandey et al. 2019).

Due to the fact that the precursor of the product may be obtained in large quantities both naturally and with the help of residents of the locality, the method is both beneficial to the environment and economically viable on a large scale. In addition, the nano clay employed in the entire procedure as a degrading and converting agent is easily accessible and has essentially no negative effects on the economy or nature.

11.6.5 Roll to Roll Production

Graphene is produced continuously on a roll of substrate material in a manufacturing technique known as "roll-to-roll production of graphene." With the help of this technique, graphene can be produced on a huge scale, making it appropriate for use in a variety of products, including electronics, energy storage, sensors, and more (Bhavsar et al. 2023).

Flexible transparent conductive film (FTCF) is a highly crucial element for creating technologies including organic light-emitting diodes, photovoltaic cells, and touchscreen displays. Here, a quick technical procedure has been established that effectively fabricates large-area rGO film with high quality and is ideally suited for the bulk roll-to-roll manufacture of FTCFs. The $Sn^{2+/}$ethanol which is a solution was used for the reductant for the first time, and an effective coating, drying, and washing procedure is used on a self-built lab-level manufacturing line. Roll-to-roll manufacturing of rGO films supported by PET (polyethylene terephthalate) has proven successful in producing rolls with a width of 25 cm (Ning et al. 2017).

These FTCFs are among the finest findings for rGO-based FTCFs to date, with transmittances of 82.9% and resistances of 800 or 3840 sq. in one case and 91.9% with resistances of 800 or 3840 sq. in the other. These FTCFs are able to meet the requirements of the majority of flexible photoelectronic devices since they can be bent 500 times without losing any of their features. By using just the as-prepared FTCF as the electrodes, an illuminating bracelet that is transparent and a highly flexible electrochromic device with complete functionality has been successfully shown. More intriguingly, the method put out here works quite well along with the silk-screen printing technique, making it possible to easily generate a variety of intricate designs and circuits in large quantities.

Researchers Liu, Yingjun, and colleagues describe in this paper a quick method for fabricating continuous graphene film by using Joule heating to heat up the rGO film combined with a high-throughput technique of roll-to-roll. The resulting graphene layer has an electrical conductivity of 4.21 S/m and a thermal conductivity of 128,520 W/mK, along with exceptional thermal and electrical properties and excellent flexibility. Roll-to-roll manufacturing uses considerable Joule heating, which is significantly more time, energy, and cost-effective than the traditional reduction method that uses an electrical furnace (Liu et al. 2019).

In order to produce vast areas of graphene on a wide scale using a straightforward technique, adhesive-assisted transfer is the best one. The roll-to-roll process has been coupled and metal etching technology to allow continuous manufacture of highly flexible graphene transparent conductive films. Nevertheless, the large metal consumption results in high production costs as well as major environmental pollution and contamination caused by the manufacturing process (Hong 2011).

Graphene may be transferred as intact and uniformly across a broad area without etching using target substrate-supported mechanical delamination, however, this method is still difficult to implement (Na et al. 2015).

A potential solution to the abovementioned problems is the newly discovered electrochemical bubbling delamination, in which the metal substrate is not damaged and the bubbles of hydrogen produced by the electrolysis reaction of water easily separate the graphene layer from the metal substrate (Gao et al. 2012). In particular, the technique of roll-to-roll with the PET support coated with the adhesive in conjunction with bubbling delamination offers great potential for the continuous production of graphene sheets and potentially other two-dimensional materials such as semiconducting monolayer films (Deng et al. 2015).

Researchers have described a UV-epoxy used as an adhesive to be a reliable multifunctional layer for the industrially scalable, etch-free manufacturing of transparent flexible graphene conductive films with outstanding performance (Ma et al.

2018). Due to its exceptional solvent stability, sufficient adhesive strength, and conformal contact with graphene, these UV-epoxy adhesives enable the bubble transfer of graphene layers without any cracks. Moreover, the UV-epoxy activated in situ produces an efficient p-type dopant called superacid $HSbF_6$, that significantly lowers the resistance of the sheet of graphene by about 95% while sustaining high stability. The etch-free constant synthesis of transparent conductive graphene films is superior to those made by using dopants or conventional transfer methods. The transparent conductive films are then used as the sensitive multi-touch panel in tablet computers.

11.6.6 Green Synthesis Methods

Reducing graphene oxide nanosheets involves using chemical reductants, like hydrazine. However, it is understood that hydrazine consumption is bad for both the environment and people. Physical issues include the hydrazine's strong corrosivity and explosiveness as well as the "doping" issues that arise with other elements when it is lowered in Graphene Oxide sheets. Because there are several N-containing species in hydrazined rGO, the electrical characteristics of graphene are changed.

The green rGO nanotechnology has been thoroughly studied to address the issues mentioned above by utilizing natural reducing agents such phyto-extracts, and biomolecules. In place of hydrazine, Zhao et al. (2014) used harmless antioxidants including melatonin and methanol to reduce the graphene oxide nanosheets. Numerous bioagents such as phytochemicals derived from various natural sources were used as agents to reduce GO (Yu et al. 2016), providing a safe, affordable, and environmentally friendly method for the creation of rGO nanosheets.

In a study (Shen et al. 2017), lignin and hemicellulose extracts from bagasse were employed as environmentally friendly reduction agents to create various nanomaterials like (Ag) silver and (Pt) platinum nanoparticles. Therefore, if successful, bagasse has a significant potential to be used as a reduction as well as a stabilizing agent in the synthesis of reduced Graphene Oxide. Researchers in a study suggested utilizing bagasse to create reduced graphene oxide sheets in a simple and sustainable manner. GO sheets created using a chemical exfoliation process were reduced by bagasse in an acidic state. The synthetic bagasse-based rGO was next tested to see how well it could eliminate dye effluent, with methyl blue being used as the standard dye. To the greatest extent of our information, this is the first instance in which bagasse and its subsequent use to remove dye effluent were used to create graphene (Gan et al. 2019).

The fabrication of graphene quantum dots from glucose using a straightforward and high-yielding hydrothermal method has been described (Gu et al. 2014); the resulting graphene quantum dots had a fluorescence quantum yield of 44.3% with considerable excitation-independent emission and green photoluminescence characteristics.

The production of water-soluble graphene quantum dots from cow's milk for simultaneous medication administration and cancer imaging has been shown as an efficient and environmentally friendly process. These multi-fluorescent, spherical quantum dots sized about 5 nm were produced using a one-pot, microwave-assisted heating technique (Thakur et al. 2016). It was discovered that their photoluminescence properties were considerably influenced by their heating duration and ionic strength. It shown in vitro theranostic applicability in cancer therapy due to the prepared complex's drug transport and bio-imaging properties.

Another, with starch as a precursor, graphene quantum dots have also been produced using a simple, environmentally friendly, one-pot hydrothermal reaction; the initial step in the preparation process is hydroxylation, followed by ring-closure condensation. The created graphene quantum dots were used for cell imaging because of their high photoluminescence emission, suitable hydrophilicity, and minimal cytotoxicity (Chen et al. 2018).

Latest research has demonstrated that some microbes, vitamin C, botanicals, albumin from bovines, and metal extracts might potentially reduce GO, opening up new avenues for the creation of graphene. The GO reduction method can make use of phytochemicals that can be found in many natural sources as reducing agents. Dextran was used as a multipurpose reduction agent along with a cellulose ester membrane (Kim et al. 2011) to create graphene (Thakur and Karak 2015). Successfully synthesized graphene by reducing GO using leaf extracts from *Mesua ferrea and Colocasia esculenta*. At very high temperatures, Feng et al. (2013) utilized starch powder as a suitable reducing agent in order to produce graphene. Pomegranate juice was used to make graphene (Tavakoli et al. 2015; Khojasteh et al. 2017). Graphene was synthesized using urea as the reducing agent (Khanam and Hasan 2019). Natural *Allium cepa* (onion) extract, which is widely available for a very inexpensive price practically everywhere in the globe, was effectively used to manufacture graphene. Additionally, this technique provides a different, quick, affordable, and environmental approach for producing graphene.

11.6.7 Hybrid Approaches

In order to produce graphene, different methods that combine several synthesis processes or integrate graphene with other substances or structures are referred to as hybrid approaches. These hybrid strategies seek to enhance the usefulness, scalability, and quality of graphene by combining the best aspects of other technologies (Jia et al. 2021). It also provides different ways to get over the restrictions of various fabrication techniques and modify graphene for particular uses. These technologies make it possible to combine various procedures, precursors, composites, or growing processes to give graphene its characteristics, scalability, and usefulness.

11.6.8 CVD (Chemical Vapor Deposition) Using Mechanical Techniques

This novel process brings together the benefits of mechanical exfoliation techniques with CVD, a popular technique for producing large-area graphene sheets.

In this method, a thin layer of graphene is grown using the CVD system on a metal catalyst, usually copper or nickel. In a high-temperature furnace, hydrocarbon gases like methane or ethylene are injected, where they react and break down on the metal surface. A continuous layer of graphene is created when the carbon atoms organize into a lattice. Physical techniques are used to exfoliate or transfer the graphene from the metal substrate once the graphene film has been created (Jia et al. 2021).

Thin layers of graphene may be peeled off mechanically by using sticky tape or a micro-fabricated stamp. Although this process consumes a lot of time and not suitable for mass manufacturing, it can produce high-quality graphene with perfect control over the number of layers.

Exfoliation using liquid phase is an alternative. On a metal substrate, graphene is first produced in a thin layer using CVD. The graphene-coated substrate is then submerged in a liquid medium while being exfoliated by sonication or another exfoliation method. To separate the layers and get individual graphene flakes, the manufactured graphene sheet is immersed in a solvent and subjected to sonication or other mechanical forces.

In this hybrid approach, CVD and physical techniques are used to produce large-area film made of graphene with better quality while maintaining control over the size of layers. High-quality graphene is produced on a metal substrate via CVD, and for various purposes, the graphene may then be transferred or separated using physical processes. This hybrid technique enables customization of the fabrication of graphene's characteristics and scalability, making it appropriate for a range of commercial and academic uses.

11.6.9 Electrochemical Exfoliation Followed by Chemical Vapor Deposition

Electrochemical exfoliation and chemical vapor deposition (CVD) are two separate processes that may be combined in a hybrid strategy for graphene synthesis in very large quantities. When a graphite electrode is immersed in an electrolyte solution, electrochemical exfoliation is used to remove graphene using an electric field. A method for growing thin sheets of graphene on a substrate surface is chemical vapor deposition.

The two methods are used successively in the hybrid method that combines electrochemical exfoliation and CVD to produce high-quality graphene sheets (Hofmann et al. 2015). An initial sample of graphene is obtained by the electrochemical exfoliation process. An electric field is applied between the counter electrode and a graphite electrode that is immersed in an electrolyte solution. The exfoliation of graphene sheets off the graphite electrode allows them to be collected in the electrolyte solution. After electrochemical exfoliation, the resulting graphene sheets must be transported to a suitable substrate for further processing and development. To apply the exfoliated graphene to the chosen substrate surface, a variety of transfer techniques, including wet transfer and dry transfer, can be used. The established substrate is subsequently given a CVD growth treatment. The substrate is put inside a chemical vapor deposition reactor, and the container is filled with carbon-containing gas. The carbon molecules from the source of gas settle and aggregate on the substrate covered with graphene under regulated pressure and temperature conditions, enabling the formation of new graphene layers. The growth parameters may be changed to regulate the thickness of the layers as well as the quality of the achieved graphene, including pressure, time, gas distribution rate, and temperature.

The manufacturing of graphene films with improved quality, scalability, and growth process control is made possible by this hybrid technique. While CVD offers a technique for controlled development and subsequent improvement of the graphene sheets, electrochemical exfoliation serves as to produce graphene from a supply of bulk graphite. Applications which require for large-area graphene sheets with certain characteristics may benefit via this hybrid technique.

In order to advance and scale up the high-quality synthesis and bulk production of graphene and its derivatives, various other combinations, such as CVD with Laser Irradiation, Template-Assisted Growth with CVD, Chemical Reduction of Graphene Oxide with Thermal Exfoliation, and Plasma-Enhance CVD with Liquid-Phase Exfoliation techniques, should be further studied by the researchers.

11.7 Global Graphene Market

A two-dimensional allotrope of carbon called graphene has extraordinary qualities like high strength, flexibility, thermal and electrical conductivity. Numerous sectors, including electronic devices, energy-storing, material composites, and healthcare, might be completely transformed by it. The current market demand for cutting-edge materials has led to consistent expansion in the graphene market. Recent years have seen a considerable increase in the graphene market as research and development activities have continued to investigate novel uses and production processes. Graphene has the ability to improve the functionality of current technology and open the door to the creation of new products.

In 2020, the market share of graphene was estimated to be worth USD 650 million, and by 2025, it is believed to be worth USD 1,459 million (Ren and Cheng 2014). The types, geography, and applications are the three divisions of the worldwide graphene market. Due to increased demand for graphene from a number of applications worldwide, the sector is expanding. The market is segmented into four types: CVD graphene, graphene oxide, few-layer graphene, and graphene nanoplatelets.

Graphene nanoplatelets held the highest market share in 2022 and are expected to continue doing so through 2025 (Kong et al. 2019).

Governments and commercial businesses are making significant investments in improving product characteristics and streamlining the manufacturing process. For instance, the Kerala government established a joint venture in February 2022 between (C-MET) and Tata Steel Limited at the Digital University of Kerala for the creation of a graphene innovation hub (Ren and Cheng 2014). Due to its exceptional qualities, including as strong electrical and thermal conduction, high electron mobility, permeability, and heat resistance, the electronics sector also dominates the graphene market. Because of this, the product is capable of creating technologically sophisticated equipment to meet changing consumer expectations. Demand will be driven by market growth indicators in the electronics and semiconductor sector, which is heavily utilized in smartphones and laptop computers. During the projected period, the graphene market from graphene oxide sector is anticipated to expand at a CAGR of 32% (Ciriminna et al. 2015). Because its oxidized state may be applied to any substrate, graphene's characteristics are adaptable. Since batteries, solar cells, and capacitors are in great demand across a variety of machinery and equipment, it finds widespread use in the electronics sector. In addition, it is used to improve the qualities of composite materials, which are widely applicable in the automotive sector to create lightweight vehicle components because of their exceptional qualities, including high tensile strength, conductivity, and elasticity.

Current R&D initiatives are offering a promising picture for industry expansion over the study period. Industry participants are concentrating on advancing technology, employing graphene for industrial applications, and expanding their market shares. In the long term, graphene is anticipated to replace a number of conventional materials, which will considerably enhance product operation and efficiency. Scientists from all around the world are researching different characteristics and modifying the materials basic qualities to enhance its functioning. For instance, a scientific paper on the ability to create reversible defects in graphene grain boundaries was released in 2019 by a group of scientists from India and the U.S (Ren and Cheng 2014). It opened the door for a variety of graphene uses by purposefully changing the manufacturing flaws.

11.8 Commercialization of Graphene Products and Their Applications

Although graphene technology has existed since 2006, there are presently growing applications in the nanotechnology industry. The journey from small-scale industrial production of graphene to large-scale industrial manufacturing followed a nonlinear pattern. In terms of emerging technology, graphene has an advantage due to its low cost as a beginning material. This is because graphene is primarily made from graphite, which is already mined and sold for about one cent per gram (Jia et al. 2021). However, a sheet of premium graphene these days costs about $100 per gram. Due to the difficulties in getting high-quality graphene as well as the high cost of manufacture, it is expensive. In addition, only a limited quantity of graphene is produced by exfoliation, CVD, epitaxial growth, etc.

In the past decade, investigations and reports of monolayer graphene's remarkable physical characteristics have been made. Since then, interest in graphene has grown significantly, leading to constant advancements in its commercialization. From 14 tonnes in 2009 to approximately 120 tonnes in 2015, graphene production has grown, and it is anticipated to reach about 1,800 tonnes by 2025 (Ren and Cheng 2014). Small businesses were required to test commercialized uses of this ground-breaking material in order to take advantage of its extraordinary characteristics. As a passive additive, graphene first found success in sporting goods and in anti-corrosion paints used in houses and buildings. Ever since the scientific knowledge of graphene's electrical as well as optical characteristics as an atomically thin material has grown, graphene has become a hot topic in both commercial electronic devices and optical technology. The graphene-based device industry has been steadily expanding since there are so many market segments that might be transformed or greatly enhanced by using graphene and similar materials. The majority of the excellent physical features of graphene have been examined at this point in the study, but attention still has to be paid to the creation of technology that connects scholarly research with commercial manufacture.

Intensive attempts have been undertaken to integrate graphene in items commercially sold with the quick development of graphene fabrication technology. Based on the special characteristics of graphene, several application areas have been found, including structural, environmental, electrical, and optoelectronic technologies. Significant progress has been made in these areas, and production levels are now within reach.

Energy is an important sector that necessitates the use of defect-free graphene in large quantities. The use of graphene in supercapacitors, batteries, and hydrogen-producing technologies and storing it provides a solution to solve the major issue of the ever-growing demand for global energy supply. The Li-ion battery electrodes can employ graphene as an active material. Two layers of monolayer graphene utilized for the adsorption of Li have been suggested to have a potential specific capacity that comes to 744 mAh g^{-1}, which is double compared to the present electrodes of graphite (Kong et al. 2019). Due to its electrical and thermal conductivity, graphene is primarily used as a component that is conductive in electrodes to increase the rate of capability and durability. For instance, $LiFePO_4$, $LiMn_2O_4$, and $LiCoO_2$ are popular cathode materials that are combined with graphene. The transmission path of the electric charge and heat transfer is greatly reduced by graphenes increased thermal and electrical conductivity compared to traditional modifications like carbon black. Using electrodes modified with graphene, the discharge capacity can increase up to 160% (Kucinskis et al. 2013).

The development of a commercial battery with graphene and a graphene-Si composite electrode is now underway. In order to increase charging speeds by five times more compared to conventional lithium-ion batteries, researchers have been working on battery technology that involves the use of graphene ball materials (Son et al. 2017). This piece of technology is anticipated to be incorporated into mobile cell phones or electronic devices in the very near future. One of the first industries to commercially utilize graphene technology is the supercapacitor industry. Elevator energy usage has been reported to be cut in half using graphene ultracapacitors, which are already in use across Europe. Supercapacitors with a mesoporous structure and graphene-based electrodes had extremely specific energy densities of 85.6 Wh kg^{-1} and 136 Wh kg^{-1}, respectively, at an ambient temperature of 80 °C. They also displayed current densities of 1 A g^{-1}. Without the need for a binding agent, conductive additives, or metallic charge collectors, laser-scribed graphene sheets have demonstrated energy and power densities of up to 1.36 mWh cm^{-3} and 20 W cm^{-3} respectively in an ionic liquid electrolyte solution. Additionally, to close the distance between the batteries and their electric double-layer capacitors, graphene may be paired with pseudocapacitive materials that preserve excellent energy density and power performance (Kong et al. 2019).

Ion or molecule filtering is one new use of graphene in the environment. Monolayer graphene sheets may also be used to filter water, with the water passing through clearly defined nanopores. Pore diameter and the functional groups that exist at the pores have a role in selectivity. The highest range of diffusion in the filtration of water is anticipated to surpass 1,000 l m^{-2} h^{-1} bar^{-1} and is dependent on the porous density (Surwade et al. 2015). Gas filtration and energy-efficient water desalination can both benefit from the high selective ability and permeability of graphene membranes.

Graphene's rapid field velocity offers a significant advantage over conventional semiconductors like Si. Due to a bandgap of zero, which causes a terrible on-and-off ratio, it has severely limited the performance of graphene-based transistors that are used as logic switches (Lin et al. 2010). Radiofrequency field effect transistors (FETs) based on graphene have advanced fast over the past ten years, breaking the previous record established by their Si counterpart by swiftly reaching a cut-off frequency of 100 GHz while maintaining a uniform gate length. Radio frequency electrical circuitry based on graphene is still being developed. Graphene-based field-effect transistors have thus far mostly been provided by suppliers for use in technological advancement. Instead of threatening Si entrenched position, graphene is gaining greater attention in fields of traditional electronic materials. For instance, graphene interconnects have a thermal conductivity of up to 5,000 W m^{-1} K^{-1} and can handle current densities of 108 A cm^{-2}, which is 100 times and 4 times, respectively, larger than that of copper (Yu et al. 2018).

Due to its exceptional qualities, graphene has the potential to be used as heat or current dispersers in the silicon integrated circuits' backend. Commercialization of graphene-based thermal management has been successful, particularly in the cooling of solid-state drives, which enables regular cooling without the need for large heat sinks. Transparent electrodes made of graphene have the added advantage of being flexible, which is beneficial in high-tech devices including radio-frequency (RF) connection, solar energy, display panels, and electronic devices (Kim et al. 2017a, b). Commercial electronics have utilized graphene as transparent electrodes in touch displays for mobile devices. Recent innovative applications of graphene make use of the distinct van der Waals forces that are present on its surface. The semiconductor thin films with poor van der Waals bonds with graphene formed on this material which mechanically removed from substrates. Further research revealed that the closer proximity of atomically thin graphene to its substrate may considerably enhance the crystallinity of deposited semiconductor thin films (Kim et al. 2017a, b). High-value non-silicon devices may experience considerable cost reductions resulting from graphene's novel capabilities.

Optoelectronics has also benefited from the unique characteristics of graphene. Theoretically encompassing the ultraviolet to infrared frequency range and extending into terahertz and even radio frequency, graphene's electrons interact significantly with photons because of its singular linear energy–momentum dispersion relation that makes it applicable in optoelectronics. Graphene is currently used to create a range of high-performing photonic parts, such as polarizers, optics sensors, saturable absorbers, and more (Romagnoli et al. 2018). These components may one day be incorporated into photonic circuits for future generations of wireless communication systems. To achieve device production in large quantities with unique features and roughness, appropriate techniques for production must also be improved. Graphene commercialization for optical communications is accelerating, as seen by the recent demonstration of an all-graphene optical communication connection prototype that, has an incredible data throughput of 25 Gbs^{-1}, and there is a huge rise in the reports on the creation of products (Kong et al. 2019).

Graphene and its derivatives have the potential to develop into potent materials for wastewater treatment and other forms of environmental detoxification. Many traditional techniques for environmental detoxification can be improved by their wonderful characteristics. Filtration, adsorption, and photocatalytic degradation are the three basic methods that can use graphenes capabilities for environmental applications. Other techniques, such as electrochemical separation, can be utilized to extract contaminants from water or the air using graphene (Panwar et al. 2022). Here, we demonstrated the most current advancements in the investigation of filtration, adsorption, and photocatalysis, and employing materials based on graphene and its derivatives. Reverse osmosis, forward osmosis, membrane distillation, and nanofiltration can all be done utilizing graphene membranes. The main benefit of graphene is its extremely thin layer, which directly influences how permeable it is to water. Pure graphene does not let water penetrate through it on its own. Any element is unable to penetrate through the pi orbitals that are delocalized on either side of the structure of graphene, but this issue may be resolved through the creation of cavities in the graphene sheets (Ge et al. 2020). For the purpose of filtering river water, researchers used graphene oxide that was added to the filter bed and discovered a significant difference in the performance with the filter and without the filter. The effectiveness of filtering out the iron particles was 84.5 and 70.7%, respectively, with and without the GO filter. Without a GO filter, the hardness removal efficiency was 61.34% and with one, it was 86.55%. Without graphene oxide, 82% of the arsenic was removed, while 90.9% was eliminated with the use of a graphene oxide filter. In the absence of the graphene oxide filter in the solution, the effective removal of fluoride was 58%, but it was 73.1% with the filter (Kumar et al. 2023). In the area of filtration, graphene has encouraging results. It can also be applied to the filtering of salt water in reverse osmosis (Dave et al. 2021). When scientists synthesized mixed matrix membranes using graphene nanoplatelets that were polydopamine-crosslinked, they discovered that graphene significantly increased the filtration of the salts NaCl and $MgSO_4$ to 99.85 and 99.95%, respectively (Ndlwana et al. 2020). All of these results suggest that graphene will soon play a crucial role in several filtering processes.

The use of graphene as a new material in adsorption for applications like sewage water treatment has a lot of potential. Due to its large surface area, very high porosity, functional groups, and π bonding, graphene has a high capacity for adsorption. Many academics have been working on various ways to increase the adsorption capacity of graphene since adsorption generally has considerable potential for the treatment of wastewater. For instance, polyvinyl alcohol-supported graphene oxide aerogel was created (Dai et al. 2016) for the treatment of wastewater. According to measurements made using UV–Vis spectroscopy on the produced sample, Congo red (CR) and methylene blue (MB) had an adsorption capacity of 96%. Using human nails to create graphene nanoplatelets (Nandi et al. 2021) were able to adsorb 99% of the arsenic in wastewater.

An improved oxidation method is photocatalytic degradation because it is capable of breaking down organic compounds like colors, which are challenging to break down using earlier techniques; it has become a superior way for cleaning up wastewater. As photocatalysts, semiconductors like ZnO, TiO_2, ZnS, and MnO_2 are frequently utilized. They exhibit band gaps between 3.0 and 3.5 (Gopinath et al. 2020). They still show certain disadvantages, such as a poor rate of visible ray absorption and a rapid recombination of the electron and holes produced during recombination. Graphene may prove to be a practical answer in this situation. The rapid electron transit, high electron uptake (due to π conjugation), high surface area, adsorption, mechanical strength, chemical, and thermal durability, as well as other remarkable features of graphene, make it trusted materials with a semiconductor to enhance their capabilities. Semiconductors may be held very securely in graphene-porous frameworks with a great number of sites for trapping. Due to its numerous capabilities and qualities, all these instances demonstrate how graphene may significantly improve and offers a hopeful future for environmental restoration.

11.9 Conclusion

The broad and prospective uses of these materials have been greatly aided by the improvements in methods for the mass manufacture and commercialization of graphene and its derivatives. In order to overcome the difficulties involved with the large-scale manufacture of graphene and its various derivatives, scientists and researchers have made significant progress in the development of scalable and affordable manufacturing procedures. The incorporation of graphene into flexible and wearable electronics has been assisted by cutting-edge techniques including roll-to-roll production, green synthesis methods, flash graphene synthesis, and hybrid approaches, bringing up new opportunities for commercial applications. The commercialization of graphene and its derivatives has also been greatly aided by the development of scalable production techniques and the optimization of manufacturing processes. The yield and cost of producing graphene have been improved by researchers, making it much more commercially feasible for incorporation in big applications. Due to this development, graphene is now being used in a variety of sectors, including electronics, biomedicine, energy storage, composites, sensors, and electronics.

In order to overcome several obstacles to the mass manufacturing and commercialization of graphene and its derivatives, more study and development are still required. These difficulties include creating effective and environmentally friendly synthesis procedures, enhancing the repeatability and quality control of large-scale synthesis, and investigating novel techniques for the functionalization of graphene to obtain desired features. These materials are anticipated to continue transforming a variety of industries with further study and innovation, allowing the creation of cutting-edge technologies and applications that take use of graphene and its derivatives special features.

Acknowledgements Kalpesh V. Bhavsar; Uday P. Jagtap; Hardik S. Churi and Kevin Jose gratefully acknowledges the support provided by National center for Nanosciences and Nanotechnology, University of Mumbai, India, and Sonopant Dandekar Shikshan Mandali's (S.D.S.M) College, Palghar, India.

References

Abhervé, A., Manas-Valero, S., Clemente-León, M., Coronado, E.: Graphene related magnetic materials: micromechanical exfoliation of 2D layered magnets based on bimetallic anilate complexes with inserted [Fe III (acac 2-trien)]+ and [Fe III (sal 2-trien)]+ molecules. Chem. Sci. **6**(8), 4665–4673 (2015)

Advincula, P.A., Luong, D.X., Chen, W., Raghuraman, S., Shahsavari, R., Tour, J.M.: Flash graphene from rubber waste. Carbon **178**, 649–656 (2021)

Algozeeb, W.A., Savas, P.E., Luong, D. X., Chen, W., Kittrell, C., Bhat, M., et al.: Flash graphene from plastic waste. ACS Nano **14**(11), 15595-15604 (2020)

Ameen, S., Akhtar, M.S., & Shin, H.S. (eds.).: Graphene Production and Application. BoD–Books on Demand (2020)

Bae, S., Kim, H., Lee, Y., Xu, X., Park, J.S., Zheng, Y., et al.: Roll-to-roll production of 30-inch graphene films for transparent electrodes. Nature Nanotechnol. **5**(8), 574-578 (2010)

Berktas, I., Hezarkhani, M., Haghighi Poudeh, L., Saner Okan, B.: Recent developments in the synthesis of graphene and graphene-like structures from waste sources by recycling and upcycling technologies: a review. Graphene Technol. **5**, 59–73 (2020)

Bhavsar, K.V., Jagtap, U.P.: Enzymes and their applications in medical sciences. Updates Intern Sci**2021**, 265 (2021)

Bhavsar, K. V., Churi, H.S., Jagtap, U.P.: Aptamer-based point-of-care diagnostic devices for infectious diseases. In: Aptasensors for Point-of-Care Diagnostics, pp. 99–126. CRC Press (2023)

Bointon, T.H., Barnes, M.D., Russo, S., Craciun, M.F.: High quality monolayer graphene synthesized by resistive heating cold wall chemical vapor deposition. Adv. Mater. **27**(28), 4200–4206 (2015)

Bourlinos, A.B., Georgakilas, V., Zboril, R., Steriotis, T.A., Stubos, A.K., Trapalis, C.: Aqueous-phase exfoliation of graphite in the presence of polyvinylpyrrolidone for the production of water-soluble graphenes. Solid State Commun. **149**(47–48), 2172–2176 (2009)

Boyd, D.A., Lin, W.H., Hsu, C.C., Teague, M.L., Chen, C.C., Lo, Y.Y., et al.: Single-step deposition of high-mobility graphene at reduced temperatures. Nature Commun. **6**(1), 662 (2015)

Brückner, F.U., Schwerdtfeger, K.: Single crystal growth with the Czochralski method involving rotational electromagnetic stirring of the melt. J. Cryst. Growthcryst. Growth **139**(3–4), 351–356 (1994)

Chen, W., Li, D., Tian, L., Xiang, W., Wang, T., Hu, W., et al.: Synthesis of graphene quantum dots from natural polymer starch for cell imaging. Green Chem. **20**(19), 4438-4442 (2018)

Churi, H. S., Dave, S.: Strategic Synthesis of Diagnostic Novel Materials Against Infectious Diseases, pp. 209–233 (2023). https://doi.org/10.1002/9783527837946.ch10

Ciriminna, R., Zhang, N., Yang, M.Q., Meneguzzo, F., Xu, Y.J., Pagliaro, M.: Commercialization of graphene-based technologies: a critical insight. Chem. Commun. **51**(33), 7090–7095 (2015)

Coleman, J.N.: Liquid exfoliation of defect-free graphene. Acc. Chem. Res. **46**(1), 14–22 (2013)

Dai, J., Huang, T., Tian, S.Q., Xiao, Y.J., Yang, J.H., Zhang, N., et al.: High structure stability and outstanding adsorption performance of graphene oxide aerogel supported by polyvinyl alcohol for waste water treatment. Mater. Des. **107**, 187–197 (2016)

Dave, S., Churi, H., Litoria, P., David, P., Das, J.: Biofilms, filtration, microbial kinetics and mechanism of degradation: a revolutionary approach (2021). https://doi.org/10.1016/B978-0-12-823804-2.00018-5

Deng, B., Liu, Z., Peng, H.: Toward mass production of CVD graphene films. Adv. Mater. **31**(9), 1800996 (2019)

Deng, B., Hsu, P. C., Chen, G., Chandrashekar, B. N., Liao, L., Ayitimuda, Z., et al.: Roll-to-roll encapsulation of metal nanowires between graphene and plastic substrate for high-performance flexible transparent electrodes. Nano Lett. **15**(6), 4206-4213 (2015)

Dave, S., Churi, H., Litoria, P., David, P., Das, J.: Biofilms, filtration, microbial kinetics and mechanism of degradation: a revolutionary approach (2021).

Dreyer, D.R., Ruoff, R.S., Bielawski, C.W.: From conception to realization: an historial account of graphene and some perspectives for its future. Angew. Chem. Int. Ed.. Chem. Int. Ed. **49**(49), 9336–9344 (2010)

Edwards, R.S., Coleman, K.S.: Graphene synthesis: relationship to applications. Nanoscale **5**(1), 38–51 (2013)

Feng, Y., Feng, N., Du, G.: A green reduction of graphene oxide via starch-based materials. RSC Adv. **3**(44), 21466–21474 (2013)

Gamo, Y., Nagashima, A., Wakabayashi, M., Terai, M., Oshima, C.: Atomic structure of monolayer graphite formed on Ni (111). Hyomen Kagaku **17**(12), 745–749 (1996)

Gamo, Y., Nagashima, A., Wakabayashi, M., Terai, M., Oshima, C.J.S.S.: Atomic structure of monolayer graphite formed on Ni (111). Surf. Sci. **374**(1–3), 61–64 (1997)

Gan, L., Li, B., Chen, Y., Yu, B., Chen, Z.: Green synthesis of reduced graphene oxide using bagasse and its application in dye removal: a waste-to-resource supply chain. Chemosphere **219**, 148–154 (2019)

Gao, L., Ren, W., Xu, H., Jin, L., Wang, Z., Ma, T., et al.: Repeated growth and bubbling transfer of graphene with millimetre-size single-crystal grains using platinum. Nat. Commun. **3**(1), 699 (20120

Ge, X., Su, G., Che, W., Yang, J., Zhou, X., Wang, Z., et al.: Atomic filtration by graphene oxide membranes to access atomically dispersed single atom catalysts. ACS Catal. **10**(18), 10468-10475 (2020)

Geim, A.K., MacDonald, A.H.: Graphene: exploring carbon flatland. Phys. Today **60**(8), 35–41 (2007)

Gong, J., Liu, J., Wen, X., Jiang, Z., Chen, X., Mijowska, E., Tang, T.: Upcycling waste polypropylene into graphene flakes on organically modified montmorillonite. Ind. Eng. Chem. Res. **53**(11), 4173–4181 (2014)

Gopinath, K.P., Madhav, N.V., Krishnan, A., Malolan, R., Rangarajan, G.: Present applications of titanium dioxide for the photocatalytic removal of pollutants from water: a review. J. Environ. Manag. **270**, 110906 (2020)

Gu, J., Hu, M.J., Guo, Q.Q., Ding, Z.F., Sun, X.L., Yang, J.: High-yield synthesis of graphene quantum dots with strong green photoluminescence. RSC Adv. **4**(91), 50141–50144 (2014)

Gutzwiller, M.C.: Effect of correlation on the ferromagnetism of transition metals. Phys. Rev. Lett. **10**(5), 159 (1963)

Hofmann, M., Chiang, W.Y., Nguyễn, T.D., Hsieh, Y.P.: Controlling the properties of graphene produced by electrochemical exfoliation. Nanotechnology **26**(33), 335607 (2015)

Hong, B.H.: Roll-to-roll production of 30-inch graphene films for transparent electrodes. In APS March Meeting Abstracts, pp. W2–002 (2011)

Hämäläinen, S.K., Boneschanscher, M.P., Jacobse, P.H., Swart, I., Pussi, K., Moritz, W., et al.: Structure and local variations of the graphene moiré on Ir (111). Phys. Rev. B **88**(20), 201406 (20130

Jia, K., Zhang, J., Zhu, Y., Sun, L., Lin, L., Liu, Z.: Toward the commercialization of chemical vapor deposition graphene films. Appl. Phys. Rev. **8**(4) (2021)

Jobst, J., Waldmann, D., Speck, F., Hirner, R., Maude, D.K., Seyller, T., Weber, H.B.: How graphene-like is epitaxial graphene? quantum oscillations and quantum hall effect (2009). arXiv:0908.1900

Kamali, A.R., Fray, D.J.: Molten salt corrosion of graphite as a possible way to make carbon nanostructures. Carbon **56**, 121–131 (2013)

Khan, U., O'Neill, A., Lotya, M., De, S., Coleman, J.N.: High-concentration solvent exfoliation of graphene. Small **6**(7), 864-871 (2010)

Khanam, P.N., Hasan, A.: Biosynthesis and characterization of graphene by using non-toxic reducing agent from Allium Cepa extract: anti-bacterial properties. Int. J. Biol. Macromol. **126**, 151–158 (2019)

Khojasteh, H., Salavati-Niasari, M., Safajou, H., Safardoust-Hojaghan, H.: Facile reduction of graphene using urea in solid phase and surface modification by N-doped graphene quantum dots for adsorption of organic dyes. Diam. Relat. Mater. Mater. **79**, 133–144 (2017)

Kim, Y.K., Kim, M.H., Min, D.H.: Biocompatible reduced graphene oxide prepared by using dextran as a multifunctional reducing agent. Chem. Commun. **47**(11), 3195–3197 (2011)

Kim, H.W., Ko, W., Ku, J., Kim, Y., Park, S., Hwang, S.: Evolution of graphene growth on Pt (111): From carbon clusters to nanoislands. J. Phys. Chem. C **121**(45), 25074–25078 (2017a)

Kim, Y., Cruz, S.S., Lee, K., Alawode, B.O., Choi, C., Song, Y., et al.: Remote epitaxy through graphene enables two-dimensional material-based layer transfer. Nature **544**(7650), 340–343 (2017)

Kobayashi, T., Bando, M., Kimura, N., Shimizu, K., Kadono, K., Umezu, N., et al.: Production of a 100-m-long high-quality graphene transparent conductive film by roll-to-roll chemical vapor deposition and transfer process. Appl. Phys. Lett. **102**(2) (2013)

Kong, W., Kum, H., Bae, S.H., Shim, J., Kim, H., Kong, L., et al.: Path towards graphene commercialization from lab to market. Nat. Nanotechnol. **14**(10), 927–938 (2019)

Kucinskis, G., Bajars, G., Kleperis, J.: Graphene in lithium ion battery cathode materials: a review. J. Power. Sources **240**, 66–79 (2013)

Kumar, S., Himanshi, Prakash, J., Verma, A., Suman, Jasrotia, R., et al.: A Review on properties and environmental applications of graphene and its derivative-based composites. Catalysts **13**(1), 111 (2023)

Lee, J.H., Lee, E.K., Joo, W.J., Jang, Y., Kim, B.S., Lim, J.Y., et al.: Wafer-scale growth of single-crystal monolayer graphene on reusable hydrogen-terminated germanium. Science **344**(6181), 286–289 (2014)

Lee, E., Ryu, J., Jang, J.: Fabrication of graphene quantum dots via size-selective precipitation and their application in upconversion-based DSSCs. Chem. Commun. **49**(85), 9995–9997 (2013)

Li, M., Liu, D., Wei, D., Song, X., Wei, D., Wee, A.T.S.: Controllable synthesis of graphene by plasma-enhanced chemical vapor deposition and its related applications. Adv. Sci. **3**(11), 1600003 (2016)

Li, X., Cai, W., An, J., Kim, S., Nah, J., Yang, D., et al.: Large-area synthesis of high-quality and uniform graphene films on copper foils. Science **324**(5932), 1312–1314 (2009)

Lin, Y.M., Dimitrakopoulos, C., Jenkins, K.A., Farmer, D.B., Chiu, H.Y., Grill, A., Avouris, P.: 100-GHz transistors from wafer-scale epitaxial graphene. Science **327**(5966), 662–662 (2010)

Liu, Y., Li, P., Wang, F., Fang, W., Xu, Z., Gao, W., Gao, C.: Rapid roll-to-roll production of graphene films using intensive joule heating. Carbon **155**, 462–468 (2019)

Lotya, M., Hernandez, Y., King, P.J., Smith, R.J., Nicolosi, V., Karlsson, L.S., et al.: Liquid phase production of graphene by exfoliation of graphite in surfactant/water solutions. J. Am. Chem. Soc. **131**(10), 3611–3620 (2009)

Lowe, S.E., Zhong, Y.L.: Challenges of Industrial Scale Graphene Oxide Production. In: Graphene Oxide: Fundamentals and Applications, pp. 410–431 (2016)

Lu, J., Yang, J.X., Wang, J., Lim, A., Wang, S., Loh, K.P.: One-pot synthesis of fluorescent carbon nanoribbons, nanoparticles, and graphene by the exfoliation of graphite in ionic liquids. ACS Nano **3**(8), 2367–2375 (2009)

Ma, L.P., Dong, S., Chen, M., Ma, W., Sun, D., Gao, Y., et al.: UV-epoxy-enabled simultaneous intact transfer and highly efficient doping for roll-to-roll production of high-performance graphene films. ACS Appl. Mater. Interfaces **10**(47), 40756–40763 (2018)

Mannix, A.J., Ye, A., Sung, S.H., Ray, A., Mujid, F., Park, C., et al.: Robotic four-dimensional pixel assembly of van der Waals solids. Nat. Nanotechnol. **17**(4), 361–366 (2022)

Mattevi, C., Kim, H., Chhowalla, M.: A review of chemical vapour deposition of graphene on copper. J. Mater. Chem. **21**(10), 3324–3334 (2011)

Mindivan, F.: The synthesis and charecterization of graphene oxide (GO) and reduced graphene oxide (rGO). Mach. Technol. Mater. **10**(6), 32–35 (2016)

Na, S.R., Suk, J.W., Tao, L., Akinwande, D., Ruoff, R.S., Huang, R., Liechti, K.M.: Selective mechanical transfer of graphene from seed copper foil using rate effects. ACS Nano **9**(2), 1325–1335 (2015)

Nandamuri, G., Roumimov, S., Solanki, R.: Remote plasma assisted growth of graphene films. Appl. Phys. Lett. **96**(15) (2010)

Nandi, D., Ghosh, S.K., Ghosh, A., Siengchin, S., Roy, A., Gupta, K., et al.: Arsenic removal from water by graphene nanoplatelets prepared from nail waste: a physicochemical study of adsorption based on process optimization, kinetics, isotherm and thermodynamics. Environ. Nanotechnol. Monit. Manag. **16**, 100564 (2021)

Ndlwana, L., Motsa, M.M., Mamba, B.B.: A unique method for dopamine-cross-linked graphene nanoplatelets within polyethersulfone membranes (GNP-pDA/PES) for enhanced mechanochemical resistance during NF and RO desalination. Eur. Polymer J. **136**, 109889 (2020)

Ning, J., Hao, L., Jin, M., Qiu, X., Shen, Y., Liang, J., et al.: A facile reduction method for roll-to-roll production of high performance graphene-based transparent conductive films. Adv. Mater. **29**(9), 1605028 (2017)

Novoselov, K.S., Jiang, D., Schedin, F., Booth, T.J., Khotkevich, V.V., Morozov, S.V., Geim, A.K.: Two-dimensional atomic crystals. Proc. Natl. Acad. Sci. **102**(30), 10451–10453 (2005)

Novoselov, K.S., Fal′ko, V.I., Colombo, L., Gellert, P.R., Schwab, M.G., Kim, K.: A roadmap for graphene. Nature **490**(7419), 192–200 (2012)

Novoselov, K.S., Geim, A.K., Morozov, S.V., Jiang, D.E., Zhang, Y., Dubonos, S.V., et al.: Electric field effect in atomically thin carbon films. Science **306**(5696), 666–669 (2004)

Ohta, T., Bostwick, A., McChesney, J.L., Seyller, T., Horn, K., Rotenberg, E.: Interlayer interaction and electronic screening in multilayer graphene investigated with angle-resolved photoemission spectroscopy. Phys. Rev. Lett. **98**(20), 206802 (2007)

Pandey, S., Karakoti, M., Dhali, S., Karki, N., SanthiBhushan, B., Tewari, C., et al.: Bulk synthesis of graphene nanosheets from plastic waste: an invincible method of solid waste management for better tomorrow. Waste Manag. **88**, 48–55 (2019)

Panwar, R., Churi, H., Dave, S.: Point-of-care electrochemical biosensors using CRISPR/Cas for RNA analysis (2022). https://doi.org/10.1016/B978-0-323-88464-8.00003-8

Park, S., Ruoff, R.S.: Chemical methods for the production of graphenes. Nat. Nanotechnol. **4**(4), 217–224 (2009)

Parvez, K., Li, R., Puniredd, S.R., Hernandez, Y., Hinkel, F., Wang, S., et al.: Electrochemically exfoliated graphene as solution-processable, highly conductive electrodes for organic electronics. ACS Nano **7**(4), 3598–3606 (2013)

Paton, K.R., Varrla, E., Backes, C., Smith, R.J., Khan, U., O'Neill, A., et al.: Scalable production of large quantities of defect-free few-layer graphene by shear exfoliation in liquids. Nat. Mater. **13**(6), 624–630 (2014)

Peng, K.J., Wu, C.L., Lin, Y.H., Liu, Y.J., Tsai, D.P., Pai, Y.H., Lin, G.R.: Hydrogen-free PECVD growth of few-layer graphene on an ultra-thin nickel film at the threshold dissolution temperature. J. Mater. Chem. C **1**(24), 3862–3870 (2013)

Peng, Q., Dearden, A. K., Crean, J., Han, L., Liu, S., Wen, X., De, S.: New materials graphyne, graphdiyne, graphone, and graphane: review of properties, synthesis, and application in nanotechnology. In: Nanotechnology, Science and Applications, pp. 1–29 (2014)

Qi, Y., Deng, B., Guo, X., Chen, S., Gao, J., Li, T., et al.: Switching vertical to horizontal graphene growth using faraday cage-assisted PECVD approach for high-performance transparent heating device. Adv. Mater. **30**(8), 1704839 (2018)

Ray, S.C.: Application and uses of graphene oxide and reduced graphene oxide. Appl. Graphene Graphene-Oxide Based Nanomater. **6**(8), 39–55 (2015)

Ren, W., Cheng, H.M.: The global growth of graphene. Nat. Nanotechnol. **9**(10), 726–730 (2014)

Romagnoli, M., Sorianello, V., Midrio, M., Koppens, F.H., Huyghebaert, C., Neumaier, D., et al.: Graphene-based integrated photonics for next-generation datacom and telecom. Nat. Rev. Mater. **3**(10), 392–414 (2018)

Rudrapati, R.: Graphene: Fabrication Methods, Properties, and Applications in Modern Industries, vol. 1. IntechOpen, London, UK (2020)

Ruths, M., Israelachvili, J.N.: Surface forces and nanorheology of molecularly thin films. In: Nanotribology and Nanomechanics II: Nanotribology, Biomimetics, and Industrial Applications, pp. 107–202 (2011)

Ryu, J., Kim, Y., Won, D., Kim, N., Park, J.S., Lee, E.K., et al.: Fast synthesis of high-performance graphene films by hydrogen-free rapid thermal chemical vapor deposition. ACS Nano **8**(1), 950–956 (2014)

Shen, Z., Han, G., Wang, X., Luo, J., Sun, R.: An ultra-light antibacterial bagasse–AgNP aerogel. J. Mater. Chem. B **5**(6), 1155–1158 (2017)

Son, I.H., Park, J.H., Park, S., Park, K., Han, S., Shin, J., et al.: Graphene balls for lithium rechargeable batteries with fast charging and high volumetric energy densities. Nat. Commun. **8**(1), 1561 (2017)

Srivastava, A., Galande, C., Ci, L., Song, L., Rai, C., Jariwala, D., et al.: Novel liquid precursor-based facile synthesis of large-area continuous, single, and few-layer graphene films. Chem. Mater. **22**(11), 3457–3461 (2010)

Stanford, M.G., Bets, K.V., Luong, D.X., Advincula, P.A., Chen, W., Li, J.T., et al.: Flash graphene morphologies. ACS Nano **14**(10), 13691–13699 (2020)

Surwade, S.P., Smirnov, S.N., Vlassiouk, I.V., Unocic, R.R., Veith, G.M., Dai, S., Mahurin, S.M.: Water desalination using nanoporous single-layer graphene. Nat. Nanotechnol. **10**(5), 459–464 (2015)

Sutter, P.: How silicon leaves the scene. Nat. Mater. **8**(3), 171–172 (2009)

Sutter, P.W., Flege, J.I., Sutter, E.A.: Epitaxial graphene on ruthenium. Nat. Mater. **7**(5), 406–411 (2008)

Tavakoli, F., Salavati-Niasari, M., Mohandes, F.: Green synthesis and characterization of graphene nanosheets. Mater. Res. Bull. **63**, 51–57 (2015)

Thakur, S., Karak, N.: Alternative methods and nature-based reagents for the reduction of graphene oxide: a review. Carbon **94**, 224–242 (2015)

Thakur, M., Mewada, A., Pandey, S., Bhori, M., Singh, K., Sharon, M., Sharon, M.: Milk-derived multi-fluorescent graphene quantum dot-based cancer theranostic system. Mater. Sci. Eng. C **67**, 468–477 (2016)

Tiwari, S.K., Kumar, V., Huczko, A., Oraon, R., Adhikari, A.D., Nayak, G.C.: Magical allotropes of carbon: prospects and applications. Crit. Rev. Solid State Mater. Sci. **41**(4), 257–317 (2016)

Varykhalov, A., Rader, O.: Graphene grown on Co (0001) films and islands: Electronic structure and its precise magnetization dependence. Phys. Rev. B **80**(3), 035437 (2009)

Verdejo, R., Bernal, M.M., Romasanta, L.J., Lopez-Manchado, M.A.: Graphene filled polymer nanocomposites. J. Mater. Chem. **21**(10), 3301–3310 (2011)

Warner, J.H., Schaffel, F., Rummeli, M., Bachmatiuk, A.: Graphene: fundamentals and emergent applications. Newnes (2012)

Wei, J., Atif, R., Vo, T., Inam, F.: Graphene nanoplatelets in epoxy system: dispersion, reaggregation, and mechanical properties of nanocomposites. J. Nanomater. **16**(1), 374–374 (2016)

Woltornist, S.J., Oyer, A.J., Carrillo, J.M.Y., Dobrynin, A.V., Adamson, D.H.: Conductive thin films of pristine graphene by solvent interface trapping. ACS Nano **7**(8), 7062–7066 (2013)

Wyss, K.M., Beckham, J.L., Chen, W., Luong, D.X., Hundi, P., Raghuraman, S., et al.: Converting plastic waste pyrolysis ash into flash graphene. Carbon **174**, 430–438 (2021)

Yamada, Y., Yasuda, H., Murota, K., Nakamura, M., Sodesawa, T., Sato, S.: Analysis of heat-treated graphite oxide by X-ray photoelectron spectroscopy. J. Mater. Sci. **48**, 8171–8198 (2013)

Yu, H., Ye, D., Butburee, T., Wang, L., Dargusch, M.: Green synthesis of porous three-dimensional nitrogen-doped graphene foam for electrochemical applications. ACS Appl. Mater. Interfaces **8**(4), 2505–2510 (2016)

Yu, C., He, Z., Song, X., Liu, Q., Gao, L., Yao, B., et al.: High-frequency flexible graphene field-effect transistors with short gate length of 50 nm and record extrinsic cut-off frequency. Physica Status Solidi (RRL)–Rapid Res. Lett. **12**(5), 1700435 (2018)

Zhao, J., Wang, Z., White, J.C., Xing, B.: Graphene in the aquatic environment: adsorption, dispersion, toxicity and transformation. Environ. Sci. Technol. **48**(17), 9995–10009 (2014)

Zurutuza, A., Marinelli, C.: Challenges and opportunities in graphene commercialization. Nat. Nanotechnol. **9**(10), 730–734 (2014)

Chapter 12
Electrochemical Exfoliation Techniques for Graphene Synthesis: Challenges and Future Opportunities

Fadeela Chundekatt Ummer and Nandakumar Kalarikkal

Abstract The synthesis of graphene derivatives through electrochemical intercalation and exfoliation is significant because of its green nature and cost-effectiveness. However, the tradeoff between the yield and properties of synthesized graphene derivatives is necessary to be considered when choosing electrochemical exfoliation as a synthesis method. Since the exfoliation takes place in an electrolyte solution, there is a possibility of side chemical reactions that adversely affect the quality of the graphene sheets in terms of defect densities, carbon/oxygen ratios, lateral size, and layer numbers, which are crucial for many of the promising applications of the material. Various process parameters significantly determine the quantity and quality of graphene derivatives produced by electrochemical exfoliation. In this chapter, we are addressing the challenges and future research directions in the electrochemical exfoliation synthesis of graphene and its functional variants for the desired application. The chapter is organized in such a way that the readers can understand the essentials of electrochemical exfoliation -its importance and features -to get an insight into the challenges to be addressed when going for electrochemical exfoliation synthesis.

Keywords Electrochemical exfoliation · Graphene · Graphene derivatives · Raman spectroscopy · Challenges · Synthesis

F. C. Ummer · N. Kalarikkal (✉)
International and Inter University Centre for Nanoscience and Nanotechnology (IIUCNN), Mahatma Gandhi University, Kottayam 686 560, Kerala, India
e-mail: nkkalarikkal@mgu.ac.in

N. Kalarikkal
School of Pure and Applied Physics, Mahatma Gandhi University, Kottayam 686 560, Kerala, India

R. Khan et al. (eds.), *Electrochemical Exfoliation of Graphene and its Derivatives*, Engineering Materials, https://doi.org/10.1007/978-981-97-2128-3_12

12.1 Introduction

The discovery of the exciting properties of the super thinnest two-dimensional sp^2 network of carbon atoms, aka graphene, created a significant research momentum in the scalable production of this material and its derivatives. Since then, researchers worldwide have tried various techniques for the scalable production of high-quality graphene sheets. The fabrication techniques of graphene are broadly classified into two categories bottom-up and top-bottom methods (Gutiérrez-Cruz et al. 2022; Lim et al. 2018; Singh et al. 2016). In the bottom-up approach, the sp^2 carbon sheets are grown from small carbon-rich molecules called precursors (Lee et al. 2016; Moreno et al. 2018; Narita et al. 2015; Son and Ham 2017). In contrast, in the top-bottom approach, the various allotropes of carbon are laterally cleaved to produce graphene sheets. The lateral cleaving is achieved by breaking the Van der Waals forces between adjacent layers of graphite. Techniques such as mechanical exfoliation, chemical exfoliation, and electrochemical exfoliation fall into the top-down approach (Yi and Shen 2015; Saleem et al. 2018; Gebreegziabher et al. 2019; Liu and Aziz 2022; Toh et al. 2014). Even though there are plenty of synthesis techniques, scalable production always faces a tradeoff with the quality of graphene sheets and vice versa. The term quality here implies 'free of defects and free of other atoms than carbon'. Among the top-down methods listed above, only mechanical exfoliation produces superior-quality graphene sheets (Whitener and Sheehan 2014). The quality of sheets significantly affects the properties of graphene, as demonstrated by Novoselov and Geim (2004), Geim and Novoselov (2007). However, mechanical exfoliation is not convenient for the scalable production of graphene. Graphene production through chemical exfoliation is achieved through the harsh oxidation of graphite. It facilitates the scalable production of graphene oxide (GO), which can be subsequently reduced to form reduced graphene oxide (rGO) with a higher carbon-to-oxygen ratio. However, chemical exfoliation suffers drawbacks such as the use of strong acids and oxidizing agents, the evolution of toxic gases such as nitric oxides, and the formation of graphene sheets with defects and other functionalities (Marcano et al. 2010; Alkhouzaam et al. 2020). Electrochemical exfoliation allows the exfoliation of graphite electrodes to produce graphene sheets in the electrolytic solution. In a typical electrochemical exfoliation, graphite is used as one of the electrodes, and the exfoliation takes place when the potential is applied. A schematic of typical electrochemical exfoliation is given in Fig. 12.1. The yield, efficiency, and quality of graphene sheets produced changes with various electrochemical experiment parameters such as voltage, electrolyte, and type of electrode. Graphite forms, such as graphite rods, flakes, and HOPG (Highly Oriented Pyrolytic Graphite), are the commonly used graphite electrodes for electro chemical exfoliation synthesis. Electrochemical exfoliation occurs through the intercalation of anions present in the electrolyte between the graphitic layers in the presence of electricity. The intercalation can break the weak van der Waals force between the adjacent graphitic layers, subsequently facilitating the exfoliation of graphitic layers. Hence, Electrochemical exfoliation can be described as electrochemical intercalation and exfoliation. The

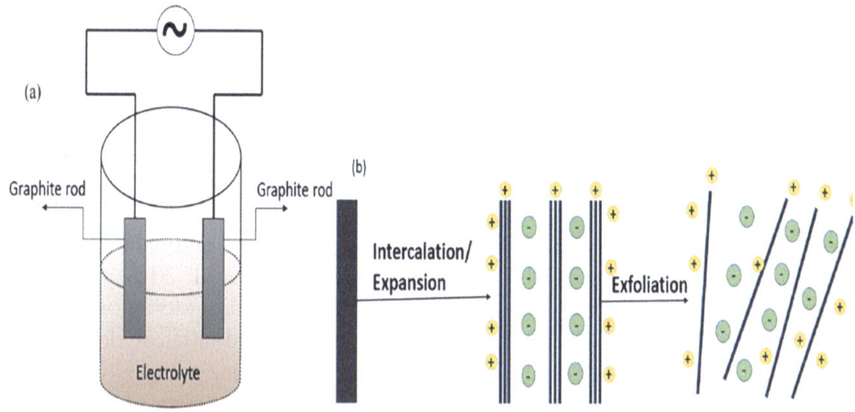

Fig. 12.1 **a** Schematic of a simple electrochemical setup **b** Schematic of graphite intercalation and exfoliation

electrochemical exfoliation is classified as cathodic or anodic exfoliation according to the bias applied. In the cathodic exfoliation, the positively charged ions intercalated into the adjacent graphitic layers because of the negative bias at the electrode. In anodic exfoliation, negatively charged ionic species are intercalated into the graphitic layers due to the positive bias at the electrode. Graphite in different structural forms is used in electrochemical exfoliation to synthesize graphene derivatives such as graphite powders, foils, rods, flakes, and plates. At the applied potential, the ions or electrolyte molecules can intercalate into the graphitic layers by breaking the Van der Waals force existing between the layers leading to the structural expansion of graphite(Liu and Aziz 2022; Abdelkader et al. 2015; Aghamohammadi et al. 2020; Wu et al. 2022).

Cao et al. showed a two-step electrochemical intercalation and oxidation for the synthesis of graphene oxide (Cao et al. 2017). Even though electrochemical exfoliation is a simple method for producing graphene and graphene derivatives, it faces several challenges that affect the quality, yield, and efficiency of electrochemical exfoliation. In this chapter, we are trying to give a brief idea to the readers about the challenging factors affecting the quality, yield, and efficiency of electrochemical exfoliation.

12.2 Electrochemically Synthesized Graphene Derivatives

The primary focus of any synthesis method is to produce graphene with minimum defects and few layers to harness all the exciting properties of graphene. Like every other wet chemical method, electrochemical exfoliation is also susceptible to the formation of defects in the graphene basal plane by incorporating oxygen and other chemical functional entities due to the side reaction that happens during electrochemical synthesis. Hence electrochemical exfoliation of graphite electrodes is generally used to synthesize graphene oxide with different C/O ratios. However, graphene sheets with a high carbon-to-oxygen ratio can be synthesized with adequate control over the various synthesis parameters (discussed in Section 12.5). Liu et al. reviewed various research articles on the electrochemical synthesis of graphene and listed the C/O ratio of the products (Liu et al. 2019). In an attempt, Munuera et al. obtained graphene with a C/O ratio of 33.3–50 up on anodic exfoliation of graphite foil with table salt as an electrolyte at 10V DC potential (Munuera et al. 2018). Munuera et al. obtained graphene oxide with a C/O ratio of 50 with anodic exfoliation of high-purity graphene foil with 10 V DC potential using sulfonate and sulfate hydrocarbons (Munuera et al. 2016). Ejigu et al. in a study, prepared graphene with a C/O ratio of 36 with the anodic exfoliation of graphene foil at 20 V in the presence of CO^{2+} ions, which act as antioxidants of graphene sheets (Ejigu et al. 2018). Chen et al. obtained graphene with a C/O ratio 26.17 with exfoliation of graphite powder with 20V potential using sulfuric acid electrolyte with melamine additives (Chen et al. 2015). Thus, it is evident that a proper selection of parameters such as exfoliation potential, electrolytes, and the graphite electrode is necessary to get graphene with a higher C/O ratio. Graphene oxide with higher oxygen content shows better dispersibility in different solvents and hence has utilities in the scalable synthesis of various graphene-based materials. Use of oxidative electrolytes and pretreated graphite electrodes allows the formation of graphene oxide with more oxygen content. The pretreatment of electrodes means the treatment of graphite electrodes with H_2SO_4 solution in the presence or absence of potential before electrochemical exfoliation. The potential applied, pretreatment time, and the concentration of acid used affect the oxidation degree and the synthesis yield (Liu et al. 2019). The use of oxidative electrolytes such as H_2SO_4 and $HClO_4$ are used in electrochemical exfoliation synthesis as it facilitates the intercalation of anions such as sulfate, chlorate, and nitrate into the graphitic layers under applied potentials. The decomposition of these anions under the reaction conditions generates radicals such as $O^{\cdot}$ and $OH^{\cdot}$ radicals will subsequently introduce oxygen functional groups in graphene oxide (Liu et al. 2019). Proper control over the pretreatment and the concentration of oxidative electrolytes are required to get the graphene oxide with desired properties for the applications.

The electrochemical technique can also be used for the synthesis of luminescent graphene oxide quantum dots (Ahirwar et al. 2017; Li et al. 2011, 2017; Kumar et al. 2021; Tan et al. 2015; Kalita et al. 2020). Some of the works in this field are discussed below. He et al. prepared graphene quantum dots with tailored emission properties by varying the water content of the electrolyte solution and current density (He et al. 2018). Tan et al. prepared red emitted graphene quantum dots from graphite by electrochemical exfoliation in $K_2S_2O_8$ solution (Tan et al. 2015). Huang et al. synthesized crystalline graphene quantum dots (GQD) by electrochemical cutting of graphene sheets in weak ammonia electrolyte with a yield of 28% more than with strong electrolytes. They have prepared the amino-functionalized graphene quantum dots by adjusting the electrolyte concentration. The weak electrolyte ensures the oxidation

Fig. 12.2 The electrolytic process of graphene paper in (**a**) ammonia and **b** NaOH solutions under different reaction times. Reprint with permission from Huang et al. (2018). Copyright © 2018, American Chemical Society

and cutting of graphene sheets by decreasing the rate of intercalation assisted exfoliation. Hence weak electrolyte favors the formation of quantum dots rather than graphene sheets. Moreover, the added advantage of using ammonia is that it can be removed from the product solution simply by evaporation (Huang et al. 2018). Figure 12.2 shows the progress of electrolysis of graphene sheets with both ammonia and NaOH. From the figure it is clear that the oxidation of graphene sheet is high with ammonia electrolyte which results in the higher dispersibility of graphene oxide sheets and the solution turns brown within 90 min.

Graphene derivatives with hetero atom dopants, functionalized graphene, hybrids, or composites can also be synthesized using electrochemical exfoliation techniques (Liu et al. 2019; Dong et al. 2015; Chang et al. 2012; Sayah et al. 2018; Xie et al. 2014). Xie et al. used a direct electrochemical process to fabricate reduced graphene oxide -copper composite films (rGO/Cu) using graphene oxide and copper sulfate solutions using cyclic voltammetry deposition (Xie et al. 2014). Figure 12.3 shows the SEM images of the rGO/Cu composite deposited at −1.2 V for different time durations.

12.3 Importance of Electrochemical Exfoliation

The importance of electrochemical exfoliation lies in the processing limitations of other synthesis methods of graphene derivatives. Wet chemical synthesis of graphene oxide using Hummers and modified Hummer's method uses harsh oxidizing agents,

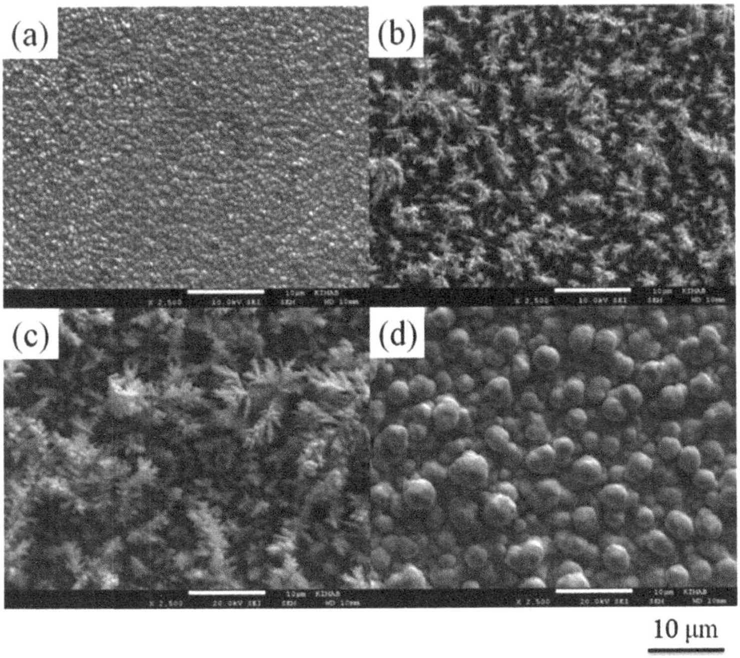

Fig. 12.3 SEM morphologies of Cu^{2+} 0.01 M/GO 0.5 mg·mL^{-1} solutions deposited under −1.2 V at different deposition times: **a** 10 min; **b** 30 min; **c** 60 min, and **d** under −0.4 V for 180 min(Xie et al. 2014). Reprint with permission from Xie et al. (2017). Copyright © 2014, American Chemical Society

leading to excessive oxide sheets (Abdelkader et al. 2015). However, electrochemical exfoliation allows the mass production of graphene materials with excellent property control. Electrochemical exfoliation is considered to be a simple and clean method and hence significant due to its environmental importance. It is a relatively quick synthesis method that requires less time and effort for purification. Another clean aspect of the electrochemical method is that no toxic reductant is required to get graphene derivatives with a high C/O ratio. As the wet chemical synthesized graphene sheets are susceptible to a high degree of oxidation, the subsequent reduction of it is required to get property control. For reduction, either thermal or chemical reduction methods are usually employed. The chemical reduction is carried out with toxic reducing agents such as hydrazine hydrate, sodium borohydride, and other potential reducing agents (Feng et al. 2020; Kuila et al. 2013). The electrochemical exfoliation synthesis allows the production of graphene sheets with a high C/O ratio; by adjusting the exfoliation parameters, and hence the use of toxic reducing agents can be avoided. The electrochemical parameters, like high negative bias, help overcome the chemical barrier for regulating oxygen-containing functional groups. The potential advantages of electrochemical exfoliation methods are the ease of operation, and direct control over the exfoliation process, as the parameters can be adjusted easily.

Electrochemical exfoliation is a fast process compared to other synthesis methods, and hence commercial production is viable cost-wise. It is considered to be environmentally benign as the used electrolyte can be recycled for further use (Parvez et al. 2015). Electrochemical exfoliation is a type of wet chemical method in which it uses electrochemical activation rather than the use of chemicals to overcome the Van der Waals barrier between the graphitic layers, and the product requires only more straightforward purification steps. In general, electrochemical exfoliation allows the tunability of material properties, including quality, defect level, and functionality, in a one-pot method (Yu et al. 2015). Su et al. showed a one-pot method for successfully synthesizing high-quality graphene sheets with 30 μm lateral dimension with better property control, including transparency and conductivity (Su et al. 2011). The authors have achieved a fast and simple exfoliation of highly oriented pyrolytic graphite using the experimental technique, as shown in Fig. 12.4. They have obtained graphene sheets with a higher 2D/G band ratio in the Raman spectrum, indicating that retaining the sp^2 C = C network in the sheets gives rise to excellent electrical properties.

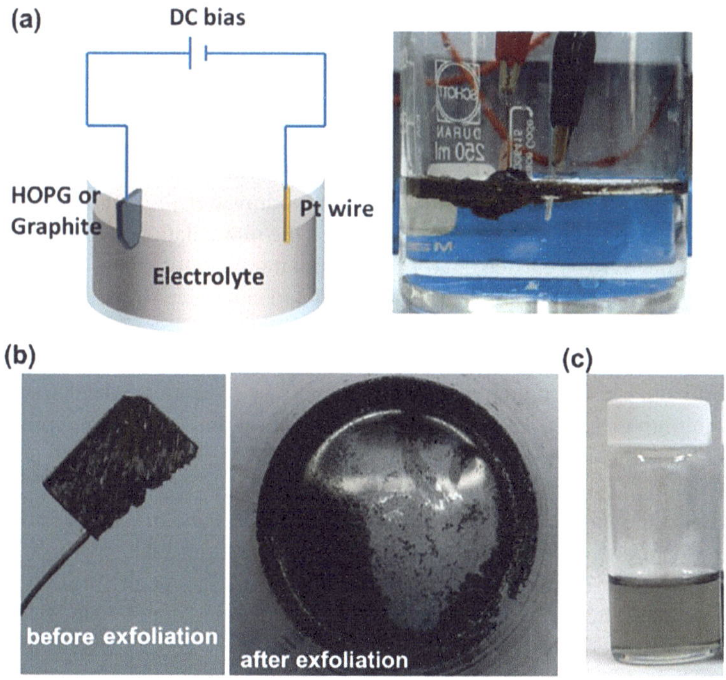

Fig. 12.4 **a** Schematic illustration and photo for electrochemical exfoliation of graphite. **b** Photos of the graphite flakes before and after electrochemical exfoliation. **c** Photo of the dispersed graphene sheets in a DMF solution (Su et al. 2011). Reprint with permission from Su et al. (2011). Copyright © 2011, American Chemical Society

12.4 Features of Electrochemical Exfoliation

The history of graphene synthesis of electrochemical exfoliation goes back to graphite intercalation chemistry (Yu et al. 2015). The first step in exfoliation synthesis is intercalating graphitic layers with ions or molecules. The intercalating species can either donate or accept an electron to or from the graphitic layers and form an ionic compound with graphite called graphite salt. Intercalation can take place in the presence or absence of a potential difference. The intercalation in the presence of electricity is called electro-intercalation, and it was first discovered in 1938. This graphene pre-era electro-intercalation forms the basis of the electrochemical exfoliation of graphene (Yu et al. 2015). The main feature of electrochemical synthesis is the intercalation of ions rather than oxidation using a harsh oxidizing reagent (Berktas et al. 2020). The exfoliation is carried out in an electrochemical cell with graphite material as one of the electrodes. The electrolytic cell for the exfoliation uses either a two-electrode or a three-electrode system with aqueous or non-aqueous electrolytes. Some studies use graphite as both electrodes, which gives a higher exfoliation rate than a single graphite electrode setup (Yu et al. 2015). However, the products formed are inhomogeneous, containing graphene sheets of different lateral dimensions. The exfoliation is classified into anodic and cathodic, depending on the potential applied. Anodic exfoliation of graphite electrodes with aqueous electrolytes is more susceptible to oxidation. Cathodic exfoliation using non-ionic electrolytes provides better control over the defect levels in the graphene sheets. As mentioned earlier in this chapter, exfoliation occurs after ions and electrolyte molecules are intercalated between the adjacent graphitic layers (Prekodravac et al. 2021); the efficiency of the exfoliation depends on the extent to which the intercalation takes place. Thus, pretreatment needs to be carried out for an efficient intercalation of the electrode. The electrochemical conditions are simple and mild and can be controlled conveniently by adjusting the parameters. According to the applications, the right control of the processing parameters ensures graphene sheets with optimum oxygen content, defect density, and electrical conductivity (Yang et al. 2019). We can synthesize graphene sheets with high or less C/O content in a single pot by playing with the processing parameters, such as interchanging electrodes from anode to cathode or vice versa, changing the electrolyte rather than using harsh oxidizing agents or toxic reducing agents. Hence the purification steps remained simple in the electrochemical exfoliation rather than the chemical methods. The most important feature of this method is the cost reduction of producing graphene derivatives. In wet chemical synthesis, $KMnO_4$ is used as the oxidizing agent (Yu et al. 2016; Lavin-Lopez et al. 2016; Shuai et al. 2020). However, $KMnO_4$ is industrially produced by electrochemical synthesis. Thus, direct electrochemical exfoliation with simple electrolytes is more effective. The procedure for electrochemical synthesis is simple as we do not have to ensure the complete removal of any oxidizing agent present in the reaction mixture to get the optimum product quality (Yu et al. 2015). The graphene derivatives prepared by electrochemical exfoliation found applications in all the areas where the graphene derivatives produced by other methods found applications. Since the electrochemically synthesized graphene has fewer defects and good conductivity, it can be used as conductive fillers and electrodes in applications such as supercapacitors, sensors, batteries, organic field effect transistors, conductive inks, and electrocatalysts (Yu et al. 2015; Maccaferri et al. 2019; Zhai et al. 2020; Li et al. 2019; Kim et al. 2018).

12.5 Process Parameters Affecting the Electrochemical Exfoliation

The efficiency of electrochemical exfoliation efficiency implies the quality as well as the yield of graphene sheets synthesized. The various process parameters, such as applied potential, current, exfoliation time, composition/concentration of electrolytes, and the quality of graphite electrodes, influence the quality and yield of graphene derivatives. Precise control over these parameters is required to get graphene sheets with controlled defect densities, oxygen content, layer numbers, and lateral dimensions according to the potential application of the material. Since the electrochemical exfoliation yields graphene sheets with oxygen content, the discussion in this section is mostly focused on graphene oxide. For the improved yield of graphene oxide, the parameters such as type and concentration of electrolytes, voltage/current, type of electrode, and operating temperature must be optimized (Liu and Aziz 2022). Fang et al. reviewed the parameters influencing electrochemical exfoliation and the mechanism of exfoliation in a simple approach (Fang et al. 2019). A detailed understanding of these parameters is necessary to develop a scalable and facile electrochemical exfoliation for the synthesis of GO (Table 12.1).

The type of electrolyte is crucial since its oxidization capabilities may affect the yield of graphene oxide. The most commonly used electrolytes are molten salts and inorganic solutions such as $(NH4)_2SO_4$, Na_2SO_4, K_2SO_4, H_2SO_4, $HClO_4$, and HNO_3. H_2SO_4 is the most popular electrolyte as it shows a better exfoliation rate as the size of the sulfate ion is similar to the interlayer distance of graphitic layers. The SO_2 and O_2 gas evolved by the decomposition of sulfate ions enhances the exfoliation rate. When the exfoliation is carried out using $(NH_2)_2SO_2$, the nitrogen and sulfur dopants are attached to the graphene oxide sheets (Liu and Aziz 2022). In addition to the type of electrolyte, its concentration also plays a crucial role in determining the yield of graphene oxide. As the concentration of electrolytes increases, the quantity

Table 12.1 List of typical electrochemical exfoliation parameters and graphene quality parameters

Typical electrochemical parameters	Typical graphene quality parameters
• Applied potential and current • Exfoliation time, composition/concentration of electrolytes • Type of graphite electrodes • The quality of graphite electrodes	• Controlled defect densities • Oxygen content • Layers numbers • Lateral dimensions

of intercalating ions increases. Hence the expansion, exfoliation, and oxidization of the graphene sheets increase. The increased concentration is reflected in the improved dispersing ability and decreased lateral dimensions. However, too much concentrated electrolyte decreases GO formation as it contains a small amount of water, water being the source of oxygen-rich functional groups. The higher concentration favors higher intercalation and exfoliation rather than the oxidization of graphene sheets. Thus, the optimum electrolyte concentration must be selected to get appropriate graphene derivatives. Munuera et al., in a study, showed the effect of several multifunctional electrolytes for synthesizing high-quality graphene sheets in the anodic exfoliation of graphite foil (Munuera et al. 2016). Similar to the type and concentration of electrolyte, the anodic potential or current is influential in determining the oxidizing potential and yield. As the applied current or voltage increases, more intercalating species are attracted towards the electrode, and the degrees of intercalation, exfoliation, and oxidization increase. A constant potential is used in all the electrochemical exfoliation for a specific time duration of several minutes to hours. It is reported that the potential between 1 and 2 V is required for pre-intercalation and is applied for a short duration of 1–10 min. The oxidization of graphite intercalation compounds is then facilitated at a higher voltage or current to form graphene oxide. Thus, higher voltage favors the yield of graphene oxide. The oxidization degree is an interplay of different factors such as voltage, electrolyte, electrode, and operating temperatures. The electrode type also affects the exfoliation and oxidization of graphene sheets; however, the spectrum of graphite electrodes available is narrow. Hence the control over the GO synthesis is limited by playing with different types of graphite electrodes. Among the available graphite electrodes, the commonly used electrodes for electrochemical exfoliation synthesis are highly oriented pyrolytic graphite (HOPG), graphite rods, graphite flakes, and graphite foils (Coroş et al. 2016; Liu et al. 2013; Roscher et al. 2019; Cooper et al. 2014; Chen et al. 2019). The mechanical strength of electrodes and their expansion are crucial in electrochemical exfoliation. The graphite foil is thin, so exfoliation is completed quickly and thus reduces the yield of graphene oxide. Similarly, with expanded graphite, exfoliation takes place easily, but GO yields less. Thus, exfoliation efficiency does not connote the high yield (Cooper et al. 2014).

12.6 Challenges of Electrochemical Exfoliation

In general, all the features of electrochemical exfoliation are paused as a challenge for electrochemical exfoliation for the mass production of defect-free graphene sheets. Since electrochemical exfoliation involves several parameters, it is necessary to know how the material quality changes with the variation of a single parameter. The efficiency of exfoliation depends on the intercalation. Intercalation requires pretreatment with oxidizing agents. It may introduce oxygen-rich functionalization and defect densities and adversely affect conductivity (Ambrosi et al. 2016); however, it is suitable for synthesizing highly dispersible graphene oxide. However, the yield of graphene oxide decreased when exfoliation took place at a faster rate. Thus, the real challenge of exfoliation with thin electrodes (graphite foils) is the tradeoff between the material property and the synthesis yield. The graphite electrode should be thick enough to slow the exfoliation and give time for the side reaction to yield oxygen-rich derivatives. The fabrication of an optimum electrode that simultaneously provides desirable material properties and yield is tedious and minimizes the cost-effectiveness of the electrochemical method. In addition to this, the reactions are not fully optimized for a commercially scalable synthesis (Maccaferri et al. 2019). The selection of electrolytes plays a significant role in determining the success of the exfoliation synthesis. An electrolyte is considered best for exfoliation if it plays two critical roles, (i) the electrolyte should provide enough ionic species to intercalate and separate the graphitic layers (ii) the electrolyte should be able to sufficiently disperse the exfoliated sheets and prevent them from reaggregation by providing functionalities to the graphene sheets. The dispersibility is associated with the functionality of the graphene sheets and is possible with aqueous electrolytes and anodic exfoliation. However, introducing oxygen functionalities creates defects in the graphene basal sheets, making them inferior for applications that expect defect-controlled graphene sheets and conductivity. The cathodic exfoliation produces graphene sheets with a higher C/O ratio (pristine graphene-like sheets). However, they are found to be less dispersible, and their dispersibility depends on the surface energy of the solvents. The dispersibility is also affected by the presence of electrolytes associated with the graphene. Hence it must be carefully removed before dispersing it in solvents. Thus, washing the product to remove the ions to prevent agglomeration poses a challenge in mass production (Ferrari et al. 2006). The equipment engineering, including the power supply (AC/DC), and the selection of parameters like potential, exfoliation time, and operating temperature for electrochemical exfoliation, is challenging. Thus, we can understand that exfoliation is not as simple as we think.

12.7 Factors Affecting the Quality of Graphene Derivatives

The quality of graphene sheets is expressed in terms of parameters such as the number of graphene layers, degree of oxidation (C/O ratio), number of defects, and lateral dimension of the graphene sheets (Munuera et al. 2017). The anodic exfoliation generates more hydroxyl and oxygen radicals from the aqueous electrolyte. These radicals attack the graphene sheets to form oxygen functionalities. The C/O ratio measures the degree of oxidization and can be determined from experimental techniques such as XPS. The XPS spectrum of graphene derivatives shows a typical band at binding energy 284.5 eV due to the C–C stretching of the sp^2 moieties and the new bands of oxygen functional groups increase (hydroxyl, carboxyl, and epoxy groups). The intensity of this XPS band at 284.5 eV decreases as the oxidation level increases (Krishnamoorthy et al. 2013). Krivenko et al. determined the surface concentration of elements in the exfoliation products of graphite at different exfoliation conditions and found that the low-voltage exfoliation results in products with a higher carbon-to-oxygen ratio. The lowest C/O ratio is obtained for cathodic and anodic plasma exfoliation among which the least for anodic plasma exfoliation (Krivenko et al. 2019). In addition to the peak intensity, the binding energies of C1's vary with the chemical environment, like when it binds to oxygen atoms using single or double bonds. Thus, XPS not only gives the C/O ratio but it gives an idea about the structural disorder of the graphene sheets. The structural quality and defect degrees of graphene derivatives can also be assessed using Raman spectroscopic studies.

The typical Raman bands are G, D, and 2D. The G and D band is located at 1595 cm^{-1} and 1360 cm^{-1} respectively. The D band denotes the structural defects in the graphene layers, and the G band is the characteristic of the vibration of sp^2 hybridized carbon atoms of the graphene basal plane. The ratio of intensities of the characteristic D and G bands (I_D/I_G) gives insight into the defect levels and number of layers in graphene sheets. The higher the I_D/I_G ratio, the higher the defect densities in the graphene sheets. The presence of the 2D band indicates the structural change of graphite to graphene (Coroş et al. 2016; López-Díaz et al. 2017; Eckmann et al. 2012; Ferrari et al. 2006). The D band indicates the presence of various disorders in the graphene basal plane, such as defects due to edge positions, vacancies, functional groups, and irregular structure. The origin of the G band is due to the E_{2g} phonon vibration of sp^2 hybridized carbon atoms in the Brillouin zone, and the 2D band arises due to the second-order vibrations of two phonons scattering in the opposite wave vectors (Han et al. 2021). In nitrogen-dopped graphene, the D band position at 1585 cm^{-1} is changed into 1555 cm^{-1}. This shift is attributed to the superposition of C = C and C = N bond vibrations. The G band is upshifted due to the increased carrier concentration due to defects and dopants (Lou et al. 2016). Munuera et al. prepared a low oxygen content graphene using exfoliation of graphite using aqueous sodium halide electrolytes (NaCl, NaBr, and NaI) in which halide ions act as sacrificial agents to prevent the extensive oxidation of graphene sheets. The I_D/I_G ratio in the Raman spectrum (Fig. 12.5) positively correlates with the oxygen-to-carbon ratio, which differs in graphene samples (Munuera et al. 2017). Figure 12.5 shows the Raman spectrum of graphene sheets prepared with three halide electrolytes. Nagyte et al. systematically studied the Raman fingerprints of graphene produced by anodic electrochemical exfoliation. They evaluated the quality of graphene sheets regarding the number of defects and sample uniformity. They have found that the quality of graphene sheets is affected by all the steps involved in the electrochemical exfoliation. They also noticed that the defect arises in the graphene sheets due to electrolyte, temperature, and other experimental conditions (Nagyte et al. 2020).

In a study, Tian et al. synthesized high-quality and crystalline graphene sheets using multiple electrolytes with the D band (I_D/I_G = 0.022) and a 2D band at 2718 cm^{-1}, indicating the low defects in the graphene sheets. The XPS spectrum shows that the graphene sheet contains 87.5% C = C/C–C, 9.6% C–O groups, and 2.9% O = C–O groups (Tian et al. 2017). Thus, XPS and Raman spectral features give an idea of the quality of graphene sheets produced with different electrochemical exfoliation parameters.

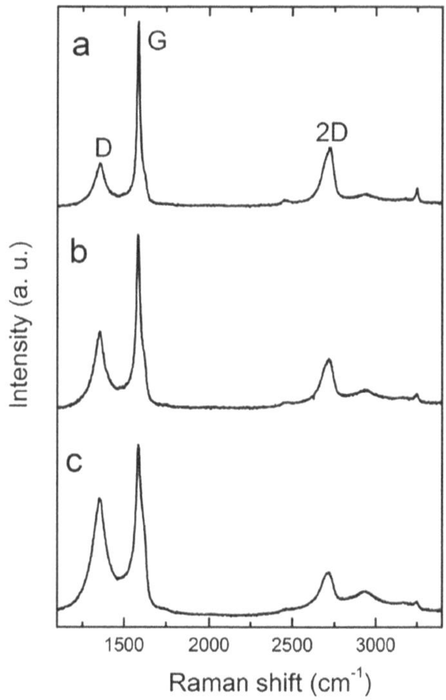

Fig. 12.5 Raman spectra of graphene samples prepared by aqueous anodic exfoliation of graphite foil in 0.05 M NaCl (**a**), 0.1 M NaBr (**b**), and 0.1 M NaI (**c**). Reprint with permission from Munuera et al. (2017). Copyright © 2017, American Chemical Society

12.8 Future of Electrochemical Exfoliation

Electrochemical exfoliation is a superior technique for synthesizing graphene derivatives as the synthesis parameters can be controlled to get graphene sheets with property control. Graphene sheets with controlled defects and functionality can be synthesized with precise control of the exfoliation parameters. A quick exfoliation with less oxidative electrolyte will produce graphene sheets with less oxidative damage and a high Carbon to oxygen ratio. However, there is no simple and well-established method for electrochemical exfoliation synthesis. For the future research direction for the scalable synthesis of graphene derivatives, it is necessary to optimize the process parameters, including the exfoliation potential, suitable graphite electrode, and electrolyte for the potential application of the material. A standard synthesis procedure will also help to compare the final device performance. Electrochemical exfoliation provides a large avenue not only for the synthesis of defect-controlled graphene sheets and functionalization with different lateral dimensions but also for the fabrication of various hybrids, including graphene-metal nanoparticle hybrids and graphene-polymer hybrids. Therefore, in the coming years, more research articles can be expected in this field.

12.9 Conclusion

Electrochemical exfoliation of graphite-based electrodes is a promising synthesis route for the commercial synthesis of graphene derivatives owing to its attractive features like simplicity, versatility, quick reaction time, and environmental friendliness. However, several factors affect the quality of the final products. There is a compendium of research articles on the electrochemical synthesis of graphene and its derivatives, which mainly focus on the effect on multiple exfoliation parameters. However, future research directions should address the principles and mechanism of electrochemical exfoliation as a separate discipline so that future researchers can get an insight into addressing all the challenges that happen during the exfoliation synthesis of graphene derivatives.

References

Abdelkader, A.M., Cooper, A.J., Dryfe, R.A.W., Kinloch, I.A.: How to get between the sheets: a review of recent works on the electrochemical exfoliation of graphene materials from bulk graphite. Nanoscale **7**, 6944–6956 (2015). https://doi.org/10.1039/C4NR06942K

Aghamohammadi, H., Eslami-Farsani, R., Torabian, M., Amousa, N.: Recent advances in one-pot functionalization of graphene using electrochemical exfoliation of graphite: a review study. Synth. Met. **269**, 116549 (2020). https://doi.org/10.1016/j.synthmet.2020.116549

Ahirwar, S., Mallick, S., Bahadur, D.: Electrochemical method to prepare graphene quantum dots and graphene oxide quantum dots. ACS Omega **2**, 8343–8353 (2017). https://doi.org/10.1021/acsomega.7b01539

Alkhouzaam, A., Qiblawey, H., Khraisheh, M., Atieh, M., Al-Ghouti, M.: Synthesis of graphene oxides particle of high oxidation degree using a modified Hummers method. Ceram. Int. **46**, 23997–24007 (2020). https://doi.org/10.1016/j.ceramint.2020.06.177

Ambrosi, A., Chua, C.K., Latiff, N.M., Loo, A.H., Wong, C.H.A., Eng, A.Y.S., Bonanni, A., Pumera, M.: Graphene and its electrochemistry–an update. Chem. Soc. Rev. **45**, 2458–2493 (2016). https://doi.org/10.1039/C6CS00136J

Berktas, I., Hezarkhani, M., Haghighi Poudeh, L., Saner Okan, B.: Recent developments in the synthesis of graphene and graphene-like structures from waste sources by recycling and upcycling technologies: a review. Graphene Technol. **5**, 59–73 (2020). https://doi.org/10.1007/s41127-020-00033-1

Cao, J., He, P., Mohammed, M.A., Zhao, X., Young, R.J., Derby, B., Kinloch, I.A., Dryfe, R.A.W.: Two-step electrochemical intercalation and oxidation of graphite for the mass production of graphene oxide. J. Am. Chem. Soc. **139**, 17446–17456 (2017). https://doi.org/10.1021/jacs.7b08515

Chang, H.-H., Chang, C.-K., Tsai, Y.-C., Liao, C.-S.: Electrochemically synthesized graphene/polypyrrole composites and their use in supercapacitor. Carbon N. Y. **50**, 2331–2336 (2012). https://doi.org/10.1016/j.carbon.2012.01.056

Chen, C.-H., Yang, S.-W., Chuang, M.-C., Woon, W.-Y., Su, C.-Y.: Towards the continuous production of high crystallinity graphene via electrochemical exfoliation with molecular in situ encapsulation. Nanoscale **7**, 15362–15373 (2015). https://doi.org/10.1039/C5NR03669K

Chen, D., Wang, F., Li, Y., Wang, W.-W., Huang, T.-X., Li, J.-F., Novoselov, K.S., Tian, Z.-Q., Zhan, D.: Programmed electrochemical exfoliation of graphite to high quality graphene. Chem. Commun.Commun. **55**, 3379–3382 (2019). https://doi.org/10.1039/C9CC00393B

Cooper, A.J., Wilson, N.R., Kinloch, I.A., Dryfe, R.A.W.: Single stage electrochemical exfoliation method for the production of few-layer graphene via intercalation of tetraalkylammonium cations. Carbon N. Y. **66**, 340–350 (2014). https://doi.org/10.1016/j.carbon.2013.09.009

Coroş, M., Pogăcean, F., Roşu, M.-C., Socaci, C., Borodi, G., Mageruşan, L., Biriş, A.R., Pruneanu, S.: Simple and cost-effective synthesis of graphene by electrochemical exfoliation of graphite rods. RSC Adv. **6**, 2651–2661 (2016). https://doi.org/10.1039/C5RA19277C

Dong, S., Xi, J., Wu, Y., Liu, H., Fu, C., Liu, H., Xiao, F.: High loading MnO_2 nanowires on graphene paper: facile electrochemical synthesis and use as flexible electrode for tracking hydrogen peroxide secretion in live cells. Anal. Chim. Acta **853**, 200–206 (2015). https://doi.org/10.1016/j.aca.2014.08.004

Eckmann, A., Felten, A., Mishchenko, A., Britnell, L., Krupke, R., Novoselov, K.S., Casiraghi, C.: Probing the nature of defects in graphene by raman spectroscopy. Nano Lett. **12**, 3925–3930 (2012). https://doi.org/10.1021/nl300901a

Ejigu, A., Fujisawa, K., Spencer, B.F., Wang, B., Terrones, M., Kinloch, I.A., Dryfe, R.A.W.: On the role of transition metal salts during electrochemical exfoliation of graphite: antioxidants or metal oxide decorators for energy storage applications. Adv. Funct. Mater.Funct. Mater. **28**, 1804357 (2018). https://doi.org/10.1002/adfm.201804357

Fang, S., Lin, Y., Hu, Y.H.: Recent advances in green, safe, and fast production of graphene oxide via electrochemical approaches. ACS Sustain. Chem. Eng. **7**, 12671–12681 (2019). https://doi.org/10.1021/acssuschemeng.9b02794

Feng, J., Ye, Y., Xiao, M., Wu, G., Ke, Y.: Synthetic routes of the reduced graphene oxide. Chem. Pap. **74**, 3767–3783 (2020). https://doi.org/10.1007/s11696-020-01196-0

Ferrari, A.C., Meyer, J.C., Scardaci, V., Casiraghi, C., Lazzeri, M., Mauri, F., Piscanec, S., Jiang, D., Novoselov, K.S., Roth, S., Geim, A.K.: Raman spectrum of graphene and graphene layers. Phys. Rev. Lett. **97**, 187401 (2006). https://doi.org/10.1103/PhysRevLett.97.187401

Gebreegziabher, G.G., Asemahegne, A.S., Ayele, D.W., Dhakshnamoorthy, M., Kumar, A.: One-step synthesis and characterization of reduced graphene oxide using chemical exfoliation method. Mater. Today Chem. **12**, 233–239 (2019). https://doi.org/10.1016/j.mtchem.2019.02.003

Geim, A.K., Novoselov, K.S.: The rise of graphene. Nat. Mater. **6**, 183–191 (2007). https://doi.org/10.1038/nmat1849

Gutiérrez-Cruz, A., Ruiz-Hernández, A.R., Vega-Clemente, J.F., Luna-Gazcón, D.G., Campos-Delgado, J.: A review of top-down and bottom-up synthesis methods for the production of graphene, graphene oxide and reduced graphene oxide. J. Mater. Sci. **57**, 14543–14578 (2022). https://doi.org/10.1007/s10853-022-07514-z

Han, Z., Gan, C., Li, X., Feng, P., Ma, X., Fan, X., Zhu, M.: Electrochemical preparation of modified-graphene additive towards lubrication requirement. Tribol. Int.. Int. **161**, 107057 (2021). https://doi.org/10.1016/j.triboint.2021.107057

He, M., Guo, X., Huang, J., Shen, H., Zeng, Q., Wang, L.: Mass production of tunable multi-color graphene quantum dots from an energy resource of coke by a one-step electrochemical exfoliation. Carbon N. Y. **140**, 508–520 (2018). https://doi.org/10.1016/j.carbon.2018.08.067

Huang, H., Yang, S., Li, Q., Yang, Y., Wang, G., You, X., Mao, B., Wang, H., Ma, Y., He, P., Liu, Z., Ding, G., Xie, X.: Electrochemical cutting in weak aqueous electrolytes: the strategy for efficient and controllable preparation of graphene quantum dots. Langmuir **34**, 250–258 (2018). https://doi.org/10.1021/acs.langmuir.7b03425

Kalita, H., Palaparthy, V.S., Baghini, M.S., Aslam, M.: Electrochemical synthesis of graphene quantum dots from graphene oxide at room temperature and its soil moisture sensing properties. Carbon N. Y. **165**, 9–17 (2020). https://doi.org/10.1016/j.carbon.2020.04.021

Kim, Y., Kwon, Y.J., Hong, J.-Y., Park, M., Lee, C.J., Lee, J.U.: Spray coating of electrochemically exfoliated graphene/conducting polymer hybrid electrode for organic field effect transistor. J. Ind. Eng. Chem. **68**, 399–405 (2018). https://doi.org/10.1016/j.jiec.2018.08.015

Krishnamoorthy, K., Veerapandian, M., Yun, K., Kim, S.-J.: The chemical and structural analysis of graphene oxide with different degrees of oxidation. Carbon N. Y. **53**, 38–49 (2013). https://doi.org/10.1016/j.carbon.2012.10.013

Krivenko, A.G., Manzhos, R.A., Kotkin, A.S., Kochergin, V.K., Piven, N.P., Manzhos, A.P.: Production of few-layer graphene structures in different modes of electrochemical exfoliation of graphite by voltage pulses. Instrum. Sci. Technol.. Sci. Technol. **47**, 535–544 (2019). https://doi.org/10.1080/10739149.2019.1607750

Kuila, T., Mishra, A.K., Khanra, P., Kim, N.H., Lee, J.H.: Recent advances in the efficient reduction of graphene oxide and its application as energy storage electrode materials. Nanoscale **5**, 52–71 (2013). https://doi.org/10.1039/C2NR32703A

Kumar, N., Abubakar Sadique, M., Khan, R.: Electrochemical exfoliation of graphene quantum dots from waste dry cell battery for biosensor applications. Mater. Lett. **305**, 130829 (2021). https://doi.org/10.1016/j.matlet.2021.130829

Lavin-Lopez, M. del P., Romero, A., Garrido, J., Sanchez-Silva, L., Valverde, J.L.: Influence of different improved hummers method modifications on the characteristics of graphite oxide in order to make a more easily scalable method. Ind. Eng. Chem. Res. **55**, 12836–12847 (2016). https://doi.org/10.1021/acs.iecr.6b03533

Lee, H.C., Liu, W.-W., Chai, S.-P., Mohamed, A.R., Lai, C.W., Khe, C.-S., Voon, C.H., Hashim, U., Hidayah, N.M.S.: Synthesis of single-layer graphene: a review of recent development. Procedia Chem. **19**, 916–921 (2016). https://doi.org/10.1016/j.proche.2016.03.135

Li, R., Yue, W., Chen, X.: Fabrication of porous carbon-coated ZnO nanoparticles on electrochemical exfoliated graphene as an anode material for lithium-ion batteries. J. Alloys Compd. **784**, 800–806 (2019). https://doi.org/10.1016/j.jallcom.2019.01.117

Li, Y., Hu, Y., Zhao, Y., Shi, G., Deng, L., Hou, Y., Qu, L.: An electrochemical avenue to green-luminescent graphene quantum dots as potential electron-acceptors for photovoltaics. Adv. Mater. **23**, 776–780 (2011). https://doi.org/10.1002/adma.201003819

Li, Y., Li, S., Wang, Y., Wang, J., Liu, H., Liu, X., Wang, L., Liu, X., Xue, W., Ma, N.: Electrochemical synthesis of phosphorus-doped graphene quantum dots for free radical scavenging. Phys. Chem. Chem. Phys. **19**, 11631–11638 (2017). https://doi.org/10.1039/C6CP06377B

Lim, J.Y., Mubarak, N.M., Abdullah, E.C., Nizamuddin, S., Khalid, M.: Inamuddin: Recent trends in the synthesis of graphene and graphene oxide based nanomaterials for removal of heavy metals—a review. J. Ind. Eng. Chem. **66**, 29–44 (2018). https://doi.org/10.1016/j.jiec.2018.05.028

Liu, F., Wang, C., Sui, X., Riaz, M.A., Xu, M., Wei, L., Chen, Y.: Synthesis of graphene materials by electrochemical exfoliation: recent progress and future potential. Carbon Energy. **1**, 173–199 (2019). https://doi.org/10.1002/cey2.14

Liu, J., Poh, C.K., Zhan, D., Lai, L., Lim, S.H., Wang, L., Liu, X., Gopal Sahoo, N., Li, C., Shen, Z., Lin, J.: Improved synthesis of graphene flakes from the multiple electrochemical exfoliation of graphite rod. Nano Energy **2**, 377–386 (2013). https://doi.org/10.1016/j.nanoen.2012.11.003

Liu, W.-W., Aziz, A.: Review on the effects of electrochemical exfoliation parameters on the yield of graphene oxide. ACS Omega **7**, 33719–33731 (2022). https://doi.org/10.1021/acsomega.2c04099

López-Díaz, D., López Holgado, M., García-Fierro, J.L., Velázquez, M.M.: Evolution of the raman spectrum with the chemical composition of graphene oxide. J. Phys. Chem. C **121**, 20489–20497 (2017). https://doi.org/10.1021/acs.jpcc.7b06236

Lou, F., Buan, M.E.M., Muthuswamy, N., Walmsley, J.C., Rønning, M., Chen, D.: One-step electrochemical synthesis of tunable nitrogen-doped graphene. J. Mater. Chem. A **4**, 1233–1243 (2016). https://doi.org/10.1039/C5TA08038J

Maccaferri, G., Terzi, F., Xia, Z., Vulcano, F., Liscio, A., Palermo, V., Zanardi, C.: Highly sensitive amperometric sensor for morphine detection based on electrochemically exfoliated graphene oxide. Application in screening tests of urine samples. Sens. Actuators B Chem. **281**, 739–745 (2019). https://doi.org/10.1016/j.snb.2018.10.163

Marcano, D.C., Kosynkin, D.V., Berlin, J.M., Sinitskii, A., Sun, Z., Slesarev, A., Alemany, L.B., Lu, W., Tour, J.M.: Improved synthesis of graphene oxide. ACS Nano **4**, 4806–4814 (2010). https://doi.org/10.1021/nn1006368

Moreno, C., Vilas-Varela, M., Kretz, B., Garcia-Lekue, A., Costache, M.V., Paradinas, M., Panighel, M., Ceballos, G., Valenzuela, S.O., Peña, D., Mugarza, A.: Bottom-up synthesis of multifunctional nanoporous graphene. Science (80-.). **360**, 199–203 (2018). https://doi.org/10.1126/science.aar2009

Munuera, J.M., Paredes, J.I., Enterría, M., Pagán, A., Villar-Rodil, S., Pereira, M.F.R., Martins, J.I., Figueiredo, J.L., Cenis, J.L., Martínez-Alonso, A., Tascón, J.M.D.: Electrochemical exfoliation of graphite in aqueous sodium halide electrolytes toward low oxygen content graphene for energy and environmental applications. ACS Appl. Mater. Interfaces **9**, 24085–24099 (2017). https://doi.org/10.1021/acsami.7b04802

Munuera, J.M., Paredes, J.I., Villar-Rodil, S., Ayán-Varela, M., Martínez-Alonso, A., Tascón, J.M.D.: Electrolytic exfoliation of graphite in water with multifunctional electrolytes: en route towards high quality, oxide-free graphene flakes. Nanoscale **8**, 2982–2998 (2016). https://doi.org/10.1039/C5NR06882G

Munuera, J.M., Paredes, J.I., Villar-Rodil, S., Castro-Muñiz, A., Martínez-Alonso, A., Tascón, J.M.D.: High quality, low-oxidized graphene via anodic exfoliation with table salt as an efficient oxidation-preventing co-electrolyte for water/oil remediation and capacitive energy storage applications. Appl. Mater. Today **11**, 246–254 (2018). https://doi.org/10.1016/j.apmt.2018.03.002

Nagyte, V., Kelly, D.J., Felten, A., Picardi, G., Shin, Y.Y., Alieva, A., Worsley, R.E., Parvez, K., Dehm, S., Krupke, R., Haigh, S.J., Oikonomou, A., Pollard, A.J., Casiraghi, C.: Raman fingerprints of graphene produced by anodic electrochemical exfoliation. Nano Lett. **20**, 3411–3419 (2020). https://doi.org/10.1021/acs.nanolett.0c00332

Narita, A., Feng, X., Müllen, K.: Bottom-up synthesis of chemically precise graphene nanoribbons. Chem. Rec. **15**, 295–309 (2015). https://doi.org/10.1002/tcr.201402082

Novoselov, K.S., Geim, A.K., Morozov, S.V., Jiang, D., Zhang, Y., Dubonos, S.V., Grigorieva, I.V., Firsov, A.A.: Electric field effect in atomically thin carbon films. Science (80-.). **306**, 666–669 (2004). https://doi.org/10.1126/science.1102896

Parvez, K., Yang, S., Feng, X., Müllen, K.: Exfoliation of graphene via wet chemical routes. Synth. Met. **210**, 123–132 (2015). https://doi.org/10.1016/j.synthmet.2015.07.014

Prekodravac, J.R., Kepić, D.P., Colmenares, J.C., Giannakoudakis, D.A., Jovanović, S.P.: A comprehensive review on selected graphene synthesis methods: from electrochemical exfoliation through rapid thermal annealing towards biomass pyrolysis. J. Mater. Chem. C **9**, 6722–6748 (2021). https://doi.org/10.1039/D1TC01316E

Roscher, S., Hoffmann, R., Prescher, M., Knittel, P., Ambacher, O.: High voltage electrochemical exfoliation of graphite for high-yield graphene production. RSC Adv. **9**, 29305–29311 (2019). https://doi.org/10.1039/C9RA04795F

Saleem, H., Haneef, M., Abbasi, H.Y.: Synthesis route of reduced graphene oxide via thermal reduction of chemically exfoliated graphene oxide. Mater. Chem. Phys. **204**, 1–7 (2018). https://doi.org/10.1016/j.matchemphys.2017.10.020

Sayah, A., Habelhames, F., Bahloul, A., Nessark, B., Bonnassieux, Y., Tendelier, D., El Jouad, M.: Electrochemical synthesis of polyaniline-exfoliated graphene composite films and their capacitance properties. J. Electroanal. Chem.Electroanal. Chem. **818**, 26–34 (2018). https://doi.org/10.1016/j.jelechem.2018.04.016

Shuai, S., Liu, Y., Zhao, C., Zhu, H., Li, Y., Zhou, K., Ge, W., Hao, J.: Improved synthesis of graphene oxide with controlled oxidation degree by using different dihydrogen phosphate as intercalators. Chem. Phys. **539**, 110938 (2020). https://doi.org/10.1016/j.chemphys.2020.110938

Singh, R.K., Kumar, R., Singh, D.P.: Graphene oxide: strategies for synthesis, reduction and frontier applications. RSC Adv. **6**, 64993–65011 (2016). https://doi.org/10.1039/C6RA07626B

Son, M., Ham, M.-H.: Low-temperature synthesis of graphene by chemical vapor deposition and its applications. FlatChem. **5**, 40–49 (2017). https://doi.org/10.1016/j.flatc.2017.07.002

Su, C.-Y., Lu, A.-Y., Xu, Y., Chen, F.-R., Khlobystov, A.N., Li, L.-J.: High-quality thin graphene films from fast electrochemical exfoliation. ACS Nano **5**, 2332–2339 (2011). https://doi.org/10.1021/nn200025p

Tan, X., Li, Y., Li, X., Zhou, S., Fan, L., Yang, S.: Electrochemical synthesis of small-sized red fluorescent graphene quantum dots as a bioimaging platform. Chem. Commun.Commun. **51**, 2544–2546 (2015). https://doi.org/10.1039/C4CC09332A

Tian, S., He, P., Chen, L., Wang, H., Ding, G., Xie, X.: Electrochemical fabrication of high quality graphene in mixed electrolyte for ultrafast electrothermal heater. Chem. Mater. **29**, 6214–6219 (2017). https://doi.org/10.1021/acs.chemmater.7b00567

Toh, S.Y., Loh, K.S., Kamarudin, S.K., Daud, W.R.W.: Graphene production via electrochemical reduction of graphene oxide: synthesis and characterisation. Chem. Eng. J. **251**, 422–434 (2014). https://doi.org/10.1016/j.cej.2014.04.004

Whitener, K.E., Sheehan, P.E.: Graphene synthesis. Diam. Relat. Mater.Relat. Mater. **46**, 25–34 (2014). https://doi.org/10.1016/j.diamond.2014.04.006

Wu, J., Wang, H., Qiu, J., Zhang, K., Shao, J., Yan, L.: Efficient preparation of high-quality graphene via anodic and cathodic simultaneous electrochemical exfoliation under the assistance of microwave. J. Colloid Interface Sci. **608**, 1422–1431 (2022). https://doi.org/10.1016/j.jcis.2021.10.098

Xie, G., Forslund, M., Pan, J.: Direct electrochemical synthesis of reduced graphene oxide (rGO)/copper composite films and their electrical/electroactive properties. ACS Appl. Mater. Interfaces **6**, 7444–7455 (2014). https://doi.org/10.1021/am500768g

Yang, Y., Hou, H., Zou, G., Shi, W., Shuai, H., Li, J., Ji, X.: Electrochemical exfoliation of graphene-like two-dimensional nanomaterials. Nanoscale **11**, 16–33 (2019). https://doi.org/10.1039/C8NR08227H

Yi, M., Shen, Z.: A review on mechanical exfoliation for the scalable production of graphene. J. Mater. Chem. A **3**, 11700–11715 (2015). https://doi.org/10.1039/C5TA00252D

Yu, H., Zhang, B., Bulin, C., Li, R., Xing, R.: High-efficient synthesis of graphene oxide based on improved hummers method. Sci. Rep. **6**, 36143 (2016). https://doi.org/10.1038/srep36143

Yu, P., Lowe, S.E., Simon, G.P., Zhong, Y.L.: Electrochemical exfoliation of graphite and production of functional graphene. Curr. Opin. Colloid Interface Sci.Opin. Colloid Interface Sci. **20**, 329–338 (2015). https://doi.org/10.1016/j.cocis.2015.10.007

Zhai, R., Xiao, Y., Ding, T., Wu, Y., Chen, S., Wei, W.: Construction of $NiCo_2S_4$ heterostructure based on electrochemically exfoliated graphene for high-performance hybrid supercapacitor electrode. J. Alloys Compd. **845**, 156164 (2020). https://doi.org/10.1016/j.jallcom.2020.156164

SPRINGER NATURE

GPSR Compliance

The European Union's (EU) General Product Safety Regulation (GPSR) is a set of rules that requires consumer products to be safe and our obligations to ensure this.

If you have any concerns about our products, you can contact us on ProductSafety@springernature.com

In case Publisher is established outside the EU, the EU authorized representative is:

Springer Nature Customer Service Center GmbH
Europaplatz 3
69115 Heidelberg, Germany

The manufacturer's authorised representative in the EU is Springer Nature Customer Service Centre GmbH, Europaplatz 3, 69115 Heidelberg, Germany. If you have any concerns regarding our products, please contact ProductSafety@springernature.com

Printed and bound by CPI Group (UK) Ltd, Croydon, CR0 4YY
01/10/2026
02225863-0008